Integral Equations

Integral Equations: Theories, Approximations and Applications

Editors

Samad Noeiaghdam
Denis N. Sidorov

MDPI • Basel • Beijing • Wuhan • Barcelona • Belgrade • Manchester • Tokyo • Cluj • Tianjin

Editors
Samad Noeiaghdam
Baikal School of BRICS
Irkutsk National Research
Technical University
Irkutsk
Russia

Denis N. Sidorov
Energy Systems Institute
Siberian Branch of Russian
Academy of Sciences
Irkutsk
Russia

Editorial Office
MDPI
St. Alban-Anlage 66
4052 Basel, Switzerland

This is a reprint of articles from the Special Issue published online in the open access journal *Symmetry* (ISSN 2073-8994) (available at: www.mdpi.com/journal/symmetry/special_issues/ Integral_Equations_Theories_Applications_Approximations).

For citation purposes, cite each article independently as indicated on the article page online and as indicated below:

LastName, A.A.; LastName, B.B.; LastName, C.C. Article Title. *Journal Name* **Year**, *Volume Number*, Page Range.

ISBN 978-3-0365-2240-1 (Hbk)
ISBN 978-3-0365-2239-5 (PDF)

© 2021 by the authors. Articles in this book are Open Access and distributed under the Creative Commons Attribution (CC BY) license, which allows users to download, copy and build upon published articles, as long as the author and publisher are properly credited, which ensures maximum dissemination and a wider impact of our publications.

The book as a whole is distributed by MDPI under the terms and conditions of the Creative Commons license CC BY-NC-ND.

Contents

About the Editors vii

Preface to "Integral Equations: Theories, Approximations and Applications" ix

Samad Noeiaghdam and Denis Sidorov
Integral Equations: Theories, Approximations, and Applications
Reprinted from: *Symmetry* **2021**, *13*, 1402, doi:10.3390/sym13081402 1

Sanda Micula
Numerical Solution of Two-Dimensional Fredholm–Volterra Integral Equations of the Second Kind
Reprinted from: *Symmetry* **2021**, *13*, 1326, doi:10.3390/sym13081326 5

Vagif Ibrahimov and Mehriban Imanova
Multistep Methods of the Hybrid Type and Their Application to Solve the Second Kind Volterra Integral Equation
Reprinted from: *Symmetry* **2021**, *13*, 1087, doi:10.3390/sym13061087 17

Alexander A. Minakov and Christoph Schick
Integro-Differential Equation for the Non-Equilibrium Thermal Response of Glass-Forming Materials: Analytical Solutions
Reprinted from: *Symmetry* **2021**, *13*, 256, doi:10.3390/sym13020256 41

Qingfeng Zhu, Yufeng Shi, Jiaqiang Wen and Hui Zhang
A Type of Time-Symmetric Stochastic System and Related Games
Reprinted from: *Symmetry* **2021**, *13*, 118, doi:10.3390/sym13010118 59

Jing Zhu, Ye Liu and Jiahui Cao
Effects of Second-Order Velocity Slip and the Different Spherical Nanoparticles on Nanofluid Flow
Reprinted from: *Symmetry* **2020**, *13*, 64, doi:10.3390/sym13010064 79

Elham Hashemizadeh, Mohammad Ali Ebadi and Samad Noeiaghdam
Matrix Method by Genocchi Polynomials for Solving Nonlinear Volterra Integral Equations with Weakly Singular Kernels
Reprinted from: *Symmetry* **2020**, *12*, 2105, doi:10.3390/sym12122105 95

Sanda Micula
A Numerical Method for Weakly Singular Nonlinear Volterra Integral Equations of the Second Kind
Reprinted from: *Symmetry* **2020**, *12*, 1862, doi:10.3390/sym12111862 111

Young Hee Geum, Arjun Kumar Rathie and Hwajoon Kim
Matrix Expression of Convolution and Its Generalized Continuous Form
Reprinted from: *Symmetry* **2020**, *12*, 1791, doi:10.3390/sym12111791 127

Sarkhosh S. Chaharborj, Zuhaila Ismail and Norsarahaida Amin
Detecting Optimal Leak Locations Using Homotopy Analysis Method for Isothermal Hydrogen-Natural Gas Mixture in an Inclined Pipeline
Reprinted from: *Symmetry* **2020**, *12*, 1769, doi:10.3390/sym12111769 141

Samad Noeiaghdam, Aliona Dreglea, Jihuan He, Zakieh Avazzadeh, Muhammad Suleman, Mohammad Ali Fariborzi Araghi, Denis N. Sidorov and Nikolai Sidorov
Error Estimation of the Homotopy Perturbation Method to Solve Second Kind Volterra Integral Equations with Piecewise Smooth Kernels: Application of the CADNA Library
Reprinted from: *Symmetry* **2020**, *12*, 1730, doi:10.3390/sym12101730 . . . . **163**

Eskandar Ameer, Hassen Aydi, Hasanen A. Hammad, Wasfi Shatanawi and Nabil Mlaiki
On (,)-Metric Spaces with Applications
Reprinted from: *Symmetry* **2020**, *12*, 1459, doi:10.3390/sym12091459 . . . . **179**

About the Editors

Samad Noeiaghdam

Samad Noeiaghdam acquired his Ph.D. degree in Applied Mathematics at Central Tehran Branch of Islamic Azad University in Iran. He is Associate Professor of Irkutsk National Research Technical University and senior researcher of South Ural State University. Through the years, he has published many papers in the field of numerical analysis, solving integral equations, ordinary differential equations, partial differential equations, solving ill-posed problems, solving fuzzy problems and bio-mathematical models.

Additionally, he has published a book and some chapters to show the applications of stochastic arithmetic to control the accuracy of mathematical methods for solving various problems. He is member of editorial and reviewer board of several Scopus-WOS journals and he has selected as a winner of outstanding reviewer 2020 by *Mathematics* (MDPI). He also has presented numerous communications in international conferences.

Denis N. Sidorov

Denis Sidorov received the Ph.D. and Dr. Habil. degrees in applied mathematics from Irkutsk State University, Irkutsk, in 2000 and 2014, respectively. He is currently PI with Melentiev Energy Systems Institute, Siberian Branch of the Russian Academy of Sciences, Irkutsk, and Head of Industrial Math Lab in Irkutsk National Research Technical University, Irkutsk. From 2001 to 2007, he was with Trinity College, Dublin, Ireland, with Université de Technologie de Compiégne and Centre National de la Recherche Scientifique, Compiègne, France, as a Postdoctoral Research Fellow, and with ASTI Holding, Singapore, as a Vision Engineer. In 2013, he was with Siegen University, Siegen, Germany, as a Deutscher Akademischer Austauschdienst Professor. He served as Distinguished Guest Professor of Hunan University (China) and Queen's University Belfast (UK) in 2016-2020. He has authored more than 140 scientific papers and four monographs. His research interests include integral and differential equations, machine learning, wind energy, and inverse problems. Dr. Sidorov is the IEEE Power and Energy Society Russia (Siberia) Chapter Chair.

Preface to "Integral Equations: Theories, Approximations and Applications"

Linear and non-linear integral equations of the first and second kinds have many applications in engineering and real life problems. Thus, we try to find efficient and accurate methods to solve these problems. The aim of this editorial is to overview the content of the Special Issue "Integral Equations: Theories, Approximations and Applications".

This Special Issue collects innovative contributions addressing the top challenges in integral equations, integro-differential equations, multi-dimensional problems, and ill-posed and singular problems with modern applications. It covers linear and non-linear integral equations of the first and second kinds, singular and ill-posed kernels, system of integral equations, high-dimensional problems, and especially new numerical, analytical, and semi-analytical methods for solving the problems mentioned by focusing on modern applications.

Samad Noeiaghdam, Denis N. Sidorov
Editors

Editorial

Integral Equations: Theories, Approximations, and Applications

Samad Noeiaghdam [1,2,*,†] and Denis Sidorov [1,3,*,†]

1. Industrial Mathematics Laboratory, Baikal School of BRICS, Irkutsk National Research Technical University, 664074 Irkutsk, Russia
2. Department of Applied Mathematics and Programming, South Ural State University, Lenin Prospect 76, 454080 Chelyabinsk, Russia
3. Energy Systems Institute, Siberian Branch of Russian Academy of Science, 664033 Irkutsk, Russia
* Correspondence: snoei@istu.edu (S.N.); sidorovdn@istu.edu (D.S.)
† These authors contributed equally to this work.

Citation: Noeiaghdam, S.; Sidorov, D. Integral Equations: Theories, Approximations, and Applications. *Symmetry* 2021, *13*, 1402. https://doi.org/10.3390/sym13081402

Received: 30 June 2021
Accepted: 20 July 2021
Published: 2 August 2021

Publisher's Note: MDPI stays neutral with regard to jurisdictional claims in published maps and institutional affiliations.

Copyright: © 2021 by the authors. Licensee MDPI, Basel, Switzerland. This article is an open access article distributed under the terms and conditions of the Creative Commons Attribution (CC BY) license (https://creativecommons.org/licenses/by/4.0/).

1. Introduction

Linear and nonlinear integral equations of the first and second kinds have many applications in engineering and real life problems. Thus, we try to find efficient and accurate methods to solve these problems. The aim of this editorial is to overview the content of the special issue "Integral Equations: Theories, Approximations and Applications". This special issue collects innovative contributions addressing the top challenges in integral equations, integro-differential equations, multi dimensional problems, and ill-posed and singular problems with modern applications. In response to our call, we had 15 submissions from 16 countries (Azerbaijan, China, Egypt, Germany, India, Indonesia, Iran, Jordan, Korea, Malaysia, Romania, Russia, Saudi Arabia, Taiwan, Vietnam, and Yemen), of which 10 were accepted and five were rejected. This issue contains 10 technical articles and one editorial. It covers linear and nonlinear integral equations of the first and second kinds, singular and ill-posed kernels, system of integral equations, high-dimensional problems and especially new numerical, analytical, and semi-analytical methods for solving the problems mentioned by focusing on modern applications.

This special issue focuses on linear and nonlinear integral equations of the first and second kinds, singular and ill-posed kernels, system of integral equations, and high-dimensional problems for solving challenging and applicable problems, especially using novel numerical, analytical, and semi-analytical methods.

2. Brief Overview of the Contributions

Ibrahimov and Imanova in "Multistep Methods of the Hybrid Type and Their Application to Solve the Second Kind Volterra Integral Equation" [1] have focused on solving the integral equations with variable boundaries. For this aim, they have applied the advanced and hybrid types of multi-step methods. They have tried to show the connection between the obtained methods and some applicable methods to solve the first order initial-value problems. Applying the methods mentioned, they can change the problem to a system of algebraic equations. They have extended the methods for solving Volterra integro-differential equations. The numerical results show the accuracy of the method.

"Integro-Differential Equation for the Non-Equilibrium Thermal Response of Glass-Forming Materials: Analytical Solutions" has been published by A.A. Minakov and C. Schick [2]. In this study, they have studied the non-equilibrium thermal response of glass-forming substances with a dynamic (time-dependent) heat capacity to fast thermal perturbations based on an integro-differential equation. They have found that the heat transfer problem can be solved analytically for a heat source with an arbitrary time dependence and different geometries. In addition, they showed that the method can be used to analyze the response to local thermal perturbations in glass-forming materials, as well as temperature fluctuations during subcritical crystal nucleation and decay. The importance of this paper is related to some applications of the thermal properties of glass-forming materials, polymers, and nanocomposites.

Zhu et al. in [3] have studied the paper titled "A Type of Time-Symmetric Stochastic System and Related Games". This paper has been concerned with a type of time-symmetric stochastic system, namely the so-called forward–backward doubly stochastic differential equations, in which the forward equations are delayed doubly stochastic differential equations and the backward equations are anticipated backward doubly stochastic differential equations. Under some monotonicity assumptions, the existence and uniqueness of measurable solutions to forward–backward doubly stochastic differential equations have been obtained. The future development of many processes depends on both their current state and historical state, and these processes can usually be represented by stochastic differential systems with time delay. Therefore, a class of nonzero sum differential game for doubly stochastic systems with time delay has been studied in this paper. A necessary condition for the open-loop Nash equilibrium point of the Pontriagin-type maximum principle has been established, and a sufficient condition for the Nash equilibrium point has been obtained. Furthermore, the above results have been applied to the study of nonzero sum differential games for linear quadratic backward doubly stochastic systems with delay.

"Effects of Second-Order Velocity Slip and the Different Spherical Nanoparticles on Nanofluid Flow" have been studied by Zhu [4]. The paper theoretically has investigated the heat transfer of nanofluids with different nanoparticles inside a parallel-plate channel. The second-order slip condition has been adopted due to the microscopic roughness in the microchannels. After proper transformation, they have tried to convert the system of nonlinear partial differential equations to the ordinary differential equations with unknown constants, and they have solved the problem using the homotopy analysis method. As we know, this method has some important applications to solve the integral equations. Several graphs have been plotted to show the convergence regions. The semi-analytical expressions between Nu_B and N_{BT} are acquired. The results show that both first-order slip parameter and second-order slip parameter have positive effects on Nu_B of the MHD flow.

Hashemizadeh et al. have presented the paper "Matrix Method by Genocchi Polynomials for Solving Nonlinear Volterra Integral Equations with Weakly Singular Kernels" in [5]. In this study, they have worked on the spectral method for solving nonlinear Volterra integral equations with weakly singular kernels based on the Genocchi polynomials. Many other interesting results concerning nonlinear equations with discontinuous symmetric kernels with the application of group symmetry have remained beyond the scope of this paper. In the proposed approach, relying on the useful properties of Genocchi polynomials, they have produced an operational matrix and a related coefficient matrix to convert nonlinear Volterra integral equations with weakly singular kernels into a system of algebraic equations. This method is very fast and gives high-precision solutions with good accuracy in a low number of repetitions compared to other methods that are available. The error boundaries for this method have also been presented. Some illustrative examples have been provided to demonstrate the capability of the proposed method. In addition, the results derived from the new method are compared to Euler's method to show the superiority of the proposed method.

Micula in [6] has focused on the paper titled "A Numerical Method for Weakly Singular Nonlinear Volterra Integral Equations of the Second Kind". This paper presents a numerical iterative method for the approximate solutions of nonlinear Volterra integral equations of the second kind, with weakly singular kernels. In this study, the existence and uniqueness conditions of the solution have been proved using the unique fixed point of an integral operator. Iterative application of that operator to an initial function yields a sequence of functions converging to the true solution. Finally, an appropriate numerical integration scheme (a certain type of product integration) has been used to produce the approximations of the solution at given nodes. The convergence of the method and the error estimates have been illustrated by the author. The proposed method has been applied to some numerical examples.

"Matrix Expression of Convolution and Its Generalized Continuous Form" has been published by Y.H. Geum [7]. In this paper, they have considered the matrix expression of convolution and its generalized continuous form. The matrix expression of convolution is effectively applied in convolutional neural networks, and, in this study, we correlate the concept of convolution in mathematics to that in the convolutional neural network. Of course, convolution is one of the main processes of deep learning, the learning method of deep neural networks, as a core technology. In addition to this, the generalized continuous form of convolution has been expressed as a new variant of Laplace-type transform that encompasses almost all existing integral transforms.

Chaharborj et al. in [8] have studied the paper titled "Detecting Optimal Leak Locations Using Homotopy Analysis Method for Isothermal Hydrogen-Natural Gas Mixture in an Inclined Pipeline". The aim of this article is to use the homotopy analysis method to pinpoint the optimal location of leakage in an inclined pipeline containing hydrogen-natural gas mixture by obtaining quick and accurate analytical solutions for nonlinear transportation equations. Because of important applications of the homotopy analysis method for solving different kinds of integral equations, we have accepted to publish this paper on this issue. The homotopy analysis method utilizes a simple and powerful technique to adjust and control the convergence region of the infinite series solution using auxiliary parameters. The auxiliary parameters provide a convenient way of controlling the convergent region of series solutions. Numerical results have indicated that the approach is highly accurate, computationally very attractive, and easy to implement.

Noeiaghdam et al. have focused on "Error Estimation of the Homotopy Perturbation Method to Solve Second Kind Volterra Integral Equations with Piecewise Smooth Kernels: Application of the CADNA Library" in [9]. In this paper, they have studied the second kind of linear Volterra integral equations with a discontinuous kernel obtained from the load leveling and energy system problems. For solving this problem, they have proposed the homotopy perturbation method. They have discussed the convergence theorem and the error analysis of the formulation to validate the accuracy of the obtained solutions. In this study, the Controle et Estimation Stochastique des Arrondis de Calculs method (CESTAC) and the Control of Accuracy and Debugging for Numerical Applications (CADNA) library have been used to control the rounding error estimation. The advantage of the discrete stochastic arithmetic has been taken to find the optimal iteration, optimal error, and optimal approximation of the homotopy perturbation method. The comparative graphs between exact and approximate solutions show the accuracy and efficiency of the method.

Ameer et al. in [10] have published the paper titled "On (ϕ, ψ)-Metric Spaces with Applications". The aim of this article is to introduce the notion of a (ϕ, ψ)-metric space, which extends the metric space concept. In these spaces, the symmetry property has been preserved. They have presented a natural topology $\tau(\phi, \psi)$ in such spaces and discuss their topological properties. They also have established the Banach contraction principle in the context of (ϕ, ψ)-metric spaces, and they have illustrated the significance of their main theorem by examples. Ultimately, as applications, the existence of a unique solution of Fredholm type integral equations in one and two dimensions have been ensured.

3. Conclusions and Outlook

The Special Issue Book "Integral Equations: Theories, Approximations, and Applications" presents a collection of articles dealing with relevant topics in the field of integral equation. Various mathematical and computational techniques and approaches were presented to solve the linear and nonlinear problems. The success of this Special Issue has motivated the editors to propose a new Special Issue "Integral Equations: Theories, Approximations, and Applications II" that will complement the present one with a focus on modern applications of integral equations. We invite the research community to submit novel contributions covering numerical, analytical, and semi-analytical methods to solve the multi-dimensional linear and nonlinear integral equations.

Funding: D.S. and S.N. have been supported by a grant from the Academic Council in the direction of the scientific school of Irkutsk National Research Technical University No. 14-NSH-RAN-2020.

Institutional Review Board Statement: Not applicable.

Informed Consent Statement: Not applicable.

Data Availability Statement: Not applicable.

Acknowledgments: We are thankful for the journal Symmetry Editorial's help and support.

Conflicts of Interest: The authors declare no conflict of interest.

References

1. Ibrahimov, V.; Imanova, M. Multistep Methods of the Hybrid Type and Their Application to Solve the Second Kind Volterra Integral Equation. *Symmetry* **2021**, *13*, 1087. [CrossRef]
2. Minakov, A.A.; Schick, C. Integro-Differential Equation for the Non-Equilibrium Thermal Response of Glass-Forming Materials: Analytical Solutions. *Symmetry* **2021**, *13*, 256. [CrossRef]
3. Zhu, Q.; Shi, Y.; Wen, J.; Zhang, H. A Type of Time-Symmetric Stochastic System and Related Games. *Symmetry* **2021**, *13*, 118. [CrossRef]
4. Zhu, J.; Liu, Y.; Cao, J. Effects of Second-Order Velocity Slip and the Different Spherical Nanoparticles on Nanofluid Flow. *Symmetry* **2021**, *13*, 64. [CrossRef]
5. Hashemizadeh, E.; Ebadi, M.A.; Noeiaghdam, S. Matrix Method by Genocchi Polynomials for Solving Nonlinear Volterra Integral Equations with Weakly Singular Kernels. *Symmetry* **2020**, *12*, 2105. [CrossRef]
6. Micula, S. A Numerical Method for Weakly Singular Nonlinear Volterra Integral Equations of the Second Kind. *Symmetry* **2020**, *12*, 1862. [CrossRef]
7. Geum, Y.H.; Rathie, A.K.; Kim, H. Matrix Expression of Convolution and Its Generalized Continuous Form. *Symmetry* **2020**, *12*, 1791. [CrossRef]
8. Chaharborj, S.; Ismail, Z.; Amin, N. Detecting Optimal Leak Locations Using Homotopy Analysis Method for Isothermal Hydrogen-Natural Gas Mixture in an Inclined Pipeline. *Symmetry* **2020**, *12*, 1769. [CrossRef]
9. Noeiaghdam, S.; Dreglea, A.; He, J.; Avazzadeh, Z.; Suleman, M.; Fariborzi Araghi, M.A.; Sidorov, D.N.; Sidorov, N. Error Estimation of the Homotopy Perturbation Method to Solve Second Kind Volterra Integral Equations with Piecewise Smooth Kernels: Application of the CADNA Library. *Symmetry* **2020**, *12*, 1730. [CrossRef]
10. Ameer, E.; Aydi, H.; Hammad, H.A.; Shatanawi, W.; Mlaiki, N. On (ϕ,ψ)-Metric Spaces with Applications. *Symmetry* **2020**, *12*, 1459. [CrossRef]

Article

Numerical Solution of Two-Dimensional Fredholm–Volterra Integral Equations of the Second Kind

Sanda Micula

Department of Mathematics and Computer Science, Babeş-Bolyai University, 400084 Cluj-Napoca, Romania; sanda.micula@ubbcluj.ro

Abstract: The paper presents an iterative numerical method for approximating solutions of two-dimensional Fredholm–Volterra integral equations of the second kind. As these equations arise in many applications, there is a constant need for accurate, but fast and simple to use numerical approximations to their solutions. The method proposed here uses successive approximations of the Mann type and a suitable cubature formula. Mann's procedure is known to converge faster than the classical Picard iteration given by the contraction principle, thus yielding a better numerical method. The existence and uniqueness of the solution is derived under certain conditions. The convergence of the method is proved, and error estimates for the approximations obtained are given. At the end, several numerical examples are analyzed, showing the applicability of the proposed method and good approximation results. In the last section, concluding remarks and future research ideas are discussed.

Keywords: Fredholm–Volterra integral equations; fixed-point theorems; numerical approximations

MSC: 45H05; 47H10; 47H09; 65D32

Citation: Micula, S. Numerical Solution of Two-Dimensional Fredholm–Volterra Integral Equations of the Second Kind. *Symmetry* **2021**, *13*, 1326. https://doi.org/10.3390/sym13081326

Academic Editors: Samad Noeiaghdam and Denis N. Sidorov

Received: 28 June 2021
Accepted: 21 July 2021
Published: 23 July 2021

Publisher's Note: MDPI stays neutral with regard to jurisdictional claims in published maps and institutional affiliations.

Copyright: © 2021 by the author. Licensee MDPI, Basel, Switzerland. This article is an open access article distributed under the terms and conditions of the Creative Commons Attribution (CC BY) license (https://creativecommons.org/licenses/by/4.0/).

1. Preliminaries

Fredholm–Volterra equations are integral equations of the following type:

$$u(t,x) = \int_0^t \int_\Omega K(t,x,\tau,y,u(\tau,y)) \, dy \, d\tau + f(t,x),$$

for $(t,x) \in [0,T] \times \Omega$, Ω a closed subset of $\mathbb{R}^n$, $n = 1, 2, 3$.

One encounters these equations in many applications in areas of physics, engineering or biology. In addition, many reformulations of boundary value problems can be written as Volterra–Fredholm integral equations. They are also used to model the progress of an epidemic and various other biological and physical problems. Integral equations with symmetric kernels are of frequent occurrence in the formulation of electronic and optic problems, as well as in optimization and spectral analysis.

In this paper, we consider mixed Fredholm–Volterra integral equations of the following form:

$$u(t,x) = \int_0^t \int_a^b K(t,x,\tau,y,u(\tau,y)) \, dy \, d\tau + f(t,x), \tag{1}$$

$(t,x) \in D = [0,T] \times [a,b]$, where $K \in C(D \times D \times \mathbb{R})$ and $f \in C(D)$.

Given the wide variety of applications, there have been substantial works on the solvability of these equations and on studying their properties. Numerical approximations of their solutions have been studied via collocation methods [1–3], block-pulse functions [4,5], Adomian decomposition methods [6], wavelet-based methods [7–9], iterative methods [10–15], differential quadratures [16], meshless procedures [17], etc. A simplified, one-dimensional case was studied in [18]. More details and considerations can be found, for example, in [19–22].

The aim of the present work is to develop a simple but quite accurate numerical method for approximating the solution of such equations. We derive a method based on fixed point theory for the existence and uniqueness of the solution, and on the use of an appropriate cubature formula for the numerical approximation. As such, the advantage of this new method consists mainly in the fact that it is easy to use and implement but gives good approximations of the solution at a given set of nodes. Compared to other classical methods used for integral equations, such as projection, Nyström or decomposition methods, this procedure does not require solving in the end an algebraic system for the values of the unknown function at the grid points. Such systems can be ill-conditioned, and may require additional procedures, which increase the computational and implementation cost of the resulting method, while decreasing its area of applicability. Instead, the proposed scheme finds the approximations at the nodes iteratively, using previously found values.

The rest of the paper is organized as follows: in Section 2, we discuss the solvability of Equation (1), via fixed point theory. Altman's algorithm [23] is employed instead of the classical Banach's theorem. This uses a Mann-type iteration (see [24]), which, by means of some parameters (the sequences ε_n and y_n, respectively, from Theorem 1 below), allows better control over the speed of convergence. With an appropriate choice of those parameters, we obtain faster successive approximations than the ones provided by the Picard-type iteration. In Section 3, we present a numerical method for approximating the solution of Equation (1), using a suitable cubature formula. Then, we analyze the convergence and give error estimates for the case when the two-dimensional trapezium rule is used for the numerical approximation of the iterates. In Section 4 we apply the proposed method to several numerical examples that are discussed in detail, showing good agreement between the theoretical results and the practical ones. Section 5 contains the concluding remarks on the procedure presented, and a discussion of ideas for future research in this area.

2. Solvability of the Integral Equation

We analyze the solvability of Equation (1) via fixed point results. To this end, we define the integral operator $F : C(D) \to C(D)$ associated with Equation (1) by the following:

$$Fu(t,x) := \int_0^t \int_a^b K(t,x,\tau,y,u(\tau,y))\, dy\, d\tau + f(t,x). \qquad (2)$$

Then, we find a solution of the Equation (1) by finding a fixed point of the operator F:

$$u = Fu. \qquad (3)$$

Let $X = C(D)$, endowed with the Chebyshev norm:

$$\|u\| := \max_{(t,x)\in D} |u(t,x)|, \ u \in X.$$

Then, it is known that $(X, \|\cdot\|)$ is a Banach space and for some $\rho > 0$, the ball $B_\rho := \{u \in C(D) \mid \|u - f\| \leq \rho\} \subseteq X$ is a closed subset. The well-known contraction principle holds for $F : X \to X$. The speed of convergence can be improved by using the following result due to Mann [24], also known as Altman's algorithm [23]:

Theorem 1. *Consider* $(X, ||\cdot||)$ *a Banach space and* $T : X \to X$ *a* q–*contraction. Let* $0 < \varepsilon_n \leq 1$ *be a sequence of numbers satisfying the following:*

$$\sum_{n=0}^{\infty} \varepsilon_n = \infty. \tag{4}$$

Then, we have the following:

(a) *Equation* $u = Tu$ *has exactly one solution* $u^* \in X$.
(b) *The sequence of successive approximations*

$$u_{n+1} = (1 - \varepsilon_n)u_n + \varepsilon_n T u_n, \quad n = 0, 1, \ldots \tag{5}$$

converges to the solution u^*, *for any* $u_0 \in X$.
(c) *For every* $n \in \mathbb{N}$, *the following error estimate holds:*

$$||u_n - u^*|| \leq \frac{e^{1-q}}{1-q} e^{-(1-q)y_n} ||u_0 - T u_0|| \tag{6}$$

where $y_0 = 0$, $y_n = \sum_{i=0}^{n-1} \varepsilon_i$, *for* $n \geq 1$.

The error estimate in Equation (6) is better than the classical error $\frac{q^n}{1-q}$ given by the contraction principle. We will use this result for our integral operator F with $\varepsilon_n = \frac{1}{n+1}$, which satisfies the requirements of Theorem 1. Then, we have the following:

Theorem 2. *Let* $K \in C(D \times D \times \mathbb{R})$, $f \in C(D)$ *and* $\rho_1 := \min_{(t,x) \in D} f(t, x)$, $\rho_2 := \max_{(t,x) \in D} f(t, x)$. *Assume the following:*

(i) *there exists a constant* $L > 0$ *such that*

$$|K(t, x, \tau, y, u) - K(t, x, \tau, y, v)| \leq L||u - v||, \tag{7}$$

for all $(t, x), (\tau, y) \in D$ *and all* $u, v \in [\rho_1 - \rho, \rho_2 + \rho]$;

(ii)
$$q := LT(b - a) < 1; \tag{8}$$

(iii)
$$M_K T(b - a) \leq \rho, \tag{9}$$

where $M_K := \max |K(t, x, \tau, y, u)|$ *over all* $(t, x), (\tau, y) \in D$ *and all* $u, v \in [\rho_1 - \rho, \rho_2 + \rho]$.

Then, the operator F in Equation (2) has exactly one fixed point, i.e., Equation (3) has exactly one solution $u^* \in B_R$, which can be obtained as the limit of the sequence of successive approximations as follows:

$$u_{n+1} = \left(1 - \frac{1}{n+1}\right) u_n + \frac{1}{n+1} F u_n, \quad n = 0, 1, \ldots, \tag{10}$$

starting with any arbitrary initial point $u_0 \in B_R$. *Moreover, for every* $n \in \mathbb{N}$, *the following error estimate holds:*

$$||u_n - u^*|| \leq \frac{e^{1-q}}{1-q} e^{-(1-q)y_n} ||u_0 - F u_0|| \tag{11}$$

where the sequence $\{y_n\}$ is defined by the following:

$$y_0 = 0, \quad y_n = \sum_{i=0}^{n-1} \frac{1}{i+1}, \quad n \geq 1. \tag{12}$$

Proof. Let u be any arbitrary point in B_ρ. For a fixed $(t,x) \in D$, we have the following:

$$|Fu(t,x) - f(t,x)| \leq \int_0^t \int_a^b |K(t,x,\tau,y,u(\tau,y))|\, dy\, d\tau \leq M_K T(b-a).$$

Then, by Equation (9), $Fu \in B_\rho$ and, thus, $F(B_\rho) \subseteq B_\rho$. Now, for every fixed $(t,x) \in D$, we use Equation (7) to obtain the following:

$$|Fu(t,x) - Fv(t,x)| \leq \int_0^t \int_a^b |K(t,x,\tau,y,u(\tau,y)) - K(t,x,\tau,y,v(\tau,y))|\, dy\, d\tau$$

$$\leq L\|u-v\| \int_0^t \int_a^b dy\, d\tau$$

$$\leq q\|u-v\|.$$

Thus,

$$\|Fu - Fv\| \leq q\|u-v\|$$

and since $q < 1$, all the conclusions follow from Theorem 1. □

Remark 1. *Let us note that the Lipschitz and contraction conditions (7) and (8) can be quite restrictive if required on the entire space. This is why we use only a local existence and uniqueness result so that these conditions need only be satisfied for $u \in B_\rho$, for some $\rho > 0$, which is much more reasonable. This observation will also be important in the next section, when we discuss the numerical approximation of the solution at the nodes (see Remark 2).*

For more considerations and details on fixed points, see [21,24].

3. A Numerical Method for Solving the Integral Equation

In order to use the iterative procedure Equation (10), we have to approximate the integrals numerically. Consider the following numerical integration scheme:

$$\int_a^b \int_c^d \varphi(s,w)\, dw\, ds = \sum_{i=0}^{m_1} \sum_{j=0}^{m_2} a_{ij} \varphi(s_i, w_j) + R_\varphi, \tag{13}$$

with nodes $a = s_0 < s_1 < \cdots < s_{m_1} = b$, $c = w_0 < w_1 < \cdots < w_{m_2} = d$, coefficients $a_{ij} \in \mathbb{R}, i = 0,1,\ldots,m_1, j = 0,1,\ldots,m_2$, such that there exists $M > 0$ with the following:

$$|R_\varphi| \leq M, \tag{14}$$

where $M \to 0$ as $m_1, m_2 \to \infty$.

For our purposes, let $0 = t_0 < t_1 < \cdots < t_{m_1} = T$ and $a = x_0 < x_1 < \cdots < x_{m_2} = b$ be partitions of $[0,T]$ and $[a,b]$, respectively, and let $u_0 = \tilde{u}_0 \equiv f$ be the initial approximation. We will use the successive iterations (10) and the numerical integration

formula (13) to approximate $u_n(t_l, x_k)$ by $\tilde{u}_n(t_l, x_k)$, for $l = \overline{0, m_1}, k = \overline{0, m_2}$ and $n = 0, 1, \ldots$
Let $l \in \{0, 1, \ldots, m_1\}$ and $k \in \{0, 1, \ldots, m_2\}$ be fixed. The following approximations hold:

$$\begin{aligned}
u_1(t_l, x_k) &= Fu_0(t_l, x_k) \\
&= \int_0^{t_l} \int_a^b K(t_l, x_k, \tau, y, f(\tau, y))\, dy\, d\tau + f(t_l, x_k) \\
&= \sum_{i=0}^l \sum_{j=0}^{m_2} a_{ij} K(t_l, x_k, t_i, x_j, f(t_i, x_j)) + R_K + f(t_l, x_k) \\
&= \tilde{u}_1(t_l, x_k) + \tilde{R}_1,
\end{aligned}$$

where

$$\tilde{u}_1(t_l, x_k) = \sum_{i=0}^l \sum_{j=0}^{m_2} a_{ij} K(t_l, x_k, t_i, x_j, f(t_i, x_j)) + f(t_l, x_k).$$

We make the following notation for the maximum error at the nodes:

$$err(u_n, \tilde{u}_n) := \max_{(t_l, x_k) \in D} |u_n(t_l, x_k) - \tilde{u}_n(t_l, x_k)|.$$

Then, by Equation (14), we have the following:

$$err(u_1, \tilde{u}_1) \leq |\tilde{R}_1| \leq M. \tag{15}$$

We continue with the next iteration:

$$\begin{aligned}
u_2(t_l, x_k) &= \left(1 - \frac{1}{2}\right) u_1(t_l, x_k) + \frac{1}{2} \left(\int_0^{t_l} \int_a^b K(t_l, x_k, \tau, y, u_1(\tau, y))\, dy\, d\tau + f(t_l, x_k) \right) \\
&= \left(1 - \frac{1}{2}\right) \tilde{u}_1(t_l, x_k) + \left(1 - \frac{1}{2}\right) (u_1(t_l, x_k) - \tilde{u}_1(t_l, x_k)) \\
&\quad + \frac{1}{2} \left(\sum_{i=0}^l \sum_{j=0}^{m_2} a_{ij} K(t_l, x_k, t_i, x_j, u_1(t_i, x_j)) + R_K + f(t_l, x_k) \right) \\
&= \left(1 - \frac{1}{2}\right) \tilde{u}_1(t_l, x_k) + \left(1 - \frac{1}{2}\right) (u_1(t_l, x_k) - \tilde{u}_1(t_l, x_k)) \\
&\quad + \frac{1}{2} \left(\sum_{i=0}^l \sum_{j=0}^{m_2} a_{ij} K(t_l, x_k, t_i, x_j, \tilde{u}_1(t_i, x_j)) + R_K + f(t_l, x_k) \right) \\
&\quad + \sum_{i=0}^l \sum_{j=0}^{m_2} a_{ij} K\left(t_l, x_k, t_i, x_j, u_1(t_i, x_j) - \tilde{u}_1(t_i, x_j)\right) \Big) \\
&= \left(1 - \frac{1}{2}\right) \tilde{u}_1(t_l, x_k) + \frac{1}{2} \left(\sum_{i=0}^l \sum_{j=0}^{m_2} a_{ij} K(t_l, x_k, t_i, x_j, \tilde{u}_1(t_i, x_j)) + f(t_l, x_k) \right) \\
&\quad + \left(1 - \frac{1}{2}\right) \left(u_1(t_l, x_k) - \tilde{u}_1(t_l, x_k) \right) \\
&\quad + \frac{1}{2} \left(\sum_{i=0}^l \sum_{j=0}^{m_2} a_{ij} K\left(t_l, x_k, t_i, x_j, u_1(t_i, x_j) - \tilde{u}_1(t_i, x_j)\right) + R_K \right) \\
&= \tilde{u}_2(t_l, x_k) + \tilde{R}_2,
\end{aligned} \tag{16}$$

with

$$\tilde{u}_2(t_l, x_k) = \left(1 - \frac{1}{2}\right)\tilde{u}_1(t_l, x_k) + \frac{1}{2}\left(\sum_{i=0}^{l}\sum_{j=0}^{m_2} a_{ij} K(t_l, x_k, t_i, x_j, \tilde{u}_1(t_i, x_j)) + f(t_l, x_k)\right),$$

$$\tilde{R}_2 = \left(1 - \frac{1}{2}\right)\left(u_1(t_l, x_k) - \tilde{u}_1(t_l, x_k)\right)$$
$$+ \frac{1}{2}\left(\sum_{i=0}^{l}\sum_{j=0}^{m_2} a_{ij} K(t_l, x_k, t_i, x_j, u_1(t_i, x_j) - \tilde{u}_1(t_i, x_j)) + R_K\right).$$

The values $\tilde{u}_2(t_l, x_k)$ can be then computed from the values obtained in the previous step. For the error estimate, let $\theta := L\sum_{i=0}^{m_1}\sum_{j=0}^{m_2}|a_{ij}|$. We have, by Equation (15), the following:

$$\begin{aligned}
err(u_2, \tilde{u}_2) &\leq |\tilde{R}_2| \\
&\leq \left(1 - \frac{1}{2}\right)|\tilde{R}_1| + \frac{1}{2}\left(\sum_{i=0}^{k}\sum_{j=0}^{m}|a_{ij}|\cdot L \cdot |\tilde{R}_1| + |R_K|\right) \\
&\leq \left(1 - \frac{1}{2}\right)M + \frac{1}{2}\left(LM\sum_{i=0}^{m}\sum_{j=0}^{m}|a_{ij}| + M\right) \quad (17) \\
&= M + \frac{1}{2}M\theta \\
&\leq M(1 + \theta).
\end{aligned}$$

Again, in a similar way, denoting by

$$\tilde{u}_n(t_l, x_k) = \left(1 - \frac{1}{n}\right)\tilde{u}_{n-1}(t_l, x_k)$$
$$+ \frac{1}{n}\left(\sum_{i=0}^{l}\sum_{j=0}^{m_2} a_{ij} K(t_l, x_k, t_i, x_j, \tilde{u}_{n-1}(t_i, x_j)) + f(t_l, x_k)\right), \quad (18)$$

for $l = 0, 1, \ldots, m_1$, $k = 0, 1, \ldots, m_2$, by induction, we find the following:

$$\begin{aligned}
err(u_n, \tilde{u}_n) &\leq |\tilde{R}_n| \\
&\leq \left(1 - \frac{1}{n}\right)|\tilde{R}_{n-1}| + \frac{1}{n}\left(\theta|\tilde{R}_{n-1}| + M\right) \\
&\leq M(1 + \theta + \cdots + \theta^{n-2})\left(1 - \frac{1}{n} + \frac{1}{n}\right) + \frac{1}{n}M\theta^{n-1} \quad (19) \\
&\leq M(1 + \theta + \cdots + \theta^{n-2}) + M\theta^{n-1} \\
&= M(1 + \theta + \cdots + \theta^{n-1}).
\end{aligned}$$

Then, we have the following approximation result:

Theorem 3. *Assume the conditions of Theorem 2 hold. In addition, assume that the coefficients in the numerical integration formula (13) satisfy the following:*

$$\theta = L\sum_{i=0}^{m_1}\sum_{j=0}^{m_2}|a_{ij}| < 1. \quad (20)$$

Then, the following error estimate holds for every $n \in \mathbb{N}$:

$$err(u^*, \tilde{u}_n) \leq \frac{e^{1-q}}{1-q} e^{-(1-q)y_n} ||u_0 - Fu_0|| + \frac{M}{1-\theta} \tag{21}$$

where u^* is the true solution of Equation (3), $\tilde{u}_n$ is the approximation given by Equation (18) and the sequence $\{y_n\}$ is defined in Equation (12).

Proof. By Equations (19) and (20), for all $l = 0, 1, \ldots, m_1$ and $k = 0, 1, \ldots, m_2$

$$|u_n(t_l, x_k) - \tilde{u}_n(t_l, x_k)| \leq \frac{M}{1-\theta}. \tag{22}$$

Since

$$|u^*(t_l, x_k) - \tilde{u}_n(t_l, x_k)| \leq |u^*(t_l, x_k) - u_n(t_l, x_k)| + |u_n(t_l, x_k) - \tilde{u}_n(t_l, x_k)|,$$

the estimate in Equation (21) now follows from Equation (22) and Theorem 2. □

Remark 2. Let us discuss condition (20), which can seem to be quite restrictive, especially since it also involves the constant L. As we will see below, when the quadrature scheme used is the trapezoidal rule, this condition reduces to the contraction condition (8) (whose applicability was discussed earlier in Remark 1), and, thus, does not introduce any new restrictions. In fact, the same thing is true for other fairly easy quadrature formulas, such as the midpoint or Simpson's rule (see [25]).

A Numerical Method Based on the Trapezoidal Rule

As discussed previously, we can use any numerical integration formula to approximate the iterates $u_n(x_k)$, as long as it satisfies condition (20). In what follows, we propose one of the simplest formulas, the two-dimensional trapezoidal rule:

$$\int_a^b \int_c^d \varphi(\tau, y) \, dy \, d\tau = \frac{(b-a)(d-c)}{4m_1 m_2} \Big[\varphi(a,c) + \varphi(b,c) + \varphi(a,d) + \varphi(b,d)$$

$$+ 2 \sum_{i=1}^{m_1-1} (\varphi(\tau_i, c) + \varphi(\tau_i, d)) \tag{23}$$

$$+ 2 \sum_{j=1}^{m_2-1} (\varphi(a, y_j) + \varphi(b, y_j)) + 4 \sum_{i=1}^{m_1-1} \sum_{j=1}^{m_2-1} \varphi(\tau_i, y_j) \Big] + R_\varphi,$$

using the nodes $s_i = a + \frac{b-a}{m_1} i$, $w_j = c + \frac{d-c}{m_2} j$, $i = \overline{0, m_1}, j = \overline{0, m_2}$. The remainder is the following:

$$R_\varphi = -\Big[\frac{(b-a)^3(d-c)}{12m_1^2 m_2} \varphi^{(2,0)}(\xi, \eta_1) + \frac{(b-a)(d-c)^3}{12m_1 m_2^2} \varphi^{(0,2)}(\xi_1, \eta) \tag{24}$$

$$+ \frac{(b-a)^3(d-c)^3}{144 m_1^2 m_2^2} \varphi^{(2,2)}(\xi, \eta) \Big], \quad \xi, \xi_1 \in (a,b), \eta, \eta_1 \in (c,d),$$

where we use the notation $\varphi^{(\alpha, \beta)}(t, x) = \frac{\partial^{\alpha+\beta} \varphi}{\partial t^\alpha \partial x^\beta}(t, x)$.

For fixed m_1, m_2, we consider the nodes $t_l = \frac{T}{m_1}l$, $x_k = a + \frac{b-a}{m_2}k$, $l = \overline{0, m_1}$, $k = \overline{0, m_2}$.
For simplicity, we will use the notation $K_{l,k,i,j} = K(t_l, x_k, t_i, x_j, u_n(t_i, x_j))$. Then we have the following:

$$\int_0^{t_l}\int_a^b K(t_l, x_k, \tau, y, u_n(\tau, y))\, dy\, d\tau = \frac{t_l(b-a)}{4lm_2}\Big[K_{l,k,0,0} + K_{l,k,l,0} + K_{l,k,0,m_2}$$

$$+ K_{l,k,l,m_2} + 2\sum_{i=0}^{l-1}\left(K_{l,k,i,0} + K_{l,k,i,m_2}\right)$$

$$+ 2\sum_{j=0}^{m_2-1}\left(K_{l,k,0,j} + K_{l,k,l,j}\right) \qquad (25)$$

$$+ 4\sum_{i=0}^{l-1}\sum_{j=0}^{m_2-1} K_{l,k,i,j}\Big] + R_K,$$

for each $l = 0, 1, \ldots, m_1$, $k = 0, 1, \ldots, m_2$. Since $\frac{t_l}{l} = \frac{T}{m_1}$, in this case, $\theta \le LT(b-a) = q$, which, by Equation (8) is strictly less than 1.

Next, let us discuss the bound M from Equation (14). By Equation (24), if $K^{(2,0)}(\tau, y, u_n(\tau, y))$, $K^{(0,2)}(\tau, y, u_n(\tau, y))$ and $K^{(2,2)}(\tau, y, u_n(\tau, y))$ are bounded, then the remainder R_K is of the form $\mathcal{O}\left(\frac{1}{m_1^2}\right) + \mathcal{O}\left(\frac{1}{m_2^2}\right)$. For simplicity, we write the function K emphasizing only the variables that it is to be differentiated with respect to, i.e., $K(\tau, y, u(\tau, y))$. We have the following:

$$K^{(2,0)}(\tau, y, u_n(\tau, y)) = \frac{\partial^2 K}{\partial \tau^2}(\tau, y, u_n(\tau, y)) + 2\frac{\partial^2 K}{\partial \tau \partial u}(\tau, y, u_n(\tau, y))\frac{\partial u}{\partial \tau}(\tau, y)$$

$$+ \frac{\partial^2 K}{\partial u^2}(\tau, y, u_n(\tau, y))\left(\frac{\partial u}{\partial \tau}(\tau, y)\right)^2$$

$$+ \frac{\partial K}{\partial u}(\tau, y, u_n(\tau, y))\frac{\partial^2 u}{\partial \tau^2}(\tau, y),$$

$$K^{(0,2)}(\tau, y, u_n(\tau, y)) = \frac{\partial^2 K}{\partial y^2}(\tau, y, u_n(\tau, y)) + 2\frac{\partial^2 K}{\partial y \partial u}(\tau, y, u_n(\tau, y))\frac{\partial u}{\partial y}(\tau, y)$$

$$+ \frac{\partial^2 K}{\partial u^2}(\tau, y, u_n(\tau, y))\left(\frac{\partial u}{\partial y}(\tau, y)\right)^2$$

$$+ \frac{\partial K}{\partial u}(\tau, y, u_n(\tau, y))\frac{\partial^2 u}{\partial y^2}(\tau, y),$$

and a similar (albeit much longer) formula can be found for $K^{(2,2)}(\tau, y, u_n(\tau, y))$, involving partial derivatives of K and u_n of up to order 4. For the partial derivatives of u_n, we have the following:

$$u_n(t, x) = \int_0^t\int_a^b K(t, x, \tau, y, u_{n-1}(\tau, y))\, dy\, d\tau + f(t, x),$$

$$\frac{\partial u_n}{\partial x}(t, x) = \int_0^t\int_a^b \frac{\partial K}{\partial x}(t, x, \tau, y, u_{n-1}(\tau, y))\, dy\, d\tau + \frac{\partial f}{\partial x}(t, x),$$

$$\frac{\partial u_n}{\partial t}(t, x) = \int_a^b K(t, x, t, y, u_{n-1}(t, y))\, dy$$

$$+ \int_0^t\int_a^b \frac{\partial K}{\partial t}(t, x, \tau, y, u_{n-1}(\tau, y))\, dy\, d\tau + \frac{\partial f}{\partial t}(t, x),$$

and so on, up to the partial derivatives of order 4.

It is now obvious that if K and f are C^4 functions with bounded fourth order partial derivatives, then there exists $M > 0$, independent of n, such that

$$|R_K| \leq M, \tag{26}$$

with $M \to 0$ as $m_1, m_2 \to \infty$. Thus, under these assumptions and those in Theorem 2, we have the following error estimate:

$$err(u^*, \tilde{u}_n) \leq \frac{e^{1-q}}{1-q} e^{-(1-q)y_n} ||u_0 - Fu_0|| + \frac{M}{1-\theta}, \tag{27}$$

for all $n = 1, 2, \ldots$, and $\{y_n\}$ given in Equation (12).

4. Numerical Examples

We now illustrate the applicability of the proposed method on several numerical examples. All computations are completed in Matlab, in double precision arithmetic. In general, the number of nodes is chosen such that the mesh size is around 0.05, which is small enough to achieve good accuracy but not so small as to increase the number of operations.

Example 1. First, let us consider the linear mixed Fredholm–Volterra equation:

$$u(t,x) = \int_0^t \int_0^2 xe^{-y} u(\tau,y) \, dy \, d\tau + t(e^x - tx), \ t \in [0,1], \tag{28}$$

with exact solution $u^*(t,x) = te^x$.

We take $\rho = 15.5$. We have $K(t,x,\tau,y,u) = xe^{-y}u$, $\dfrac{\partial K}{\partial u} = xe^{-y}$ and the following:

$$LT(b-a) \approx 0.74 < 1,$$
$$M_K T(b-a) \approx 15.37 \leq \rho,$$

so all the hypotheses of Theorem 3 are satisfied. Additionally, for $\rho = 15.5$, we have that $u^* \in B_\rho$.

We consider the two-dimensional trapezoidal rule with $m_1 = 18$ and $m_2 = 36$, with the corresponding nodes $t_i = \dfrac{1}{m_1} i, i = \overline{0,m_1}$ and $x_j = \dfrac{2}{m_2} j, j = \overline{0,m_2}$. Table 1 contains the errors $err(u^*, \tilde{u}_n)$, with initial approximation $u_0(t,x) = f(t,x) = t(e^x - tx)$. The CPU time per iteration is approximately 1.01.

Table 1. Errors for Example 1, $m_1 = 18$, $m_2 = 36$.

n	$err(u^*, \tilde{u}_n)$
1	1.080492×10^{-1}
5	1.210778×10^{-4}
10	5.837723×10^{-6}

Example 2. Next, consider the following nonlinear integral equation:

$$u(t,x) = 2 \int_0^t \int_0^1 x^2 y\tau e^{-\tau} e^{u(\tau,y)} \, dy \, d\tau + x^2(1 - e^{-t}), \ t \in [0, 1/4], \tag{29}$$

whose exact solution is $u^*(t,x) = tx^2$.

Here, $K = \dfrac{\partial K}{\partial u} = 2x^2 y \tau e^{-\tau} e^u$. Thus, for $\rho = 1$, we have the following:

$$LT(b-a) \approx 0.33 < 1,$$
$$M_K T(b-a) \approx 0.33 \leq \rho,$$

thus, Theorem 3 is applicable and $u^* \in B_\rho$.

Again, we use the trapezoidal rule with $m_1 = m_2 = 18$ and nodes $t_i = \dfrac{1}{4m_1}i, i = \overline{0, m_1}$, $x_j = \dfrac{1}{m_2}j, j = \overline{0, m_2}$. The errors $err(u^*, \tilde{u}_n)$ are given in Table 2, with initial approximation $u_0(t,x) = f(t,x) = x^2(1 - e^{-t})$. The CPU time per iteration is approximately 0.89.

Table 2. Errors for Example 2, $m_1 = m_2 = 18$.

n	$err(u^*, \tilde{u}_n)$
1	2.034743×10^{-1}
5	9.354733×10^{-4}
10	3.077314×10^{-5}

Example 3. Last, consider the nonlinear mixed Fredholm–Volterra equation as follows:

$$u(t,x) = 2\int_0^t \int_0^1 x \cos\tau (u(\tau,y))^2 \, dy \, d\tau + \dfrac{x \sin t}{9}(9 - \sin^2 t), \tag{30}$$

for $t \in [0, 1/2]$. The exact solution of Equation (30) is $u^*(t,x) = x \sin t$.

We have $K(t, x, \tau, y, u) = 2xu^2 \cos \tau$ and $\dfrac{\partial K}{\partial u} = 4xu \cos \tau$. Choosing $\rho = 0.3$, we obtain the following:

$$LT(b-a) \approx 0.53 < 1,$$
$$M_K T(b-a) \approx 0.28 \leq \rho,$$

so Theorem 3 can be used and $u^* \in B_\rho$.

Again, the trapezoidal rule is used with $m_1 = m_2 = 18$ and nodes $t_i = \dfrac{1}{2m_1}i, i = \overline{0, m_1}$ and $x_j = \dfrac{1}{m_2}j, j = \overline{0, m_2}$. In Table 3 we give the errors $err(u^*, \tilde{u}_n)$ with initial approximation $u_0(t,x) = f(t,x) = \dfrac{x \sin t}{9}(9 - \sin^2 t)$. The CPU time per iteration is approximately 0.98.

Table 3. Errors for Example 3, $m_1 = m_2 = 18$.

n	$err(u^*, \tilde{u}_n)$
1	2.733605×10^{-1}
5	7.890241×10^{-4}
10	2.766358×10^{-5}

5. Conclusions

We presented a numerical method for approximating solutions of two-dimensional mixed Fredholm–Volterra integral equations of the second kind, using a combination of successive approximations for fixed points and cubature formulas. In this paper, we used Altman's algorithm and the Mann iteration for finding fixed points of an integral operator and the two-dimensional trapezium rule for the numerical integration of the iterates. This has many advantages: in the first place, the fixed point result we used not only guarantees the existence of a unique solution, but also gives a procedure for finding it by successive iterations. Moreover, Mann iterates converge faster than Picard ones (see [24]), so better accuracy is obtained with fewer iterations. In addition, by using the trapezoidal rule, the contraction condition for the integral operator also guarantees the convergence of the numerical approximations. Secondly, the choice of the trapezoidal scheme makes the method easy to use and implement since most mathematical software have this rule built-in. Last, but not least, many popular approximation methods, such as Nyström, collocation, Galerkin or Adomian decomposition methods, lead to difficult-to-solve systems of algebraic equations that are many times ill-conditioned. Such problems are avoided here since the computation of an approximate value only requires the values obtained at the previous step. This reduces the computational and implementation cost of the method. Still, the method proposed converges with order $\mathcal{O}\left(e^{-(1-q)y_n}\right) + \mathcal{O}\left(\frac{1}{m_1^2}\right) + \mathcal{O}\left(\frac{1}{m_2^2}\right)$ (with $\{y_n\}$ given in Equation (12)), producing good resulting approximations as the numerical examples show. On the downside, there are some limitations to the types of equations that this method can be applied to, due to the constraints in Theorem 2.

These ideas can be continued in studying other types of mixed integral equations, such as equations in higher dimensions ($\Omega \subseteq \mathbb{R}^2$ or $\mathbb{R}^3$), equations with singular kernels (arising, for example, in reformulations of the heat equation), or kernels with modified argument, etc. Other types of successive approximations or other numerical integration schemes can also be explored.

Funding: This research did not receive any specific grants from funding agencies in the public, commercial, or not-for-profit sectors.

Institutional Review Board Statement: Not applicable.

Informed Consent Statement: Not applicable.

Data Availability Statement: Data sharing not applicable.

Conflicts of Interest: The author declares no conflict of interest.

References

1. Brunner, H. *Collocation Methods for Volterra Integral and Related Functional Differential Equations*; Cambridge University Press: Cambridge, UK, 2004.
2. Hafez, R.M.; Doha, E.H.; Bhrawy, A.H.; Băleanu, D. Numerical Solutions of Two-Dimensional Mixed Volterra-Fredholm Integral Equations Via Bernoulli Collocation Method. *Rom. J. Phys.* **2017**, *62*, 1–11.
3. Ordokhani, Y.; Razzaghi, M. Solution of nonlinear Volterra-Fredholm-Hammerstein integral equations via a collocation method and rationalized Haar functions. *Appl. Math. Lett.* **2008**, *21*, 4–9.
4. Maleknejad, K.; Mahdiani, K. Solving nonlinear mixed Volterra-Fredholm integral equations with two dimensional block-pulse functions using direct method. *Commun. Nonlinear. Sci. Numer. Simulat.* **2011**, *16*, 3512–3519.
5. Mashayekhi, S.; Razzaghi, M.; Tripak, O. Solution of the Nonlinear Mixed Volterra-Fredholm Integral Equations by Hybrid of Block-Pulse Functions and Bernoulli Polynomials. *Sci. World J.* **2014**, *2014*, 1–8.
6. El-Kalla, I.L.; Abd-Eemonem, R.A.; Gomaa, A.M. Numerical Approach For Solving a Class of Nonlinear Mixed Volterra Fredholm Integral Equations. *Electron. J. Math. Anal. Appl.* **2016**, *4*, 1–10.
7. Micula, S.; Cattani, C. On a numerical method based on wavelets for Fredholm-Hammerstein integral equations of the second kind. *Math. Method. Appl. Sci.* **2018**, *41*, 9103–9115.
8. Aziz, I.; Islam, S.; Khan, F. A new method based on Haar wavelet for the numerical solution of two-dimensional nonlinear integral equations. *J. Comput. Appl. Math.* **2014**, *272*, 70–80.
9. Aziz, I.; Islam, S. New algorithms for the numerical solution of nonlinear Fredholm Volterra integral equations using Haar wavelets. *J. Comput. Appl. Math.* **2013**, *239*, 333–345.
10. Micula, S. A Numerical Method for Weakly Singular Nonlinear Volterra Integral Equations of the Second Kind. *Symmetry* **2020**, *12*, 1862.
11. Micula, S. On some iterative numerical methods for a Volterra functional integral equation of the second kind. *J. Fixed Point Theory Appl.* **2017**, *19*, 1815–1824.
12. Micula, S. A fast converging iterative method for Volterra integral equations of the second kind with delayed arguments. *Fixed Point Theor. RO* **2015**, *16*, 371–380.

13. Ahmadi Shali, J.; Joderi Akbarfam, A.A.; Ebadi, G. Approximate Solutions of Nonlinear Volterra-Fredholm Integral Equations. *Int. J. Nonlin. Sci.* **2012**, *14*, 425–433.
14. Wang, K.; Wang, Q.; Guan, K. Iterative method and convergence analysis for a kind of mixed nonlinear Volterra-Fredholm integral equation. *Appl. Math. Comp.* **2013**, *225*, 631–637.
15. Wazwaz, A.M. A reliable treatment for mixed Volterra-Fredholm integral equations. *Appl. Math. Comp.* **2002**, *127*, 405–414.
16. Islam, S.; Ali, A.; Zafar, A.; Hussain, I. A Differential Quadrature Based Approach for Volterra Partial Integro-Differential Equation with a Weakly Singular Kernel. *CMES Comput. Model. Eng. Sci.* **2020**, *124*, 915–935.
17. Islam, S.; Zaheer, D. Meshless methods for two-dimensional oscillatory Fredholm integral equations. *J. Comput. Appl. Math.* **2018**, *335*, 33–50.
18. Micula, S. On Some Iterative Numerical Methods for Mixed Volterra–Fredholm Integral Equations. *Symmetry* **2019**, *11*, 1200.
19. Atkinson, K.E. *The Numerical Solution of Integral Equations of the Second Kind, Cambridge Monographs on Applied and Computational Mathematics*; Cambridge University Press: Cambridge, UK, 1997.
20. Wazwaz, A.M. *Linear and Nonlinear Integral Equations, Methods and Applications*; Higher Education Press: Beijing, China; Springer: New York, NY, USA, 2011.
21. Bacoțiu, C. *Picard Operators and Applications*; Napoca Star: Cluj-Napoca, Romania, 2008.
22. Sidorov, D.N. Existence and blow-up of Kantorovich principal continuous solutions of nonlinear integral equations. *Diff. Equat.* **2014**, *50*, 1217–1224.
23. Altman, M. A Stronger Fixed Point Theorem for Contraction Mappings. preprint. 1981.
24. Berinde, V. *Iterative Approximation of Fixed Points, Lecture Notes in Mathematics*; Springer: Berlin/Heidelberg, Germany; New York, NY, USA, 2007.
25. Dobrițoiu, M. *Integral Equations with Modified Argument (in Romanian)*; Cluj University Press: Cluj-Napoca, Romania, 2009.

Article

Multistep Methods of the Hybrid Type and Their Application to Solve the Second Kind Volterra Integral Equation

Vagif Ibrahimov [1,2,*] **and Mehriban Imanova** [1,2,3]

1 Computational Mathematics, Baku State University, Baku AZ1148, Azerbaijan; mehriban.imanova@sdf.gov.az
2 Institute of Control Systems Named after Academician A. Huseynov, Baku AZ1141, Azerbaijan
3 Science Development Foundation under the President of the Republic of Azerbaijan, Baku AZ1025, Azerbaijan
* Correspondence: ibvag47@mail.ru

Abstract: There are some classes of methods for solving integral equations of the variable boundaries. It is known that each method has its own advantages and disadvantages. By taking into account the disadvantages of known methods, here was constructed a new method free from them. For this, we have used multistep methods of advanced and hybrid types for the construction methods, with the best properties of the intersection of them. We also show some connection of the methods constructed here with the methods which are using solving of the initial-value problem for ODEs of the first order. Some of the constructed methods have been applied to solve model problems. A formula is proposed to determine the maximal values of the order of accuracy for the stable and unstable methods, constructed here. Note that to construct the new methods, here we propose to use the system of algebraic equations which allows us to construct methods with the best properties by using the minimal volume of the computational works at each step. For the construction of more exact methods, here we have proposed to use the multistep second derivative method, which has comparisons with the known methods. We have constructed some formulas to determine the maximal order of accuracy, and also determined the necessary and sufficient conditions for the convergence of the methods constructed here. One can proved by multistep methods, which are usually applied to solve the initial-value problem for ODE, demonstrating the applications of these methods to solve Volterra integro-differential equations. For the illustration of the results, we have constructed some concrete methods, and one of them has been applied to solve a model equation.

Keywords: Volterra integral equation; multistep method with constant coefficients; degree and stability; advanced multistep methods; hybrid method; multistep second derivative methods; necessary condition for the convergency

Citation: Ibrahimov, V.; Imanova, M. Multistep Methods of the Hybrid Type and Their Application to Solve the Second Kind Volterra Integral Equation. *Symmetry* **2021**, *13*, 1087. https://doi.org/10.3390/sym13061087

Academic Editors: Samad Noeiaghdam, Denis N. Sidorov and Marin Marin

Received: 11 April 2021
Accepted: 12 May 2021
Published: 18 June 2021

Publisher's Note: MDPI stays neutral with regard to jurisdictional claims in published maps and institutional affiliations.

Copyright: © 2021 by the authors. Licensee MDPI, Basel, Switzerland. This article is an open access article distributed under the terms and conditions of the Creative Commons Attribution (CC BY) license (https://creativecommons.org/licenses/by/4.0/).

1. Introduction

It is known that many problems of the natural sciences are reduced to the solving of integral equations of variable boundaries, which are called integral equations of Volterra type. Vito Volterra (proud Italian) fundamentally investigated these equations and also reduced the mathematical models of many problems of the natural sciences to solve these integral equations. As is known, to solve Volterra integral equations is one of the basic directions in modern mathematics. For objectivity, let us note that scientists have met with the need to solve integral equations with variable boundaries before Vito Volterra (see, for example, [1–6]). Now, let us consider the following integral equation of Volterra type:

$$y(x) = f(x) + \int_{\alpha(x)}^{\beta(x)} K(x, s, y(s)) ds, x \epsilon [x_0, X]. \tag{1}$$

One of the popular Volterra integral equations can be written as:

$$y(x) = f(x) + \int_{x_0}^{x} K(x, s, y(s))ds, x_0 \leq s \leq x \leq X. \tag{2}$$

This equation is taken as known, if given the functions $f(x)$ and $K(x, s, z)$.

Suppose that the given functions $f(x)$ and $K(x, s, z)$ are sufficiently smooth and Equation (2) has the unique solution $y(x)$, which is defined on the segment $[x_0, X]$. It follows that the solution $y(x)$ is also a sufficiently smooth function. For the construction of numerical methods to solve Equation (2), let us divide the segment $[x_0, X]$ to N equal parts by using nodes $x_{i+1} = x_i + h$ ($i = 0, 1, \ldots, N$). Here, $0 < h$ is the step size.

As is known, for the solving of Equation (2) there are numerous methods constructed by different authors. There exists one-step and multistep methods constructed for the solving of Equation (2). Let us note that some authors, for the solving of Equation (2), have proposed to use the spline function or collocation methods (see, for example, [7–11]).

There are many works dedicated to the solving of integral equations, which have used the quadrature methods (in [12] for the calculation of definite integrals proposed to use the new way). Note that for solving Volterra integral equations, many authors constructed methods which are different from the above noted (see, for example, [13–16]). It is known that in this case the number of calculations increases when going from the current point to the next. By taking into account this property, in [13] they have constructed a method which is released from the indicated disadvantages. By generalization of this method, here we have constructed more exact stable methods, which we have applied to solve Equation (2). For the presentation of the essence of these methods, let us consider the partial case of Equation (2) which is obtained when replacing $K(x, s, y) = \varphi(s, y)$. In this case Equation (2) can be written as the following:

$$y(x) = f(x) + \int_{x_0}^{x} \varphi(s, y(s))ds. \tag{3}$$

It is not hard to understand that solving this equation is equivalent to solving the following initial-value problem for ODEs of the first order:

$$y'(x) = f'(x) + \varphi(x, y(x)), \quad y(x_0) = f(x_0). \tag{4}$$

It follows from here that the solution of the integral equation of (3) and the initial-value problem (4) can be found by one and the same method. It is not the only case, when the initial-value problem for ODEs and Volterra integral equations can be solved by one and the same methods. To show this, let us consider the following case, when the function of $K(x, s, y)$ is degenerate and can be presented in the following form:

$$K(x, s, y) = \sum_{j=1}^{m} a_j(x) b_j(s, y). \tag{5}$$

By taking this in Equation (2), we receive:

$$y(x) = f(x) + \sum_{j=1}^{m} a_j(x) v_j(x).$$

The function $v_j(x)$, $j = 0, 1, \ldots, m$ can be determined as the solution of the following problem:

$$v_j'(x) = b_j(x, y(x)), \quad v_j(x_0) = 0 \ (j = 1, 2, \ldots, m). \tag{6}$$

It is clear that by solving the system of ODEs (6), one can find the values of the function $y(x)$ at the nodes (mesh points) by using the solution of the system (6). Note that in this case this system of ODEs and the integral equation of (2) can be solved by one and the same methods. Thus, there are some domains in which the integral equation of (2) and the initial-value problem for ODEs can be solved by one and the same methods. Here, it is shown that this domain can be extended and the error received in this case can be estimated (see, for example, [17–19]).

It is not difficult to prove that by using the Lagrange interpolation polynomial the function $K(x, s, y)$ can be presented as:

$$K(x, s, y) = \sum_{j=1}^{k} l_j(x) b_j(s, y) + R_k(x), \tag{7}$$

where $l_j(x)$ ($i = 1, 2, \ldots, m$) are the basic Lagrange function and $R_m(x)$ is the remainder term. By comparison of the equality of (5) and (7) we receive some connection between them.

2. Construction of Multistep Methods to Solve Both Equations (2) and (4)

Let us note that the known multistep method with constant coefficients can be applied to solve the Volterra integral equation. In the result of which, one can constructed by the following method (see, for example, [17–19]):

$$\sum_{i=0}^{k} \alpha_i y_{n+i} = \sum_{i=0}^{k} \alpha_i f_{n+i} + h \sum_{i=0}^{k} \sum_{j=i}^{k} \beta_i^{(j)} K(x_{n+j}, x_{n+i}, y_{n+i}). \tag{8}$$

For the construction methods with the improved properties, here we have used the generalization of the multistep methods, which can be written as the following (see, for example, [20–28]):

$$\sum_{i=0}^{m} \alpha_i y_{n+i} = h \sum_{i=0}^{k} \beta_i y'_{n+i} \quad (n = 0, 1, \ldots, N-1; l = \max(m, k)). \tag{9}$$

For the value $m \geq k$ from the method of (9), it follows the known multistep methods, but for the value $m < k$ it follows advanced methods (formally), but in reality these methods do not depend on each other. Therefore, each of them is an independent object of investigation. The stable advanced method is more accurate ($p \leq k + l + 1$, for $k \geq 3l$, and $m = k - l$, here, p is the degree and k is the order of finite-difference method (9)) (see, for example, [23,24]). Let us note that by Dahlquist's laws, there exists stable methods of type (8) which have the degree $p_{max} = 2[k/2] + 2$. Here, we use the conceptions of the stability, degree, and order, defined by Dahlquist (see, for example, [21–28]). By the above-described way, we find that the stable methods of the advanced type are more exact than the stable multistep methods. However, unstable multistep methods are more exact than the advanced method. Namely, $p \leq 2k$ for method (8) and $p \leq 2k - l$ for method (9).

Advanced methods have been constructed by Kouella for the calculation of the return of Holliley's comet. Note that some advanced concrete methods have been constructed by known scientists such as Laplace, Steklov, etc. However, all stable advanced methods constructed by different specialists had the degree $p \leq 2[k/2] + 2$. That is, they obeyed the law of Dahlquist. The advantages and disadvantages of advanced methods have been shown in [23,24], and for the correction of some of the disadvantages of the advanced method there have been constructed special predictor–corrector methods (see, for example, [29]). Note that if $m > k$ in this case method (9) will be explicit and the maximal value for stable explicit methods can be found by the formula $p \leq m$. Thus, it is proven that stable advanced methods are more exact than the explicit and implicit methods of type (9). Now, let us investigate these methods by using their other properties.

If method (9) is applied to solve the problem (4), then we receive:

$$\sum_{i=0}^{m} \alpha_i(y_{n+i} - f_{n+i}) = h \sum_{i=0}^{k} \beta_i \varphi_{n+i}. \tag{10}$$

where $f_m = f(x_m)$, $\varphi_m = \varphi(x_m, y_m)$, and α_j, β_i ($j = 0, 1, \ldots, m; i = 0, 1, \ldots, k$) are the coefficients of method (10) or (9).

Let us input $m = k$. In this case, for the value $\alpha_k \neq 0$, we receive the implicit method if $\beta_k \neq 0$. If we compare these methods, then we find that the explicit ($\beta_k = 0$) methods can be applied to solve some problems. However, in the application of implicit methods arises some difficulties for elimination, of which here it is proposed to use the predictor–corrector methods. Now, let us consider the application of the advanced methods. For this, we have $k = m + l$ and $\alpha_{k-l} \neq 0$. In this case, method (10) can be written as:

$$\sum_{i=0}^{m} \alpha_i(y_{n+i} - f_{n+i}) = h \sum_{i=0}^{m-1} \beta_i \varphi_{n+i} + h \sum_{i=m}^{m+l} \beta_i \varphi_{n+i}. \tag{11}$$

In the application of this method arises some difficulties related to the calculation of the second part, which is located on the right hand side of the equality of (11).

It is evident that for the calculation of the second sum on the right hand side of equality (11), we need to define the values $y_{n+m}, y_{n+m+1}, \ldots, y_{n+m+l}$. This difficulty can be solved by using some methods for the calculation of values y_{n+m+j} ($0 \leq j \leq l$). For this aim, one can use the predictor–corrector methods. Let us note that if method (11) is implicit, then it will take place that $l = 0$. In this case, one can also use predictor–corrector methods (see [29]). For the sake of objectivity, let us note that for using method (11), one can use the same predictor–corrector method in both the cases $l = 0$ and $l \neq 0$. Therefore, in the using of method (11), additional difficulties do not arise for the case $l > 0$.

Let us note that method (10) can be applied to the solving of Equation (2), from the results of which one can receive the following:

$$\sum_{i=0}^{m} \alpha_i(y_{n+i} - f_{n+i}) = h \sum_{i=0}^{k} \sum_{j=i}^{k} \beta_i^{(j)} K(x_{n+j}, x_{n+i}, y_{n+i}). \tag{12}$$

It is clear that this method can be written as method (11), and in the application of them to solve some problems can be used in the above-described way. Therefore, let us consider the determination of the values of the coefficients α_i, $\beta_i^{(j)}$ ($i, j = 0, 1, \ldots, k$), as the basic properties of the multistep methods depend on the values of their coefficients. For this aim let us suppose that by any methods we have found the values of the coefficients α_i, $\beta_i^{(j)}$ ($i, j = 0, 1, \ldots, k$), by the choosing of which method (12) can have the degree of p. The conception of degree can be defined by the following way:

Definition 1. *The integer p is called the degree for method (12), if the following holds:*

$$\sum_{i=0}^{m} \alpha_i(y(x_{n+i}) - f(x_{n+i})) = h \sum_{i=0}^{k} \sum_{j=i}^{k} \beta_i^{(j)} K(x_{n+j}, x_{n+i}, y(x_{n+i})) + O(h^{p+1}), \; h \to 0. \tag{13}$$

where $y(x_m)$ is the exact value of the solution of the problem (2) at the point x_m ($m \geq 0$).

By the above-described way we have constructed methods (8) and (12) to solve the Volterra integral equation of the second kind, presented by the equation of (2). It is known that both theoretical and practical interests are stable methods with a high order of accuracy. Therefore, let us define the maximum values of the degree for the methods (8) and (12). For this, let us consider the following theorem:

Theorem 1. *If the methods (8) and (12) have the degree of p and p_1, respectively, then $1 \leq p \leq 2k$ and $1 \leq p_1 \leq k + m$.*

If methods (8) and (12) are stable, then the following holds:

$$1 \leq p \leq 2[k/2] + 2,\ 1 \leq p_1 \leq k + l + 1\ (if\ m = k - l\ and\ k \geq 3l).$$

Proof. It is obvious that the theorem also holds in the case $K(x, s, y) = \varphi(s, y)$. In this case, methods (8) and (12) will match with method (9) (and in the case $m = k$). By Dahlquist's rule we find that $p_1 \leq k + m$, and the method with the degree $p_{max} = m + k$ (and also for the case $m = k$) is unique. If these methods are stable, then there are methods with the degree:

$$p \leq 2[k/2] + 2,\ p_1 \leq k + l + 1,$$

for all the values of K. □

Generally speaking, there is no uniqueness for the methods of type (8) and (12) from the corresponding conditions (see [20,23]).

From here we find that the local truncation error for this method can be presented as $O(h^{p+1})$. It is not difficult to understand that the equality of (13) will also hold in the case when $K(x, s, y) = \varphi(s, y)$. In this case we find that the integral equation of (2) will be same with the equation of (3) (see, for example, [1,30–34]). By the above-described way we have proved that the solution of the integral equation of (3) coincides with the solution of the initial-value problem for ODEs of the first order, which have been written as the problem of (4). It follows that to the solving of the Equation (3) can been applied the methods constructed for solving the initial-value problem for ODEs. Taking into account that in the method of (12) one can replace the function of $K(x, s, y)$ with the function of $\varphi(s, y)$, then we find that in this case from method (12) one can receive the following:

$$\sum_{i=0}^{m} \alpha_i y_{n+i} = \sum_{i=0}^{m} \alpha_i f_{n+i} + h \sum_{i=0}^{k} \sum_{j=i}^{k} \beta_i^{(j)} \varphi(x_{n+i}, y_{n+i}). \tag{14}$$

If in the method of (14) we use the next replacement:

$$\sum_{j=i}^{k} \beta_i^{(j)} = \beta_i\ (i = 0, 1, \ldots, k), \tag{15}$$

then the receiving method will be same as method (10).

If we assume that the coefficients β_i ($0 \leq i \leq k$) are known, then it follows that (15) is the system of linear algebraic equations. Note that the solution of this system is not unique. Generally speaking, finding the solution of system (15) is not difficult. As seen from here, the value of the degree p is independent from the coefficients of $\beta_i^{(j)}$ ($i, j = 0, 1, \ldots, k$). Let us prove that the value of p depends on the values of the coefficients of α_i, β_i ($i = 0, 1, \ldots, k$). To illustrate this, here we propose to use the following Taylor series:

$$y(x + ih) = y(x) + ihy'(x) + \frac{(ih)^2}{2!}y''(x) + \cdots + \frac{(ih)^p}{p!}y^p(x) + O(h^{p+1}), \tag{16}$$

$$y'(x + ih) = y'(x) + ihy''(x) + \frac{(ih)^2}{2!}y'''(x) + \cdots + \frac{(ih)^{p-1}}{(p-1)!}y^p(x) + O(h^p). \tag{17}$$

By taking into account these series in (14), we receive the following:

$$\sum_{i=0}^{m} \alpha_i(y(x+ih) - f(x+ih)) - h \sum_{i=0}^{k} \beta_i \varphi(x+ih) = \sum_{i=0}^{m} \alpha_i((y(x) - f(x)) + h(\sum_{i=0}^{m} i\alpha_i - \sum_{i=0}^{k} \beta_i)(y(x) - f(x))'$$
$$+ h^2 \left(\sum_{i=0}^{m} \frac{i^2}{2!}\alpha_i - \sum_{i=0}^{k} \beta_i \right)(y(x) - f(x))'' + \cdots + h \left(\sum_{i=0}^{m} \frac{i^2}{2!}\alpha_i - \sum_{i=0}^{k} \frac{i^{p-1}}{(p-1)!}\beta_i \right)(y(x) - f(x))^p \quad (18)$$
$$+ O(h^{p+1}) = 0$$

where $x = x_0 + nh$ is a fixed point and $(y(x) - f(x))^{(j)} = \varphi^j(x, y)(j = 0, 1, \ldots, p)$ (it follows from here the equality which is similar to Equation (4)).

Suppose that the following equalities hold:

$$\sum_{i=0}^{m} \alpha_i = 0; \quad \sum_{i=0}^{k} \beta_i = \sum_{i=0}^{m} i\alpha_i; \quad \sum_{i=0}^{k} i\beta_i = \sum_{i=0}^{m} \frac{i^2}{2!}\alpha_i, \ldots, \sum_{i=0}^{k} \frac{i^{p-1}}{(p-1)!}\beta_i = \sum_{i=0}^{m} \frac{i^p}{p!}\alpha_i. \quad (19)$$

Then, from (18), we receive the following:

$$\sum_{i=0}^{m} \alpha_i(y(x+ih) - f(x+ih)) - h \sum_{i=0}^{k} \beta_i \varphi(x+ih) = O(h^{p+1}), h \to 0, \quad (20)$$

where $x = x_0 + nh$ is a fixed point.

In this case we find that the method of (14) has the degree of p. Now, let us prove that if the asymptotic equality of (20) holds, then the system of algebraic Equation (19) will have a solution. It is not hard to understand that, if the asymptotic equality of (20) holds, then we find that the following also holds:

$$\sum_{i=0}^{m} \alpha_i(y(x) - f(x)) + h(\sum_{i=0}^{m} i\alpha_i - \sum_{i=0}^{k} \beta_i)(y'(x) - f'(x)) + h^2(\sum_{i=0}^{m} \frac{i^2}{2!}\alpha_i - \sum_{i=0}^{k} i\beta_i)(y''(x) - f''(x)) + \cdots$$
$$+ h^p(\sum_{i=0}^{m} \frac{i^p}{p!}\alpha_i - \sum_{i=0}^{k} \frac{i^{p-1}}{(p-1)!}\beta_i)(y^{(p)}(x) - f^{(p)}(x)) = 0 \quad (21)$$

Let us consider the following notation:

$$z(x) = y(x) - f(x).$$

It is known that if $z(x)$ is a sufficiently smooth function, then $z(x), z'(x), \ldots, z^p(x)$ is the independent linear system, if $z^{(j)}0$ $(0 \leq j \leq p)$. If we take this into account in the equality of (21), then from that it follows the system of (19). It follows from here that the fulfillment of the condition (19) for the coefficients of method (14) is necessary, and is a sufficient condition for the holding of the asymptotic equality of (20). Thus, we have proved the following lemma:

Lemma 1. *In order for the method of (14) to have a degree p, the satisfaction of its coefficients by the system of algebraic Equation (19) is necessary and sufficient.*

In the system of (19) there are $k + m + 2$ unknowns and $p + 1$ equations. Equation (19) is a system of linear algebraic equations and in the case $p + 1 = k + m + 2$, the determinant of this system is nonzero (in this case receiving the Vandermond determinant). As was noted above, the condition $\alpha_m \neq 0$ must hold. By taking into account this condition, we receive $p + 1 = k + m + 2$. It follows that $p \leq k + m$. In the case $m = k$, from here we receive Dahlquist's rule, which can be written as $p \leq 2k$ ($p_{max} = 2k$ or $p_{max} = k + m$). One can prove that the method with the degree $p_{max} = k + m$ is unique (see, for example, [33–36]).

Comment 1. *In the above-described method for the comparison of the advanced and multistep methods, we usually use the values of the variables m and k. Note that the advanced methods can not be received from the multistep methods. For the proving of this, let us consider the following k-step methods:*

$$\sum_{i=0}^{k} \alpha_i y_{n+i} = h \sum_{i=0}^{k} \beta_i y'_{n+i}. \tag{22}$$

We usually suppose that $\alpha_k \neq 0$, which has a relation with finding the value y_{n+k} as the solution of the finite-difference equation of (22). From here, we find that in the case $\alpha_k = 0$, the equality (22) is transferable to the other class method. As was noted, if method (22) is stable and $\alpha_k \neq 0$, then $p \leq 2[k/2] + 2$, and there are stable methods with the degree $p_{max} = 2[k/2] + 2$ for all the values of the order k.

Note that here we used the following definition for the stability:

Definition 2. *Method (9) is called stable if the roots of the following polynomial*

$$\rho(\lambda) = \alpha_m \lambda^m + \alpha_{m-1} \lambda^{m-1} + \cdots + \alpha_1 \lambda + \alpha_0$$

lie inside the unit circle, on the boundary of which there are no multiply roots.

Comment 2. *As was noted above for method (9), the condition $p \leq k + m$ holds. If $m = k - l$ ($l > 0$) then we receive $p \leq 2k - l$. However, if method (9) is stable, then $p \leq k + l + 1$. It is not hard to understand that the linear parts of the methods, which are investigated here, are the same. Therefore, the conception of stability for them is defined in one and the same way. It follows from here that the methods (9) and (22) are independent from each other, because these methods are the independent objects of research.*

For the construction of methods with a high order of accuracy or higher degrees, the hybrid method is often used (see, for example, [37–43]). Therefore, let us consider the following paragraph.

3. Construction of a Generalized Hybrid Method and Its Application

Let us remember some of the popular methods, which have been applied to solve the problem (4). Among of them are the Euler methods (explicit and implicit) and trapezoidal and midpoint rules. The midpoint rule differs from others in that this method uses the calculation of variables of the type $y(x_n + h/2)$. This variable can be written in a more general form as $y(x_n + v_i h)$, ($|v_i| < 1$, $i = 0, 1, 2, \ldots, k$). By the generalization of the midpoint rule, one can construct the following hybrid method:

$$\sum_{i=0}^{k} \alpha_i (y_{n+i} - f_{n+i}) = h \sum_{i=0}^{k} \beta_i \varphi_{n+i+v_i} \ (|v_i| < 1, \ i = 0, 1, 2, \ldots, k). \tag{23}$$

In the work of [42], they constructed a hybrid method with the degree $p_{max} = 4$, which can be received from method (23) in the case $k = 1$. However, from the multistep method (22) for the case $k = 1$, one can receive the method with the degree $p_{max} = 2$. By simple comparison, we find that the hybrid methods can be taken as the perspective. From Equation (23) one can receive the midpoint rule, which can be taken as the explicit method. It is known that this method has the degree $p = 2$. However, as was noted above from the method of (22), one can receive the explicit method with the degree $p_{max} = 1$ for $k = 1$. This comparison shows that the hybrid methods have some advantages over all the known methods. Note that some hybrid methods have the extended region of stability and all the methods investigated here are linear multistep methods; therefore, the linear part of these methods has the same properties. It follows that the conception of stability and degree can be defined in the same way for all linear methods. Let us note that the values of the coefficients α_i ($i = 0, 1, \ldots, k$) can be different from the corresponding coefficients of the other methods.

Now, let us define the values of the coefficients α_i, β_i, $\nu_i (i = 0, 1, \ldots, k)$. To this end, let us use the following Taylor series:

$$y'(x + l_i h) = y'(x) + l_i h y''(x) + \frac{(l_i h)^2}{2!} y'''(x) + \cdots + \frac{(l_i h)^{p-1}}{(p-1)!} y^p(x) + O(h^p) \quad (24)$$

By taking into account Equations (16) and (24) in the method of Equation (23), we receive:

$$\sum_{i=0}^{k} (\alpha_i(y(x+ih) - f(x+ih)) - h\beta_i \varphi(x + (i+\nu_i)h, y(x + (i+\nu_i)h))$$
$$= \sum_{i=0}^{k} \alpha_i(y(x) - f(x)) + h \sum_{i=0}^{k} (i\alpha_i - \beta_i)(y'(x) - f'(x)) + h^2 \sum_{i=0}^{k} (\frac{i^2}{2!}\alpha_i - l_i\beta_i)(y''(x) \quad (25)$$
$$- f''(x)) + \cdots + h^p \sum_{i=0}^{k} (\frac{i^p}{p!}\alpha_i - \frac{l_i^{p-1}}{(p-1)!}\beta_i)(y^{(p)}(x) - f^{(p)}(x)) + O(h^{p+1}) = 0$$

where $x = x_0 + nh$ is a fixed point and $l_i = i + \nu_i$ $(i = 0, 1, \ldots, k)$.

By using the discussion, which we have used in the investigation of method (9), and taking into account the comparison of asymptotic equality (18) with asymptotic equality (25), we receive the following system for finding the determined values of the coefficients α_i, β_i, $\nu_i (i = 0, 1, \ldots, k)$:

$$\sum_{i=0}^{k} \alpha_i = 0; \quad \sum_{i=0}^{k} \beta_i = \sum_{i=0}^{k} i\alpha_i; \quad \sum_{i=0}^{k} (i+\nu_i)\beta_i = \sum_{i=0}^{m} \frac{i^2}{2!}\alpha_i, \ldots, \sum_{i=0}^{k} \frac{(i+\nu_i)^{p-1}}{(p-1)!}\beta_i = \sum_{i=0}^{m} \frac{i^p}{p!}\alpha_i. \quad (26)$$

This is a nonlinear system of algebraic equations. In this system there are $3k + 3$ unknowns, but the amount of equations in this system is equal to $p + 1$. Note that this is a nonlinear system of algebraic equations, because defining the exact solution of such a system is not easy. By taking this into account, scientists proposed to use some numerical methods for solving them. For this, they used Mathcard 2015. Note that by using the approximate solution of system (26), they have constructed some methods with the degree of p. The application of some of them to solving model problems has shown that in reality, the received results correspond to the results received by the methods with a degree less than p. Therefore, finding a private solution of system (26) is very important. Let us note that these results correspond to the theoretical.

The system of (26) to remember the nonlinear system of algebraic equations is used for finding the coefficients of the Gauss method. Therefore, some solutions of this system will be also solutions of the corresponding Gauss system which is used for finding Gauss nodes and coefficients (see, for example, [43–49]).

To solve system (26) is more simple than the corresponding Gauss system, and usually by the solution of (26) one can construct hybrid methods which are different from Gauss methods. By taking into account these properties, here we have proposed to construct stable methods with high degrees. For this, let us consider the following method:

$$\sum_{i=0}^{k} \alpha_i y_{n+i} = \sum_{i=0}^{k} \alpha_i f_{n+i} + h \sum_{i=0}^{k} \beta_i \varphi_{n+i} + h \sum_{i=0}^{k} \gamma_i \varphi_{n+i+\nu_i}. \quad (27)$$

One can consider this method as the linear combination of methods (10) and (23). It is easy to prove that the system of algebraic equations which has been constructed for finding the coefficients of method (27) can be constructed as the linear combination of systems (26) and (19), respectively. By the generalization of the system (26) or (19), we receive the following system of algebraic equations:

$$\sum_{i=0}^{k} \alpha_i = 0; \quad \sum_{i=0}^{k} (\beta_i + \gamma_i) = \sum_{i=0}^{k} i\alpha_i; \quad \sum_{i=0}^{k} (i\beta_i + l_i\gamma_i) = \sum_{i=0}^{m} \frac{i^2}{2!}\alpha_i, \ldots, \sum_{i=0}^{k} \left(\frac{(i)^{p-1}}{(p-1)!}\beta_i + \frac{l_i^{p-1}}{(p-1)!}\gamma_i \right)$$
$$= \sum_{i=0}^{m} \frac{i^p}{p!}\alpha_i, \ (l_i = i + \nu_i, \ 0 \le i \le k). \quad (28)$$

By taking into account the series (16), (17), and (24) in the equality of (27), one can receive the system of (28), in the case when the method will have the degree of p. Now, let us investigate the solvability of the system (28).

As was noted above, the system (28) is nonlinear, therefore to find the exact solution of that is perhaps not always possible. Hence, let us consider an investigation of the system of (28). In this system participates $4k+4$ unknowns and $p+1$ nonlinear algebraic equations. Let us consider the case $p+1 \leq 4k+4$. Note that without breaking the generality one can take $\alpha_k = 1$. By taking this into account, let us investigate the system of (28) for $p \leq 4k+2$. It is clear that in the case of $k=1$, there does not arise a question on the stability of the methods of type (27). Let us in the system of (28) take $k=1$. In this case, from system (28) we receive the following (by taking into account the condition $\alpha_1 = 1$, we receive $\alpha_0 = -1$):

$$\beta_0 + \gamma_0 + \beta_1 + \gamma_1 = 1;\ \beta_1 + l_0^j \gamma_0 + l_1^j \gamma_1 = 1/(j+1)\ (1 \leq j \leq 5). \tag{29}$$

By using the solution of system (29) one can construct the following method:

$$y_{n+1} = y_n + h(y'_{n+1} + y'_n)/12 + 5h(y'_{n+\beta} + y'_{n+1-\beta})/12,\ \beta = (5 - \sqrt{5})/10. \tag{30}$$

This method has the degree $p = 6$. Similar investigations have been given by some authors (see, for example, [32,39–42,45]). For the application of this method to solve Equation (2), it can be presented in the following form:

$$y_{n+1} - f_{n+1} = y_n - f_n + h(K(x_{n+1}, x_{n+1}, y_{n+1}) + K(x_{n+1}, x_n, y_n) + 2K(x_n, x_n, y_n))/24$$
$$+ 5h(K(x_{n+1}, x_{n+\beta}, y_{n+\beta}) + K(x_{n+\beta}, x_{n+\beta}, y_{n+\beta}) + K(x_{n+1}, x_{n+1-\beta}, y_{n+1-\beta}) \tag{31}$$
$$+ + K(x_{n+1-\beta}, x_{n+1-\beta}, y_{n+1-\beta}))/24.$$

In the construction, method (30) has used the solution of system (29) and the solution of system (15) by the addition of the following:

$$\sum_{j=1}^{k} \gamma_i^{(j)} = \gamma_i\ (i = 0, 1, \ldots, k). \tag{32}$$

By this way one can construct the following multistep hybrid method:

$$\sum_{i=0}^{k} \alpha_i (y_{n+i} - f_{n+i}) = h \sum_{i=0}^{k} \sum_{j=i}^{k} \beta_i^{(j)} K(x_{n+j}, x_{n+i}, y_{n+i}) +$$
$$+ h \sum_{i=0}^{k} \sum_{j=i}^{k} \gamma_i^{(j)} K(x_{n+i+v_j}, x_{n+i+v_i}, y_{n+i+v_i}),\ (|v_i| < 1; i = 0, 1, \ldots, k) \tag{33}$$

by taking into account the solutions of the systems (15) and (32).

Hence, note that method (31) can be received from the method of (33) as the partial case. The solution of systems (15) and (32) is not unique, and the order of accuracy of method (33) is independent from the solution of mentioned systems. As was proved above, the exactness of the methods of type (33) depends on the values of the coefficients β_i, γ_i, and v_i ($i = 0, 1, \ldots, k$), which can be found as the solution of system (28).

For the construction of methods of type (33), with high accuracy, let us consider the case of $k = 2$.

At first, let us define the values of $\alpha_i (i = 0, 1, 2)$ by taking into account that the constructed method must be stable. It is clear that in this case the roots of the following polynomial:

$$\lambda^2 + \alpha_1 \lambda + \alpha_0 = 0$$

must satisfy the condition of stability. Here, we have considered the following variant:

$$\alpha_1 = 0,\ \alpha_0 = -1,\ \alpha_2 = 1.$$

In this case, by using the solution of nonlinear system (28) and taking into account the solution of systems (15) and (32), one can construct the following method:

$$y_{i+2} = y_i + h(9y'_{i+2} + 64y'_{i+1} + 9y'_i)/90 + 49h(y'_{n+1+\alpha} + y'_{n+1-\alpha})/90, \ \alpha = \sqrt{21}/7. \quad (34)$$

For the application of this method to solve nonlinear Volterra integral equations of second order, that can be modified as following:

$$\begin{aligned} y_{i+2} = y_i + f_{i+2} - f_i &+ h(9K(x_{i+2}, x_{i+2}, y_{i+2}) + 32K(x_{i+2}, x_{i+1}, y_{i+1}) + 32K(x_{i+1}, x_{i+1}, y_{i+1}) + 5K(x_{i+1}, x_i, y_i) \\ &+ 4K(x_i, x_i, y_i))/90 + 49h(K(x_{n+2}, x_{n+1+\alpha}, y_{n+1+\alpha}) \\ &+ K(x_{n+1+\alpha}, x_{n+1+\alpha}, y_{n+1+\alpha})) + K(x_{n+1}, x_{n+1-\alpha}, y_{n+1-\alpha}) + K(x_{n+1-\alpha}, x_{n+1-\alpha}, y_{n+1-\alpha}))/180 \end{aligned} \quad (35)$$

It is not difficult to prove that the method of (34) can be presented in another form, which will be different from the formula (35). As was noted above, for the construction of more exact methods one can use advanced (forward-jumping) methods. However, some specialists, for the construction of more exact methods, proposed using the multistep second derivative methods with constant coefficients. For receiving some information about these methods, one can use the content of the following section.

4. On Some Properties of Multistep Second Derivative Methods with Constant Coefficients

In the last sections, we have given some information about advanced methods which have comparisons with multistep methods. Note that multistep second derivative methods can also be the advanced type. By taking into account this property, let us consider the following multistep second derivative methods, which are fundamentally investigated by some authors (see, for example, [46–48]):

$$\sum_{i=0}^{k} \alpha_i y_{n+i} = h \sum_{i=0}^{k} \overline{\beta_i} y'_{n+i} + h^2 \sum_{i=0}^{k} \overline{\gamma_i} y''_{n+i}. \quad (36)$$

This method, after application to the solving of problem (2), can be presented as the following:

$$\sum_{i=0}^{k} \alpha_i(y_{n+i} - f_{n+i}) = h \sum_{i=0}^{k} \sum_{j=i}^{k} \overline{\beta}_i^{(j)} K(x_{n+j}, x_{n+i}, y_{n+i}) + h^2 \sum_{i=0}^{k} \sum_{j=i}^{k} \overline{\gamma}_i^{(j)} G(x_{n+j}, x_{n+i}, y_{n+i}) \quad (37)$$

where the function of $G(x, x, y)$ is defined as: $G(x, x, y) = \frac{d}{dx} K(x, s, y(s))\big|_{s=x}$. Depending on the used way to construct method (37), the function $G(x, z, y)$ can be defined in another form, but the received results of which will be the same with the method of (37). It follows to note that if in method (37) we input $K(x, s, y) = \varphi(s, y)$, then method (36) can be received from method (37) as the partial case.

To explain the above description, it is enough to apply methods (36) and (37) to solve the following problem:

$$y' = \varphi(x, y), \ y(x_0) = y_0, \ x_0 \leq x \leq X.$$

In this case we receive:

$$\sum_{i=0}^{k} \alpha_i y_{n+i} = h \sum_{i=0}^{k} \overline{\beta_i} \varphi_{n+i} + h^2 \sum_{i=0}^{k} \overline{\gamma_i} g_{n+i}, \quad (38)$$

$$\sum_{i=0}^{k} \alpha_i y_{n+i} = h \sum_{i=0}^{k} \sum_{j=i}^{k} \overline{\beta}_i^{(j)} \varphi_{n+i} + h^2 \sum_{i=0}^{k} \sum_{j=i}^{k} \overline{\gamma}_i^{(j)} g_{n+i}, \quad (39)$$

where the function $g(x, y)$ is defined as $g(x, y) = \varphi'_x(x, y) + \varphi'_y(x, y)\varphi(x, y)$.

It is evident that if we take

$$\sum_{j=i}^{k} \overline{\beta}_i^{(j)} = \overline{\beta}_i; \quad \sum_{j=i}^{k} \overline{\gamma}_i^{(j)} = \overline{\gamma}_i \ (i = 0, 1, 2, \ldots, k), \tag{40}$$

then from the formula (39) follows method (38). Therefore, to define the values of the coefficients $\overline{\beta}_i^{(j)}$, $\overline{\gamma}_i^{(j)}$ $(i, j = 0, 1, 2, \ldots, k)$, one can use the system (40), and the system of equations, which are independent from the determination of the coefficients $\overline{\beta}_i$, $\overline{\gamma}_i$ $(i = 0, 1, 2, \ldots, k)$, participate in the formula of (36). By taking this into account, let us consider the definitions of the values of the coefficients α_i, $\overline{\beta}_i$, $\overline{\gamma}_i$ $(i = 0, 1, 2, \ldots, k)$. To this end, one can use the scheme which was used in the construction of system (28). In this case, the system of algebraic equations for finding the values of the coefficients α_i, $\overline{\beta}_i$, $\overline{\gamma}_i$ in one variant can be written in the following form:

$$\sum_{i=0}^{k} \alpha_i = 0; \ \sum_{i=0}^{k} \overline{\beta}_i = i\alpha_i; \ \sum_{i=0}^{k} i\overline{\beta}_i + \sum_{i=0}^{k} \overline{\gamma}_i = \sum_{i=0}^{m} \frac{i^2}{2!} \alpha_i, \ldots, \sum_{i=0}^{k} \frac{i^l}{l!} \overline{\beta}_i + \sum_{i=0}^{k} \frac{i^{l-1}}{(l-1)!} \overline{\gamma}_i = \sum_{i=0}^{m} \frac{i^{l+1}}{(l+1)!} \alpha_i, \ (l = 2, 3, \ldots, p). \tag{41}$$

Note that the system of (28) is nonlinear, but system (41) is linear. By taking into account that the determinant of system (41) is nonzero, we find that if the amount of the unknowns and of the equations are the same, then we find that system (41) has a unique solution. Note that the amount of unknowns in system (41) is equal to $3k + 3$, but the amount of equations is equal to $p + 1$. It follows that if $p < 3k + 1$ then the system of (41) will have any solution, but in the case $p = 3k + 1$, the corresponding solution of system (41) will be unique. It follows from here that $p_{max} = 3k + 1$. Note that the degree p for method (36) can be defined as follows (see, for example, [46–48]):

Definition 3. *The integer p is called the degree for method (36) if the following holds:*

$$\sum_{i=0}^{k} (\alpha_i y(x+ih) - h\overline{\beta}_i y'(x+ih) - h^2 \overline{\gamma}_i y''(x+ih)) = O(h^{p+1}), \ h \to 0. \tag{42}$$

Let us note that this and definition 1 are the same.

It is not difficult to prove that if method (36) is stable then the relationship between k and p (order and degree) can be presented as $p \leq 2k + 2$, and there exists stable methods with the degree $p = 2k + 2$ for all the values of k (order). For the value $k = 1$ we find that the maximal value for stable and unstable methods are the same; in other words, $3k + 1 = 2k + 2$ for $k = 1$. Note that in this case ($k = 1$) there are not any unstable methods, so the one-step method satisfies the condition of stability. By simple comparison we find that the stable methods of type (27) are more exact than the stable methods of type (39), but application of hybrid methods of type (23) or (27) is more difficult. Note that these difficulties are related to the calculation of the values $y_{n+i+\nu_i}$ $(i = 0, 1, \ldots, k)$. It is evident that for the calculation of these values arises the necessity to construct a special method for calculating them. Therefore, to give some advantages of these methods are difficult.

If there exist methods for calculations of the values $y_{n+i+\nu_i} (i = 0, 1, \ldots, k)$, then the methods of hybrid types will have some advantages. It is not difficult to prove that the hybrid methods constructed by using the methods of type (38) will be more exact than the known methods, but will have a more complex structure. As was noted above, the system (41) is linear; therefore, the system (41) can be solved by using the known methods. However, in the increasing the values of order k the values of the calculation in the determination of the values of the coefficients α_i, $\overline{\beta}_i$, $\overline{\gamma}_i$ $(i = 0, 1, 2, \ldots, k)$ also increase. For decreasing the values of the calculation works, here we have proposed a new way for the calculation of the values of the coefficients α_i, $\overline{\beta}_i$, $\overline{\gamma}_i$ $(i = 0, 1, 2, \ldots, k)$ (see [49]). To this end, we required analyticity from the solution of the considered problem. In the work of [48], they proposed a way by which the condition of analyticity of the solution could be simplified and replaced with the conditions that usually are used in other works (see, for example, [48]). For the method to have a degree of p one can use the following way, which is received by using the above-mentioned results:

$$\overline{\alpha_0} = -\rho_0 + \rho_1 - \rho_2 + \cdots + (-1)^{k-1}\rho_{k-2} + (-1)^k \rho_{k-1},$$

$$\overline{\alpha_i} = \sum_{j=i-1}^{k-1} (-1)^{j-i+1}(j+1)j(j-1)\ldots(j-i+2)\rho_j/i!; \ i = 1, 2, \ldots, k.$$

$$\overline{\beta_0} = \delta_0 - \delta_1 + \delta_2 + \cdots + (-1)^{k-1}\delta_{k-1} + (-1)^k \delta_k,$$

$$\overline{\beta_i} = \sum_{j=1}^{k} (-1)^{j-1} j(j-1) \ldots (j-i+1)\delta_j/i!; \ i = 1, 2, \ldots, k. \tag{43}$$

$$\overline{\gamma_0} = l_0 + l_1 - l_2 + \cdots + (-1)^{k-1} l_{k-1} + (-1)^k l_k,$$

$$\overline{\gamma_i} = \sum_{j=1}^{k} (-1)^{j-1} j(j-1) \ldots (j-i+1) l_j/i!; \ i = 1, 2, \ldots, k.$$

The variables ρ_i, δ_i, l_i ($i = 0, 1, 2, \ldots, k$) can be defined from the following system of linear algebraic equations.

$$\sum_{i=0}^{j} c_i \rho_{j-i} + \sum_{i=1}^{j} (-1)^{j-i+1} l_{i-1}/(j-i+1) = \delta_j, \ j = 0, 1, \ldots, k; \ \rho_k = 0, \tag{44}$$

$$\sum_{i=j+1}^{j+k} c_i \rho_{j+k-i} + \sum_{i=j}^{j} (-1)^i l_{i+k-1}/i = 0, \ j = 0, 1, \ldots, k;$$

$$\sum_{i=j+k-1}^{j+2k} c_i \rho_{j+2k-i} + \sum_{i=j+k}^{j+2k} (-1)^i l_{j+2k-i}/i = 0; j = 1, 2, \ldots, k;$$

$$\sum_{i=2k+2}^{3k+1} c_i \rho_{3k+1-i} + \sum_{i=2k+1}^{3k+1} (-1)^i l_{3k+1-i} = C,$$

where C is the constant for the coefficient of the main leading term in the expansion of the error of method (39), but the coefficients c_i ($i = 0, 1, \ldots$) are defined by the following formula:

$$c_i = \frac{1}{i!}\int_0^1 u(u-1)\ldots(u-i+1)du; (i = 1, 2, \ldots), \ c_0 = 1, \ c_1 = 1/26, \ c_2 = -\frac{1}{12}, \ldots$$

It is easy to prove that the coefficients can be calculated by the following formula:

$$c_m = \sum_{i=1}^{m}(-1)^{i-1} c_{m-i}/(i+1) \ (m \geq 1, \ c_0 = 1).$$

Note that someone may think that finding a solution to systems (43) and (44) is more difficult than finding the solution to system (41). However, it is not. As is known, scientists have mostly constructed stable methods, considering that they are convergent. By taking this into account, we find that, basically, one can assume that the values of the quantities ρ_i ($i = 0, 1, \ldots, k-1$) are known. In this case, we find that to solve the system of (44) is simplified. As a result of which, we obtain the actual solution of one system, which is system (43). By using this solution in the system of (42) we can compute the values of the unknowns δ_j ($j = 0, 1, \ldots, k$). By using these values one can find the values of the coefficients α_i, $\overline{\beta_i}$, $\overline{\gamma_i}$ ($i = 0, 1, 2, \ldots, k$). Taking into account the solution of the systems

(42) and (43), here we have constructed stable methods with the degree $p = 8$ for $k = 3$. The constructed stable method with the degree $p = 8$ can be written as the following:

$$y_{n+3} = (y_{n+2} + y_{n+1} + y_n)/3 + h(10,781y'_{n+3} + 22,707y'_{n+2} + 16,659y'_{n+1} + 4285y'_n)/27,216$$
$$-h^2(2099y''_{n+3} - 7227y''_{n+2} - 2853y''_{n+1} - 979y''_n)/45,360 + 3h^9 y_n^{(9)}/156,800 + O(h^{10}). \quad (45)$$

Let us consider a comparison of the hybrid method of type (34) with method (45). These methods have the degree $p = 8$ and are stable. Note that method (34) has the order of $k = 2$, but method of (45) has the order $k = 3$. It follows that for using method (45) we must know the values (y_n, y_{n+1} and y_{n+2}) but for using method (34) we must know two values (y_n and y_{n+1}). For the application of method (45) it is necessary to use one explicit method as the predictor formula. However, for the application of method (34) to solve some problems, it needs to use two methods for the calculation of the values of the type $y(x_m \pm vh)$ ($|v| < 1$). If in the method of (36) or (38) we input $k = 2$ then the degree for the stable methods will hold the condition of $p \leq 6$. It follows that to give some advantages of any of these methods is difficult. Each of them has its own advantages and disadvantages. Now let, us apply method (45) to solve a Volterra integral equation. In this case we receive the following:

$$y_{n+3} = (y_{n+2} + y_{n+1} + y_n)/3 + f_{n+3} - (f_{n+2} + f_{n+1} + f_n)/3 + h(10,781K(x_{n+3}, x_{n+3}, y_{n+3})$$
$$+11,707K(x_{n+3}, x_{n+2}, y_{n+2}) + 11,000K(x_{n+2}, x_{n+2}, y_{n+2}) + 8659K(x_{n+2}, x_{n+1}, y_{n+1})$$
$$+8000K(x_{n+1}, x_{n+1}, y_{n+1}) + 2185K(x_{n+2}, x_n, y_n) + 2100K(x_n, x_n, y_n)/27,216 \quad (46)$$
$$-h^2(2099G(x_{n+3}, x_{n+3}, y_{n+3}) - 7000G(x_{n+3}, x_{n+2}, y_{n+2}) - 227G(x_{n+2}, x_{n+2}, y_{n+2})$$
$$-1453G(x_{n+2}, x_{n+1}, y_{n+1}) - 1400G(x_{n+1}, x_{n+1}, y_{n+1}) - 9006(x_{n+1}, x_n, y_n)$$
$$-79G(x_n, x_n, y_n))/45,360$$

It is not easy to determine the value of y_{n+3} by method (46), so, as in this case, we receive the nonlinear algebraic equation. For solving this equation, here we propose to use the predictor–corrector methods (see, for example, [29]). To this end, one can use the stable explicit methods as the predictor methods. In this case, the degree for this method will satisfy the condition $p \leq 6$, but method (46) has the degree $p = 8$. It follows that for the construction of methods with suitable accuracy, the condition $k \geq 4$ must be held. However, one can use the suitable stable explicit method (46) as the predictor and corrector method.

It is obvious that someone can propose a way to increase the accuracy of the calculated values y_{n+3}, which differ from the above description. Note that similar difficulties arise in the application of the quadrature method to solve the nonlinear Volterra integral equations.

Thus, we have shown that by using the properties of the investigated problem one can choose a suitable method. Lately, the specialists have predominantly used hybrid methods. Here, we have a comparison of the same numerical methods by using the conception stability, degree, and the volume of the computational works on each step. However, some authors, for the comparison of numerical methods, use the conception of the region of stability. This question can be solved by taking into account the results received when using predictor–corrector methods. Usually, the values found by the predictor–corrector methods are used for the definition of the boundaries for the step size $h > 0$.

As is known, for the construction of the multistep methods are usually given the amount of mesh points. Therefore, to find some relation between the order k and the degree p for the investigated multistep methods is very important. It is known that one of the important questions in the investigation of multistep methods is defining the necessary conditions for their convergence. These conditions for method (36) and the methods which are received from that as partial cases, for example, method (22), have been investigated by Dahlquist (see, for example, [27]). By taking this into account, in the next section let us define the necessary condition for the convergence of method (33), which was received by the application of method (27) to solve Volterra integral equations.

5. The Conditions Imposed on Coefficients of Method (33)

Let us suppose that method (33) is convergence and prove that the following conditions are satisfied.

A. The coefficients α_i, β_i, γ_i, ν_i ($i = 0, 1, \ldots, k$) are real numbers and $\alpha_k \neq 0$;
B. The characteristic polynomials:

$$\rho(\lambda) \equiv \sum_{i=0}^{k} \alpha_i \lambda^i; \; \delta(\lambda) \equiv \sum_{i=0}^{k} \beta_i \lambda^i; \; \gamma(\lambda) \equiv \sum_{i=0}^{k} \gamma_i \lambda^i;$$

have no common factor different from the constant;

C. The conditions $P \geq 1$ and $\delta(1) + \gamma(1) \neq 0$ hold.

The necessity of the condition $\alpha_k \neq 0$ is proved above. Therefore, condition A is obvious. Let us consider condition B and suppose otherwise. It follows that the polynomials have a common factor, which differs from the constant. Denote that by $\varphi(\lambda)$. Then, one can write $\varphi(\lambda) const$, and by the E, denote the shift operator. In this case, the following holds: $E^i y(x) = y(x + ih)$ or $\sum_{i=0}^{k} \alpha_i y_{n+i} = \sum_{i=0}^{k} \alpha_i y(x + ih) = \sum_{i=0}^{k} \alpha_i E^i y(x)$, here $x = x_0 + nh$ is a fixed point.

It is easy to see that one can write the following:

$$E^j E^i K(x_n, x_n, y_n) = K(x_{n+j}, x_{n+i}, y_{n+i}) = K(x_{n+i} + (j - i)h, x_{n+i}, y_{n+i}) \quad (47)$$

If for the fixed point $x = x_n + ih$, passing the limit to respect the first argument, then we receive:

$$\lim_{h \to 0} K(x + (j - i)h, x, y(x)) = K(x, x, y(x)).$$

By using this in the equality of (47) we receive:

$$\lim_{h \to 0} E^j E^i K(x_n, x_n, y_n) = E^i K(x_n, x_n, y_n).$$

Hence, $x = x_n + ih$ is fixed.

If we use these properties in the following expression:

$$\lim_{h \to 0} \sum_{i=0}^{k} \sum_{j=i}^{k} \beta_i^{(j)} K(x_{n+j}, x_{n+i}, y_{n+i}) = \sum_{i=0}^{k} \sum_{j=i}^{k} \beta_i^{(j)} K(x_{n+i}, x_{n+i}, y_{n+i}). \quad (48)$$

By taking into account the systems of (15) and (32) in the equality of (48), we receive:

$$\sum_{i=0}^{k} \sum_{j=i}^{k} \beta_i^{(j)} K(x_{n+j}, x_{n+i}, y_{n+i}) = \sum_{i=0}^{k} \beta_i K(x_{n+i}, x_{n+i}, y_{n+i}) + O(h).$$

Thus, we find that method (33) can be written as:

$$\sum_{i=0}^{k} \alpha_i (y_{n+i} - f_{n+i}) - h \sum_{i=0}^{k} \beta_i K(x_{n+i}, x_{n+i}, y_{n+i}) - h \sum_{i=0}^{k} \gamma_i K(x_{n+i+\nu_i}, x_{n+i+\nu_i}, y_{n+i+\nu_i}) = O(h^r), \; (r > 1) \quad (49)$$

From here we receive:

$$\rho(E)(y_n - f_n) - h\delta(E)K(x_n, x_n, y_n) - h\gamma(E)K(x_n, x_n, y_n) = 0. \quad (50)$$

By the above-described way, we prove that the finite-difference Equations (33) and (50) are equivalents. Note that Equation (49) is homogeneous, therefore the solvability theorem for the finite-difference Equations (33) and (50) are the same. By this assumption, we find that the polynomials $\rho(\lambda)$, $\delta(\lambda)$, and $\gamma(\lambda)$ have the common factor, which is denoted by $\varphi(\lambda)$. If we use the function $\varphi(\lambda)$ in the equality of (50), then we receive:

$$\varphi(E)(\rho_1(E)(y_n - f_n) - h\delta_1(E)K(x_n, x_n, y_n) - h\gamma_1(E)K(x_n, x_n, y_n)) = 0. \quad (51)$$

By using the condition $\varphi(\lambda) \neq const$, in the equality of (51) we receive the following:

$$\rho_1(E)(y_n - f_n) - h\delta_1(E)K(x_n, x_n, y_n) - h\gamma_1(E)K(x_n, x_n, y_n) = 0, \qquad (52)$$

where $\varphi(\lambda)\rho_1(\lambda) = \rho(\lambda); \varphi(\lambda)\delta_1(\lambda) = \delta(\lambda); \varphi(\lambda)\gamma_1(\lambda) = \gamma(\lambda)$.

By the simple comparison of Equations (50) and (52) we find that these equations are equivalents. Note that the finite-difference Equation (50) has the order of k, and the order of Equation (52) satisfies the condition $k_1 < k$ (k_1- is the order of Equation (52)). It is known that for the k_1- initial values the finite-difference Equation (52) has the unique solution. It is easy to prove that in this case the solution of the Equation (51) will be unique for the k_1- initial values ($k_1 < k$ satisfies). It is known that the finite-difference equation of the order k has a unique solution if it must be given k- initial values (solvability theorem). Obtaining a contradiction shows that the condition of B takes place. Now, let us prove the validity of the condition C.

It is evident that Equation (50) can be written as:

$$\rho(E)(y(x) - f(x)) = O(h), \qquad (53)$$

where $x = x_0 + nh$ is fixed. If, here, we pass the limit for the $h \to 0$, then we receive:

$$\rho(1) = 0, \qquad (54)$$

so $y(x)f(x)$. This condition is called the necessary condition for the convergence of method (33). By taking the equality of (50), we receive the following:

$$(E - 1)\rho_1(E)(y_n - f_n) - h(\delta(E) + \gamma(E))K(x_n, x_n, y_n) = 0$$

or,

$$\rho_1(E)(y_{n+1} - y_n - f_{n+1} + f_n) - h(\delta(E) + \gamma(E))K(x_n, x_n, y_n) = 0. \qquad (55)$$

By changing the meaning of variable n from zero to m and summing the received equalities, one can find the following:

$$\rho_1(E)(y_{m+1} - y_0 - f_{m+1} + f_0) = (\delta(E) + \gamma(E))h \sum_{l=0}^{m} K(x_l, x_l, y_l). \qquad (56)$$

If, here, we pass the limit for $h \to 0$, and take into account the equality of (48), then we receive:

$$\rho_1(1)(y(x) - y_0 - f(x) + f_0) = (\delta(1) + \gamma(1)) \int_{x_0}^{x} F(x, s, y(s))ds, \qquad (57)$$

where $x = x_0 + mh$ is a fixed point.

By the comparison of Equation (57) with Equation (2) and taking into account $y(x_0) = f(x_0)$, we receive:

$$\rho_1(1) = \delta(1) + \gamma(1). \qquad (58)$$

By using $\rho(1) = 0$, one can write:

$$\rho_1(\lambda) = \frac{\rho(\lambda) - \rho(1)}{\lambda - 1}.$$

Hence, we have:

$$\rho_1(1) = \lim_{\lambda \to 1} \frac{\rho(\lambda) - \rho(1)}{\lambda - 1} \text{ or } \rho_1(1) = \rho'(1).$$

By taking this into account in the equality of (58), one can write the following:

$$\rho'(1) = \delta(1) + \gamma(1). \tag{59}$$

By comparison of equalities (54) and (58) with the first two equations of system (28), we receive the conditions that $p \geq 1$ satisfies. Now, we prove that $\delta(1) + \gamma(1) \neq 0$, and when supposing otherwise, input $\delta(1) + \gamma(1) = 0$. In this case, from Equation (59) we receive $\rho'(1) = 0$. Thus, by using Equation (54) we receive $\rho(1) = \rho'(1) = 0$. It follows from here that $\lambda=1$ is twice the root of the polynomial $\rho(\lambda)$. Now, we prove that in this case method (33) is not convergence. To this end, we use the following error of method (33):

$$\varepsilon_m = y(x_m) - y_m \ (m = 0, 1, 2, \ldots).$$

Let us in the equality of (33) change the approximate values y_m by its exact values. Then, we receive:

$$\sum_{i=0}^{k}(\alpha_i(y(x_{n+i}) - f(x_{n+i})) - h\sum_{j=i}^{k}(\beta_i^{(j)} K(x_{n+j}, x_{n+i}, y(x_{n+i}) + \gamma_i^{(j)} K(x_{n+j+v_j}, x_{n+i+v_i}, y(x_{n+i})))) = R_n. \tag{60}$$

where R_n- is the reminder term.
If we subtract equality (33) from (60), then we receive

$$\sum_{i=0}^{k}(\alpha_i \varepsilon_{n+i} - h\sum_{j=i}^{k}(\beta_i^{(j)} L(\tilde{\xi}_{n+i})\varepsilon_{n+i} + \gamma_i^{(j)} \overline{L}(\overline{\tilde{\xi}}_{n+i}))\varepsilon_{n+i+v_i}) = R_n, \tag{61}$$

where

$$L_i(\tilde{\xi}_{n+i}) = K'_y(x_{n+j}, x_{n+i}, \tilde{\xi}_{n+i}); \ \overline{L}_i(\overline{\tilde{\xi}}_{n+i}) = K'_y(x_{n+j+v_j}, x_{n+i+v_i}, \overline{\tilde{\xi}}_{n+i}),$$

Variable $\tilde{\xi}_{n+i}$ lies between the values of y_{n+i} and $y(x_{n+i})$, but $\overline{\tilde{\xi}}_{n+i}$ lies between the values of y_{n+i+v_i} and $y(x_{n+i+v_i})$, respectively. The Equation (61) is the nonhomogeneous finite-difference equation. The corresponding homogeneous equation has the following form:

$$\sum_{i=0}^{k}\alpha_i \varepsilon_{n+i} = 0. \tag{62}$$

Let us note that one can receive Equation (62) from Equation (61) by going to the limit as the $h \to 0$.

As is known, the general solution of a homogeneous finite-difference equation with constant coefficients can be written as the following:

$$\varepsilon_m = c_1 \lambda_1^m + c_2 \lambda_2^m + \cdots + c_k \lambda_k^m \ (\lambda_i \neq \lambda_j \ if \ i \neq j), \tag{63}$$

where $\lambda_l (l = 0, 1, \ldots, k)$ are the roots of the characteristic polynomial $\rho(\lambda)$. If we use the condition $\rho_1(1) = \rho(1) = 0$, we find that $\lambda = 1$ is twice the root. In this case the solution of Equation (62) can be presented in the following form:

$$\varepsilon_m = c_1 + c_2 m + c_3 \lambda_3^m + \cdots + c_k \lambda_k^m. \tag{64}$$

As follows from here, the error of the investigated method is unbounded. Hence, if in Equation (64) we pass the limit as the step size tends to zero, (i.e., $h \to 0$), then we receive $\varepsilon_m \to \infty$. By this way, we have proved that if $\delta(1) + \gamma(1) = 0$, then the method does not converge, which contradicts the assumption. Thus, we have proved that the conditions A, B, and C take place.

Note that Dahlquist's results can be received from the results received here. It follows that the results obtained here are the development of Dahlquist's rule.

6. Remark

By establishing the direct relation between ODE and Volterra integral equations, we have constructed numerical methods for the solving of both Volterra integral equations and ODE. Now, we want to illustrate that one can construct simple methods for solving ODE and the Volterra integral equation and also Volterra integro-differential equations. For simplicity, we input $f(x) \equiv 0$, and in this case we have:

$$y(x_{n+1}) = \int_{x_0}^{x_{n+1}} K(x_{n+1}, s, y(s)) ds,$$

or,

$$y(x_{n+1}) = \int_{x_0}^{x_n} K(x_{n+1}, s, y(s)) ds + \int_{x_n}^{x_{n+1}} K(x_{n+1}, s, y(s)) ds. \quad (65)$$

The first integral can be presented as the following:

$$\int_{x_0}^{x_n} K(x_{n+1}, s, y(s)) ds = \int_{x_0}^{x_n} K(x_n, s, y(s)) ds + h \int_{x_0}^{x_n} K'_x(\xi_n, s, y(s)) ds \ (x_n < \xi_n < x_n + h).$$

By taking into account that the function $K(x, s, y)$ and its derivatives are bounded, then for the fixed point $x_0 + nh$, we receive:

$$\lim_{h \to 0} \int_{x_0}^{x_n} K(x_{n+1}, s, y(s)) ds = \int_{x_0}^{x_n} K(x_n, s, y(s)) ds.$$

By taking this in Equation (65) for the calculation of the value $y_{n+1} \approx y(x_{n+1})$, we receive the following:

$$y_{n+1} = y_n + \int_{x_n}^{x_{n+1}} K(x_{n+1}, s, y(s)) ds. \quad (66)$$

From here one can obtain the following methods:

$$y_{n+1} = y_n + hK(x_{n+1}, x_{n+1}, y_{n+1}); \ y_{n+1} = y_n + hK(x_{n+1}, x_n, y_n);$$

$$y_{n+1} = y_n + hK(x_{n+1}, x_n, y_n); \ y_{n+1} = y_n + h(K(x_{n+1}, x_{n+1}, y_{n+1}) + K(x_{n+1}, x_n, y_n))/2.$$

By the generalization of these methods, we receive the following known quadrature formula:

$$y_{n+1} = y_n + h \sum_{i=0}^{1} \sum_{j=i}^{1} \beta_i^{(j)} K(x_{n+j}, x_{n+i}, y_{n+i}).$$

For the construction of methods of hybrid type it is enough to apply the midpoint rule to the calculation of integrals. In this case, we receive:

$$y_{n+1} = y_n + hK\left(x_{n+1}, x_{n+\frac{1}{2}}, y_{n+\frac{1}{2}}\right).$$

By using the exact solution of Equation (2), one can write the following:

$$y'(x) = K(x, x, y(x)) + \int_{x_0}^{x} K'_x(x, s, y(s)) ds, \ y(x_0) = 0. \quad (67)$$

If we apply the multistep method with constant coefficients to solve the problem (67), then we receive:

$$\sum_{i=0}^{k} \alpha_i y_{n+i} = h \sum_{i=0}^{k} \beta_i K(x_{n+i}, x_{n+i}, y_{n+i}) + h \sum_{i=0}^{k} \beta_i \left(\int_{x_0}^{x_n} K'_x(x_{n+i}, s, y(s)) ds + \int_{x_n}^{x_{n+i}} K'_x(x_{n+i}, s, y(s)) ds \right). \tag{68}$$

From this equality one can write the following:

$$h \sum_{i=0}^{k} \beta_i \int_{x_0}^{x_n} K'_x(x_{n+i}, s, y(s)) ds = \sum_{i=0}^{k} \alpha_i y_{n+i} - h \sum_{i=0}^{k} \beta_i (K(x_{n+i}, x_{n+i}, y_{n+i}) + \int_{x_n}^{x_{n+i}} K'_x(x_{n+i}, s, y(s)) ds. \tag{69}$$

The right hand side of this equality results in the application of the following method:

$$\sum_{i=0}^{k} \alpha_i y_{n+i} = h \sum_{i=0}^{k} \beta_i y'_{n+i}, \tag{70}$$

to solve the following initial value problem:

$$y'(x) = K(x, x, y(x)) + \int_{x_n}^{x} K'_x(x, s, y(s)) ds, \quad y(x_n) = y_n.$$

By taking into account that method (70) has the degree of p, then from the equality of (69) we find that the following holds:

$$h \sum_{i=0}^{k} \beta_i \int_{x_0}^{x_n} K'_x(x_{n+i}, s, y(s)) ds = O(h^{p+1}).$$

Thus, we prove that if method (70) has the degree of p, then method (68) has also the degree of p.

Let us approximate the function of $K'_x(x_{n+i}, s, y(s))$ in the following form:

$$h K'_x(x_{n+m}, s, y(s)) = \sum_{j=0}^{k} b_j K(x_{n+j}, s, y(s)).$$

By using this formula, one can write the following:

$$h \sum_{i=0}^{k} \beta_i \int_{x_0}^{x_{n+i}} K'_x(x_{n+i}, s, y(s)) ds = \sum_{i=0}^{k} \beta_i \int_{x_n}^{x_{n+i}} \sum_{j=0}^{k} b_j K(x_{n+j}, s, y(s)) ds.$$

By using some quadrature formulas for the calculation of the definite integral, one can write:

$$\sum_{i=0}^{k} \beta_i \int_{x_n}^{x_{n+i}} \sum_{j=i}^{k} b_j K(x_{n+j}, s, y(s)) ds = h \sum_{i=0}^{k} \sum_{j=i}^{k} \widetilde{\beta}^{(j)} K(x_{n+j}, x_{n+i}, y_{n+i}).$$

If we take this in the equality of (68), then we receive the following method:

$$\sum_{i=0}^{k} \alpha_i y_{n+i} = h \sum_{i=0}^{k} \sum_{j=i}^{k} \bar{\beta}_i^{(j)} K(x_{n+j}, x_{n+i}, y_{n+i}),$$

where $\bar{\beta}_i^{(j)} = \beta_i + \sum_{j=i}^{k} \widetilde{\beta}^{(j)}$.

This method is the same as method (8).

7. Numerical Results

For the illustration of receiving theoretical results, let us consider the following model:

$$y(x) = 1 + \lambda \int_0^x y(s)ds, \qquad (71)$$

the exact solution for which can be presented as $y(x) = \exp(\lambda x)$, where $\lambda = const$.

This example well describes the behavior properties of the errors received in the application of the method which we used to solve Equation (2). Note that this integral equation has a direct relation with the following problem:

$$y\prime = \lambda y, \ y(0) = 1,$$

describing the behavior of the solution in the following problem:

$$y' = f(x, y), \ y(x_0) = y_0,$$

which was fundamentally investigated by Dahlquist. Example (71) has been solved by using the methods (31), (34), and the following:

$$y_{n+1} = y_n + h(y'_{n+\alpha} + y'_{n+1-\alpha})/2, \ \alpha = \frac{1}{2} - \sqrt{3}/6. \qquad (72)$$

Results are tabulated in Table 1.

Table 1. The results received for the case $h = 0.05$ and $\lambda = 1, 5, 10$.

x	$\lambda = 1$		$\lambda = 5$		$\lambda = 10$		$\lambda = 15$	
	Method 34	Method 72	Method 34	Method 72	Method 34	Method 72	Method 34	Method 72
0.1	5.2×10^{-12}	5.0×10^{-9}	2.4×10^{-8}	4.1×10^{-6}	1.2×10^{-6}	1.0×10^{-4}	1.4×10^{-5}	8.4×10^{-4}
0.4	3.0×10^{-11}	2.5×10^{-8}	4.6×10^{-7}	7.4×10^{-5}	1.1×10^{-4}	8.4×10^{-3}	5.7×10^{-3}	3.0×10^{-1}
0.7	7.3×10^{-11}	6.0×10^{-8}	3.6×10^{-6}	5.8×10^{-4}	3.7×10^{-3}	3.0×10^{-1}	9.1×10^{-1}	4.8×10^{1}
1.0	1.4×10^{-10}	1.1×10^{-7}	2.3×10^{-5}	3.7×10^{-3}	1.0×10^{-1}	8.5×10^{0}	1.1×10^{2}	6.1×10^{3}

In Tables 1–5 we have tabulated the results received by the application of methods (34) and (72) to solve example (71) for the different values of step size $h > 0$ and parameter λ.

Table 2. The results received for the case $h = 0.05$ and $\lambda = -1, -5, -10$.

x	$\lambda = 1$		$\lambda = 5$		$\lambda = 10$	
	Method 34	Method 72	Method 34	Method 72	Method 34	Method 72
0.1	1.9×10^{-9}	5.5×10^{-7}	7.9×10^{-6}	4.4×10^{-4}	3.5×10^{-4}	1.0×10^{-2}
0.4	1.7×10^{-8}	3.0×10^{-6}	2.3×10^{-4}	8.0×10^{-3}	4.7×10^{-2}	7.6×10^{-1}
0.7	4.2×10^{-8}	7.1×10^{-6}	1.9×10^{-3}	6.1×10^{-2}	1.7×10^{0}	2.7×10^{1}
1.0	8.3×10^{-8}	1.4×10^{-5}	1.2×10^{-2}	4.0×10^{-1}	5.1×10^{1}	7.6×10^{2}

Table 3. The results received for the case $h = 0.01$ and $m = 1, 5, 10, 15$.

x	$\lambda = -1$		$\lambda = -5$		$\lambda = -10$	
	Method 34	Method 72	Method 34	Method 72	Method 34	Method 72
0.1	1.7×10^{-9}	5.0×10^{-7}	9.1×10^{-5}	2.9×10^{-3}	5.1×10^{-4}	1.0×10^{-2}
0.4	8.0×10^{-9}	1.5×10^{-6}	2.8×10^{-5}	5.6×10^{-4}	3.1×10^{-5}	4.4×10^{-4}
0.7	1.1×10^{-8}	1.9×10^{-6}	2.5×10^{-6}	4.9×10^{-5}	6.4×10^{-7}	8.0×10^{-6}
1.0	1.2×10^{-8}	2.0×10^{-6}	1.8×10^{-7}	3.4×10^{-6}	1.0×10^{-8}	1.2×10^{-7}

Table 4. The results received for the case $h = 0.01$ and $\lambda = -1, -5, -10, -15$.

x	$\lambda = -1$		$\lambda = -5$		$\lambda = -10$		$\lambda = -15$	
	Method 34	Method 72	Method 34	Method 72	Method 34	Method 72	Method 34	Method 72
0.1	4.3×10^{-12}	3.8×10^{-9}	9.3×10^{-9}	1.6×10^{-6}	1.9×10^{-7}	1.7×10^{-5}	8.8×10^{-7}	5.3×10^{-5}
0.4	1.4×10^{-11}	1.1×10^{-8}	8.9×10^{-9}	1.5×10^{-6}	4.0×10^{-8}	3.3×10^{-6}	4.2×10^{-8}	2.4×10^{-6}
0.7	1.9×10^{-11}	1.5×10^{-8}	3.5×10^{-9}	5.7×10^{-7}	3.5×10^{-9}	2.9×10^{-7}	8.2×10^{-10}	4.6×10^{-8}
1.0	2.0×10^{-11}	1.5×10^{-8}	1.1×10^{-9}	1.8×10^{-7}	2.5×10^{-10}	2.0×10^{-8}	1.3×10^{-11}	7.3×10^{-10}

Table 5. The results received for the case $h = 0.01$ and $\lambda = \pm 1, \pm 5, \pm 10, \pm 15$.

x	m = 1	m = 5	m = 10	m = 15	m = −1	m = −5	m = −10	m = −15
0.1	1.4×10^{-8}	1.2×10^{-5}	3.1×10^{-4}	2.5×10^{-3}	1.1×10^{-8}	5.0×10^{-6}	5.0×10^{-5}	1.6×10^{-4}
0.4	7.4×10^{-8}	2.2×10^{-4}	2.5×10^{-2}	8.9×10^{-1}	3.4×10^{-8}	4.4×10^{-6}	1.0×10^{-5}	7.2×10^{-6}
0.7	1.7×10^{-7}	1.7×10^{-3}	8.7×10^{-1}	1.4×10^{2}	4.4×10^{-8}	1.7×10^{-6}	8.8×10^{-7}	1.4×10^{-7}
1.0	3.4×10^{-7}	1.1×10^{-2}	2.5×10^{1}	1.8×10^{4}	4.6×10^{-8}	5.5×10^{-7}	6.2×10^{-8}	2.2×10^{-9}

By the simple comparison of the received results one can argue that obtaining results is justified. Note that the results received for the negative m ($m < 0$) are better than the received results for the positive ($m > 0$). It follows from the fact that the exact values of the solution will be sufficiently small for large values of the quantity of $|m|$ ($m < 0$).

According to the results in Tables 1–4, we find that the results obtained by using method (34) are better. It is naturally because method (34) is more accurate than method (72), having the degree $p = 4$. Note that these methods have the hybrid type. Now, let us compare the results obtained by the hybrid and advanced methods. For this purpose, let us solve example (71) by the application of the following method, which is received from method (10), having the degree $p = 3$:

$$y_{n+1} = y_n + h(8y'_{n+1} + 5y'_n - y'_{n+2})/12 \tag{73}$$

Results for this method are tabulated in Table 5.

The results received by method (73) can be taken as better, which is explained by the fact that in method (73) the information about the solution at the previous and the next points is used.

8. Conclusions

Here, we have considered the comparison of some numerical methods, which have been applied to solve Volterra integral equations. To this end, we used the conception of stability and degree (order of accuracy) of the investigated methods. We also constructed a formula by which one can define the maximal value for the degree of the stable and unstable multistep methods having different forms (advanced, hybrid, etc.). We prove that the multistep second derivative methods and multistep methods of hybrid type, which have been applied to solve Volterra integral equations, are more exact than the others. These results can be taken as the development of Dahlquist's results, which were received for the multistep second derivative, and show that hybrid methods have an application to solve Volterra integral equation of the second kind. Additionally, here we find the necessary condition for the convergence of the methods proposed to solve Volterra integral equations. For the investigation of the convergence of proposed methods, here we used the theory of finite-difference equations with constant coefficients. Therefore, multistep methods, here, are investigated in a very simple form. We prove that the initial value problem for ODE and the Volterra integral equation can be solved by one and the same methods. Here, algorithms have been constructed using a similar form to (50) and (51). Some of the received results are illustrated by the model equations. The methods proposed here are promising and we hope that they will find their followers.

Author Contributions: Formulation of the problem, V.I. and M.I.; methodology, V.I.; software, M.I.; validation, V.I. and M.I.; investigation, V.I. and M.I.; writing—original draft preparation, M.I.; writing—review and editing, V.I. and M.I.; funding acquisition, V.I. All authors have read and agreed to the published version of the manuscript.

Funding: This research received no external funding.

Data Availability Statement: Not Applicable.

Acknowledgments: The authors wish to express their thanks to academicians Telman Aliyev and Ali Abbasov for their suggestion to investigate the computational aspects of our problem and for their frequent valuable suggestions. This work was supported by the Science Development Foundation under the President of the Republic of Azerbaijan—Grant № EİF-MQM-ETS-2020-1(35)-08/01/1-M-01, (For Vagif Ibrahimov).

Conflicts of Interest: The authors declare no conflict of interest.

References

1. Verlan, A.F.; Sizikov, V.S. *Integral Equations, Methods, Algorithms, Programs*; Nauka Dumka: Kiev, Ukraine, 1986; p. 543.
2. Lubich, C. Runge-Kutta Theory for Volterra and Abel Integral Equations of the Second Kind. *Math. Comput.* **1983**, *41*, 87–102. [CrossRef]
3. Volterra, V. *Theory of Functional and of Integral and Integro-Differential Equations*; Dover Publications: New York, NY, USA; Nauka: Moscow, Russia, 1982; p. 304. (In Russia)
4. Manzhirov, A.V.; Polyanin, A.D. *Handbow of Integral Equations*; Factorial Press: New York, NY, USA, 1998; p. 384.
5. Baker, C.T.H. *The Numerical Treatment of Integral Equations*; Clarendon Press; Oxford University Press: Oxford, UK, 1978; p. 1034.
6. Brunner, H. Implicit Runge-Kutta Methods of Optimal Order of Volterra Integro-Differential Equations. *Math. Comput.* **1984**, *42*, 95–109. [CrossRef]
7. Brunner, H. Iterated collocation methods and their discretizations for Volterra integral equations of second kind. *SIAM J. Numer. Anal.* **1984**, *21*, 1132–1145. [CrossRef]
8. Sarah, H.H.; Mohammed Ali Murad, S.; Najeed, N. Solution of Second Kind Volterra Integral Equations Using Third order Non-Polynomial Spline Function. *Baghdad Sci. J.* **2015**, *12*, 406–411.
9. El Tom, N.E. Numerical solution of Volterra integral equations by spline functions. *BIT Numer. Math.* **1973**, *13*, 1–7. [CrossRef]
10. Noeiaghdam, S.; Sidorov, D.; Sizikov, V.; Sidorov, N. Control on Taylor-collocation method to solve the Weakly regular Volterra integral equations of the first kind by using the CESTAL by method. *Appl. Comput. Math.* **2020**, *19*, 87–105.
11. Brunner, H.; Hairer, E.; Norsett, S. Runge-Kutta Theory for Volterra Integral Equations of the Second Kind. *Math. Comput.* **1982**, *39*, 147–163. [CrossRef]
12. Trifunov, Z. Definite Integral For Calculating Volume of Revolution That is Generated By Revolving The Region About The X(Y)-Axis And Thei Visualization. *Educ. Altern.* **2020**, *18*, 178–186.
13. Amiraliyev, G.N.; Kuelu, M.; Yapman, O. A fitted approximate method for Volterra delay-integro-differential equation with initial layer. *Hacet. J. Math. Stat.* **2019**, *48*, 1417–1429. [CrossRef]
14. Amiraliyev, C.; Yilmaz, B. Fitted Difference Method for a singularly Perturbed Initial Value Problem. *Int. J. Math. Comput.* **2014**, *22*, 1–10.
15. Noeiaghdam, S.; Dreglea, A.; He, J.; Avazzadeh, Z.; Araghi, M.A.F. Error Estimation of the Homotopy Perturbation Method to Solve Second Kind Volterra With Piecewise Smooth Kernels: Application of the CADNA Library. *Symmetry* **2020**, *12*, 1730. [CrossRef]
16. Araghi, M.A.F.; Noeiaghdam, S. Valid implementation of the sine collocation method to solve linear integral equation by the CADNA library. In Proceedings of the Second National Conference of Mathematics Advanced Engineering with Mathematical Technique, Mashad Branch, Iran, 19–20 April 2017; pp. 1–8.

17. Ibrahimov, V.R.; Imanova, M.N. On a new method of solution to Volterra integral equation. *Trans. Issue Math. Mech. Ser. Phys. Tech. Math. Sci.* **2007**, *XXVII*, 197–204.
18. Mehdiyeva, G.Y.; Ibrahimov, V.R. *On the Way to Determine Coefficients of Multistep Method*; News of Baku University Physics and Mathematics Series; Baku University: Baku, Azerbaijan, 2008; Volume 2, pp. 35–39.
19. Imanova, M.N.; Ibrahimov, V.R. *On the Application of the General Quadrature Method*; News of Baku University Physics and Mathematics Series; Baku University: Baku, Azerbaijan, 2008; Volume 3, pp. 83–91.
20. Dahlquist, G. Convergence and stability in the numerical integration of ODEs. *Math. Scand.* **1956**, *4*, 33–53. [CrossRef]
21. Shura-Bura, M.R. Error estimates for numerical integration of ordinary differential equations. *Prikl Matem. Fur.* **1952**, *5*, 575–588.
22. Mehdiyeva, G.Y.; Ibrahimov, V.R.; Imanova, M. Solving Volterra Integro-Differential by the Second Derivative Methods. *Appl. Math. Inf. Sci.* **2015**, *9*, 2521–2527.
23. Ibrahimov, V.R. One non-linear method for solving Cauchy problem for ODE, Differential equations and applications. In Proceedings of the Reports. Second İnternational Conference, Rousse, Bulgaria, 1 December 1982; pp. 310–319.
24. Ibrahimov, V.R. The Estimation on k-step method under weak assumptions on the solution of the given problem. In Proceedings of the XI-International Conference on Nonlinear Ascillation, Budapest, Hungary, 6–11 August 1988; pp. 543–546.
25. Mehdiyeva, G.; Ibrahimov, V.R.; Imanova, M. Application of a second derivative multistep method to numerical solution of Volterra integral equation of second kind. *Pak. J. Stat. Oper. Res.* **2012**, *8*, 245–258.
26. Mehdiyeva, G.Y.; Ibrahimov, V.R. *On the Research of Multi-Step Methods with Constant Coefficients*; Lambert Academic Publishing: Saarbrücken, Germany, 2013; p. 314.
27. Dahlquist, G. Stability and Error Bounds in the Numerical Integration of Ordinary Differential Equations. 85 S. Stockholm 1959. K. Tekniska Högskolans Handlingar. *J. Appl. Math. Mech.* **1961**, *41*, 267–268.
28. Imanova, M.N. On the comparison of Gauss and Hybrid methods and their application to calculation of definite integrals. *J. Phys. Conf. Ser.* **2020**, *1564*. [CrossRef]
29. Ibrahimov, V.R. Convergence of the predictor-corrector methods. *Appl. Math.* **1984**, *4*, 187–197.
30. Mehdiyeva, G.Y.; Ibrahimov, V.R.; Imanova, M.N. On the construction of the advanced Hybrid Methods and application to solving Volterra Integral Equation. *WSEAS Weak Trans. Syst. Control* **2019**, *14*, 183–189.
31. Imanova, M.N. On some comparison of Computing Indefinite integrals with the solution of the initial-value problem for ODE. *WSEAS Trans. Math.* **2020**, *19*, 19. [CrossRef]
32. Bolaji, B. Fully, Implicit Hybrid Block–Predictor Corrector Method for the Numerical Integration $y''' = f(x, y, y', y'')$, $y(xo) = \eta o$, $y'(xo) = \eta 1$, $y''(xo) = \eta 3$. *J. Sci. Res. Rep.* **2015**, *6*, 165–171. [CrossRef]
33. Mehdiyeva, G.Y.; Ibrahimov, V.R. On the Computation of Double Integrals by Using Some Connection Between the Wave Equation and The System Of ODE. In Proceedings of the IAPE'20, Second Edition of the International Conference on Innovative Applied Energy, Cambridge, UK, 15–16 September 2020.
34. Ibrahimov, V.R.; Imanova, M.N. On a Research of Symmetric Equations of Volterra Type. *Int. J. Math. Models Methods Appl. Sci.* **2014**, *8*, 434–440.
35. Brunner, H. Marginal Stability and Stabilization in the Numerical Integration of Ordinary Differential Equations. *Math. Comput.* **1970**, *24*, 635–646. [CrossRef]
36. Gupta, G.K. A polynomial representation of hybrid methods for solving ordinary differential equations. *Math. Comput.* **1979**, *33*, 1251–1256. [CrossRef]
37. Butcher, J.C. A modified multistep method for the numerical integration of ordinary differential equations. *J. Assoc. Comput. Math.* **1965**, *12*, 124–135. [CrossRef]
38. Gear, C.S. Hybrid methods for initial value problems in ordinary differential equations. *SIAM J. Numer. Anal.* **1965**, *2*, 69–86. [CrossRef]
39. Makroglou, A. Hybrid methods in the numerical solution of Volterra integro-differential equations. *J. Numer. Anal.* **1982**, *2*, 21–35. [CrossRef]
40. Bulatov, M.B.; Chistakov, E.B. The numerical solution of integral-differential systems with a singular matrix at the derivative multistep methods. *Differ. Equ.* **2006**, *42*, 1218–1255. [CrossRef]
41. El-Baghdady Galal, I.; El-Azab, M.S. A New Chebyshev Spectral-collocation Method for Solving a Class of One-dimensional Linear Parabolic Partial Integro-differential Equations. *Br. J. Math. Comput. Sci.* **2015**, *6*, 172–186. [CrossRef]
42. Mehdiyeva, G.Y.; Imanova, M.N.; Ibrahimov, V.R. An application of the hybrid methods to the numerical solution of ordinary differential equations of second order. *Math. Mech. Inf.* **2012**, *4*, 46–54.
43. Ibrahimov, V.R.; Mehdiyeva, G.; Imanova, M. On the construction of the multistep Methods to solving for initial-value problem for ODE and the Volterra intego-differential equations. In Proceedings of the International Conference on Indefinite Applied Energy, Oxford, UK, 14–15 March 2019.
44. Babushka, I.; Vitasek, E.; Prager, M. *Numerical Processes for Solving Differential Equations*; Mir: Moscow, Russia, 1969; 368p.
45. Mehdiyeva, G.Y.; Ibrahimov, V.R.; Imanova, M.N. On the construction test equation and its applying to solving Volterra integral equation. In Proceedings of the 17th WSEAS International Conference on Applied Mathematics Mathematical Methods for Information Science and Economics (AMATH '12), Montreux, Switzerland, 29–31 December 2012; pp. 109–114.
46. Ibrahimov, V.R. On a relation between degree and order for the stable advanced formula. *J. Comput. Math. Math. Phys.* **1990**, *7*, 1045–1056.

47. Ehigie, J.O.; Okunuga, S.A.; Sofoluwe, A.B.; Akanbi, M.A. On generalized 2-step continuous linear multistep method of hybrid type for the integration of second order ordinary differential equations. *Arch. Appl. Res.* **2010**, *2*, 362–372.
48. Urabe, M. An implicit one-step Method of High-Order Accuracy dor the Numerical Integrations of ODE. *Numer. Math.* **1970**, *2*, 151–164. [CrossRef]
49. Mirzaev, R.R.; Mehdiyeva, G.Y.; Ibrahimov, V.R. On an application of a multistep method for solving Volterra integral equations of the second kind. In Proceedings of the International Conference on Theoretical and Mathematical Foundations of Computer Science, Orlando, FL, USA, 12–14 July 2010; pp. 46–51.

Article
Integro-Differential Equation for the Non-Equilibrium Thermal Response of Glass-Forming Materials: Analytical Solutions [†]

Alexander A. Minakov [1] and Christoph Schick [2,3,*]

1. Prokhorov General Physics Institute of the Russian Academy of Sciences, GPI RAS, Vavilov Str. 38, 119991 Moscow, Russia; minakov@nsc.gpi.ru
2. Institute of Physics and Competence Centre CALOR, University of Rostock, Albert-Einstein-Str. 23-24, 18051 Rostock, Germany
3. Alexander Butlerov Institute of Chemistry, Kazan Federal University, Kremlyovskaya Str. 18, 420008 Kazan, Russia
* Correspondence: christoph.schick@uni-rostock.de
† Dedicated to the memory of Prof. Dr. Ernst-Joachim Donth (1937–2020).

Abstract: An integro-differential equation describes the non-equilibrium thermal response of glass-forming substances with a dynamic (time-dependent) heat capacity to fast thermal perturbations. We found that this heat transfer problem could be solved analytically for a heat source with an arbitrary time dependence and different geometries. The method can be used to analyze the response to local thermal perturbations in glass-forming materials, as well as temperature fluctuations during subcritical crystal nucleation and decay. The results obtained can be useful for applications and a better understanding of the thermal properties of glass-forming materials, polymers, and nanocomposites.

Keywords: non-equilibrium heat transfer problem; time-dependent response function; second-kind integro-differential equations; Volterra integral equations

1. Introduction

The dynamic heat capacity $c_{dyn}(t)$ of glass-forming liquids and glasses, considered as a function of time t, has been intensively studied since the pioneering work of Birge and Nagel published in 1985 [1]. However, much earlier, experiments on the dispersion and absorption of ultrasonic waves in polyatomic gases and liquids were explained by the relaxation of the specific heat in these substances. The experiments were comprehensively reviewed by Herzfeld and Litovitz in 1959 [2]. The relaxation of the specific heat of polyatomic gases and liquids is caused by a slow energy exchange between the external (translational) and internal (vibrational and rotational) degrees of freedom. Thus, the energy exchange in polyatomic gases is characterized by a limited set of characteristic relaxation times τ_i [2]. The spectrum of relaxation times of glass-forming liquids and glasses is extremely broad, and it can be considered as a continuous spectrum [3–24]. Naturally, this broad spectrum is observed not only for the relaxation of the dynamic heat capacity [3–7,18–20] but also for dielectric susceptibility [8–18], light scattering [23], and viscosity [8,24–26]. Since the specific heat of glass-forming substances depends on time, it follows that the thermal response of these materials to a thermal perturbation at time t depends on the temperature distribution $T(t, \vec{r})$ in the system in previous times. This effect is especially significant for fast local thermal perturbations [27,28].

The time dependence of the heat capacity of glass-forming liquids and glasses leads to a non-equilibrium (non-Fourier) thermal response of these materials to fast thermal perturbations. This non-equilibrium thermal response can be described by the integro-differential heat equation considered in [27–29]. This equation can be solved similarly to the second-kind Volterra integral equations. Volterra integral equations have several applications in many branches of science, technology, and industry. Viscoelasticity is one of the fields of physics where the Volterra equations are often used [25]. The applications of Volterra equations in renewable energy is an example of their use in industry [30–32]. Volterra integro-differential equations are usually difficult to solve analytically; therefore, approximate solutions and numerical methods are often used [33]. An interesting method for the numerical solution to nth-order integro-differential equations has been developed based on the integral mean value theorem [34]. However, to better understand the nature of physical phenomena, it is crucial to establish qualitative relationships between physical parameters and the ongoing physical processes. Thus, we focused on finding analytical solutions of the integro-differential heat equation that describe the non-equilibrium thermal response of glass-forming materials.

The heat transfer equation considered in this article was solved analytically and may be of interest for various applications. In previous articles, we found solutions for the equation with rectangular pulsed heat sources [28,29]. In this study, we focused on analytical solutions of the equation for a heat source with arbitrary time dependence. These solutions can be useful for studying local thermal perturbations in glass-forming materials, as well as temperature fluctuations during subcritical crystal nucleation and decay; these temperature fluctuations can have a significant effect on the kinetics of crystal nucleation in glass-forming materials.

The heat transfer problem with memory was analyzed in general terms by Miller in 1978 [35]. For an external heat flux represented by a smooth function, he proved the existence, uniqueness, and continuous dependence on parameters of the solution for heat transfer with memory. In this study, we considered a special case of integro-differential heat equations with kernels corresponding to glass-forming materials with dynamic heat capacity; these equations can be solved analytically. We focused on an external heat flux $\Phi(t, \vec{r})$ acting on finite intervals in space and time. In this study, we restricted ourselves to considering the multiplicatively separable heat flux $\Phi(\vec{r}) F(t)$. We considered the heat flux as a continuous and piecewise smooth function. Thus, we assumed that both $F(t)$ and $\Phi(\vec{r})$ were continuous piecewise smooth functions. This is a sufficient condition for the absolute and uniform convergence of the Fourier series of the functions $F(t)$ and $\Phi(\vec{r})$ [36]. In addition, we focused on the dynamic behavior of substances with a dynamic heat capacity and restricted ourselves to considering only homogeneous boundary value problems.

The rest of the paper is organized as follows. Sections 2 and 3 formulate the heat transfer problem to be solved. Sections 4 and 5 discuss solutions for rectangular-pulsed, sinusoidal, and arbitrary heat sources that were considered for planar and spherical geometries. In Section 6, the effect of the relaxation time distribution on the thermal responses is discussed for different temperatures and materials. It is shown that the effect of the time dispersion of the dynamic heat capacity on the thermal response $T(t, r)$ during local heating was significant. Examples of solutions $T(t, r)$ for real glass-forming materials were considered. The temperature dependence of the relaxation-time distribution was taken into account.

2. Applicability of the Heat Equation with Dynamic Heat Capacity

In this study, we focused on the heat equation for non-metallic glass-forming materials at temperatures T outside the low-temperature range. Nonlocal effects of heat conduction [37] were not taken into account for $(\partial \ln(T)/\partial x)^{-1} \gg l_{ph}$, where l_{ph} is the phonon mean-free-path. Furthermore, for small temperature changes, we considered the thermal parameters of materials independent of T. However, the temperature dependence of the relaxation time spectrum of the dynamic heat capacity $c_{dyn}(t)$ is discussed in Section 6. This relaxation time spectrum is very broad in glass-forming materials. Therefore, the time dispersion of the dynamic heat capacity is significant over a wide range of time scales. Conversely, the thermal conductivity λ of non-metallic glass-forming materials can be considered an equilibrium (time-independent) thermal parameter, at least for $t > 1$ ns and at temperatures above the low-temperature range. In fact, thermal excitations associated with thermal conductivity come to local equilibrium much faster than in 1 ns in glass-forming liquids and glasses [38–41], and the phonon mean free path is about 1 nm or less [41–47]. Therefore, we focused on the diffusion-type Fourier heat conduction. However, we took into account the fact that the local heat absorption at a given moment of time t is determined not only by a change in the local temperature $T(t, \vec{r})$ at this moment but also at each previous moment. The temporal dispersion of the dynamic heat capacity $c_{dyn}(t)$ of glass-forming materials can be described within the framework of the linear response theory [6,48], similarly to the temporal dispersion of the dielectric constant [49].

3. Heat Equation with Dynamic Heat Capacity

Heat transfer in glass-forming materials with a dynamic heat capacity $c_{dyn}(t)$ can be described using the following integro-differential heat equation [27,28]:

$$\frac{\partial}{\partial t} \int_0^\infty \rho c_{dyn}(\tau) T'\left(t - \tau, \vec{r}\right) d\tau = \lambda \Delta T\left(t, \vec{r}\right) + \Phi\left(t, \vec{r}\right), \quad (1)$$

where $T'(t, \vec{r}) = \frac{\partial}{\partial t} T(t, \vec{r})$, Δ is the Laplacian, and $\Phi(t, \vec{r})$ is the external heat flux. Equation (1) can be used for glass-forming substances at least for $t > 1$ ns and length scales greater than 1 nm.

Consider the problem with zero initial conditions $\Phi(t, \vec{r}) = 0$ and $T(t, \vec{r}) = 0$ for $t \leq 0$. Thus, from Equation (1) we obtain:

$$\frac{\partial}{\partial t} \int_0^t \rho c_{dyn}(t - \tau) T'\left(\tau, \vec{r}\right) d\tau - \lambda \Delta T\left(t, \vec{r}\right) = \Phi\left(t, \vec{r}\right). \quad (2)$$

Equation (2) can be solved if the dynamic heat capacity $c_{dyn}(t)$ is a given function. Usually, relaxation phenomena in glass-forming materials are described by the stretched exponential Kohlrausch relaxation function $exp\left(-(t/\tau_K)^\beta\right)$ for $\beta \in (0,1]$, where β and the Kohlrausch relaxation time τ_K characterize the relaxation time spectrum (for more details, see [50,51]). As a completely monotonic function, the stretched exponent can be represented as a continuous sum of exponentials (see Bernstein's theorem [52]). It should be noted that typical relaxation functions, such as exponential, stretched exponential, and power-law relaxation functions, are completely monotonic [53]. Thus, we assumed that the dynamic heat capacity $c_{dyn}(t)$ is a completely monotonic function of time; therefore, $c_{dyn}(t)$ can be represented as a continuous sum of exponentials, as seen in Equation (3):

$$c_{dyn}(t) = c_0 - (c_0 - c_{in}) \int_0^\infty H(\tau_0) exp(-t/\tau_0) d\tau_0, \quad (3)$$

where c_0 and c_{in} are the equilibrium and initial heat capacities, that is, $c_{dyn}(t) \to c_{in}$ as $t \to 0$ and $c_{dyn}(t) \to c_0$ as $t \to \infty$. The distribution function $H(\tau_0)$ is normalized as follows: $\int_0^\infty H(\tau_0) d\tau_0 = 1$. However, for practical use, $c_{dyn}(t)$ can be given on a finite interval $[\tau_{min}, \tau_{max}]$ if this interval is wide enough (see [28] for details). The distribution function $H(\tau_0)$ can be found by using broadband heat capacity spectroscopy [19], as was done in [27–29] (see Example 3).

As a first step, consider the solution to Equation (2) for the auxiliary problem, which corresponds to the dynamic heat capacity $c_{dyn}(t)$ obeying the Debye relaxation law:

$$c_{dyn}(t) = c_0[1 - \varepsilon_0 exp(-t/\tau_0)], \quad (4)$$

where $\varepsilon_0 = (c_0 - c_{in})/c_0$. Subsequently, the final solution for any given $c_{dyn}(t)$ can be obtained using the solution for an arbitrary positive τ_0 if $H(\tau_0)$, c_0, and c_{in} are specified. This final solution can be represented as a continuous sum of solutions depending on τ_0 and distributed according to the normalized distribution function $H(\tau_0)$. Using

Equations (2) and (4), we obtain the following integro-differential equation with a difference kernel:

$$\frac{\partial}{\partial t}T(t,\vec{r}) - D_0\Delta T(t,\vec{r}) = \frac{\Phi(t,\vec{r})}{\rho c_0} + \varepsilon_0 \frac{\partial}{\partial t}\int_0^t exp\left(-\frac{t-\tau}{\tau_0}\right)T'(\tau,\vec{r})d\tau, \quad (5)$$

where $D_0 = \lambda/\rho c_0$ is the equilibrium thermal diffusivity of the glass-forming substance, τ_0 is a positive parameter, and $0 < \varepsilon_0 < 1$. Usually, in glass-forming materials, ε_0 is in the range 0.2–0.3 [19,54] and sometimes even more [55].

Next, consider a one-dimensional example for a sample with a flat geometry. After that, the spherically symmetric problem in three dimensions can be reduced to the one-dimensional problem mentioned above by using $T(t,r) = U(t,r)/r$.

4. Heat Equation with Dynamic Heat Capacity: Plane Geometry

As a basic example, consider a one-dimensional problem with homogeneous boundary and initial conditions. Consider an infinite plate of thickness d. Let the x-axis be directed along the normal to the plate surface. In many practically important cases, the heat flux $\Phi(t,x)$ can be considered as a multiplicatively separable function $\Phi(x)F(t)$, for example, in laser [56,57] and Joule heating [58], as well as local heating due to crystal nucleation [59–61]. Thus, we consider the heat flux in the form $\Phi(x)F(t)$ with continuous piecewise smooth $F(t)$ and $\Phi(x)$ functions. Let the heat flux $\Phi(x)F(t)$ be distributed on the domain $[0,d]$ and act during the time interval $[0,\tau_p]$, i.e., $F(t) = 0$ for $t \leq 0$ and $\tau_p \leq t$. Suppose the dynamic heat capacity is described by Equation (4). Then, from Equation (5), we obtain:

$$\frac{\partial}{\partial t}T(t,x) - D_0\partial^2 T(t,x)/\partial x^2 = \frac{\Phi(x)F(t)}{\rho c_0} + \varepsilon_0\frac{\partial}{\partial t}\int_0^t exp\left(-\frac{t-\tau}{\tau_0}\right)T'(\tau,x)d\tau. \quad (6)$$

We focused on the dynamic behavior of the thermal response $T(t,x)$ to the external heat flux $\Phi(x)F(t)$ in materials with a dynamic heat capacity. Suppose the temperature increases from the initial (thermostat) temperature. Thus, consider a homogeneous boundary value problem for the boundary conditions $T(t,0) = 0$, $T(t,d) = 0$, and the zero initial condition $T(t,x) = 0$ for $t \leq 0$. The solution $T(t,x)$ can be represented as a series:

$$T(t,x) = \sum_{n=1}^{\infty} \psi_n(t) sin\left(\frac{\pi n x}{d}\right), \quad (7)$$

where the functions $\psi_n(t)$ must satisfy Equation (8):

$$\psi'_n(t) + \frac{\psi_n(t)}{\tau_n} = \frac{\Phi_n F(t)}{\rho c_0} + \varepsilon_0\frac{\partial}{\partial t}\int_0^t exp\left(-\frac{t-\tau}{\tau_0}\right)\psi'_n(\tau)d\tau, \quad (8)$$

where $\psi'_n(t) = \frac{\partial}{\partial t}\psi_n(t)$, $\Phi_n = \frac{2}{d}\int_0^d \Phi(x)sin(\pi n x/d)dx$, and $\tau_n^{-1} = D_0(\pi n/d)^2$.

Let $\tilde{T}(t,x)$ and $\tilde{\psi}_n(t)$ denote the solutions of the conventional Equations (6) and (8) with $\varepsilon_0 = 0$, respectively. The solution $\tilde{\psi}_n(t)$ can be represented by Equation (9) [62]:

$$\tilde{\psi}_n(t) = \frac{\Phi_n}{\rho c_0}\int_0^t F(\tau)exp\left(-\frac{t-\tau}{\tau_n}\right)d\tau. \quad (9)$$

For example, consider the solution to Equation (8) with $\varepsilon_0 = 0$ for $F(t) = \theta(t)$, where $\theta(t)$ is the Heaviside unit step function (with the condition $\theta(0) = 0$). Denote this solution by $\tilde{\chi}_n(t)$. Then, we have $\tilde{\chi}_n(t) = \frac{\Phi_n}{\rho c_0}\tau_n(1 - exp(-t/\tau_n))$. Next, for a sinusoidal heat source $F_m(t)$ of duration τ_p, where $F_m(t) = sin(\pi m t/\tau_p)$ for $t \in [0,\tau_p]$ and $F_m(t) = 0$

outside $(0, \tau_p)$, we obtain the solution $\tilde{\psi}_{n,m}(t)$ of Equation (8) with $\varepsilon_0 = 0$, as seen in Equations (10) and (11):

$$\tilde{\psi}_{n,m}(t) = \frac{\Phi_n}{\rho c_0} \frac{\frac{1}{\tau_n}\sin(\pi m t/\tau_p) + \frac{\pi m}{\tau_p}[\exp(-t/\tau_n) - \cos(\pi m t/\tau_p)]}{\left(\frac{1}{\tau_n}\right)^2 + \left(\frac{\pi m}{\tau_p}\right)^2} \quad \text{for } 0 \le t \le \tau_p, \quad (10)$$

$$\tilde{\psi}_{n,m}(t) = \frac{\Phi_n}{\rho c_0} \frac{\frac{\pi m}{\tau_p}\exp(-t/\tau_n)[1 - \exp(\tau_p/\tau_n)\cos(\pi m)]}{\left(\frac{1}{\tau_n}\right)^2 + \left(\frac{\pi m}{\tau_p}\right)^2} \quad \text{for } \tau_p \le t. \quad (11)$$

Thus, for the conventional Equation (6) with $\varepsilon_0 = 0$ and the sinusoidal heat source $F_m(t)$, the solution is:

$$\check{T}_m(t, x) = \sum_{m=1} \tilde{\psi}_{n,m}(t)\sin\left(\frac{\pi n x}{d}\right). \quad (12)$$

To find the solutions $T(t, x)$ and $\psi_n(t)$ of Equations (6) and (8) with nonzero positive ε_0 and τ_0, we first solve an auxiliary problem with $F(t) = \theta(t)$, where $\theta(t)$ is the Heaviside unit step function, as seen in Equation (13). The solution to this auxiliary problem is denoted by $\varphi_n(t)$:

$$\varphi'_n(t) + \frac{\varphi_n(t)}{\tau_n} = \frac{\Phi_n}{\rho c_0}\theta(t) + \varepsilon_0 \frac{\partial}{\partial t}\int_0^t \exp\left(-\frac{t-\tau}{\tau_0}\right)\varphi'_n(\tau)d\tau. \quad (13)$$

Equation (13) can be solved using the Laplace transform method (see Appendix A for details). In fact, the solution is obtained similarly to the Volterra integral equations with a difference kernel [63]. Thus, we have:

$$\varphi_n(t) = \frac{\Phi_n}{\rho c_0}\tau_n\left[1 + \frac{\tau_0\gamma_n\mu_n(\exp(-\mu_n t) - \exp(-\gamma_n t))}{(\gamma_n - \mu_n)} + \frac{\mu_n\exp(-\gamma_n t) - \gamma_n\exp(-\mu_n t)}{(\gamma_n - \mu_n)}\right], \quad (14)$$

$$\varphi'_n(t) = \frac{\Phi_n}{\rho c_0}\frac{\tau_n\gamma_n\mu_n}{(\gamma_n - \mu_n)}[(\tau_0\gamma_n - 1)\exp(-\gamma_n t) + (1 - \tau_0\mu_n)\exp(-\mu_n t)], \quad (15)$$

where $-\gamma_n$ and $-\mu_n$ are the roots of the polynomial $(1 - \varepsilon_0)p^2 + p\left(\tau_n^{-1} + \tau_0^{-1}\right) + \tau_n^{-1}\tau_0^{-1}$ (see Appendix A). The parameters γ_n and μ_n are real-valued and positive. Moreover, $\gamma_n - \mu_n > 0$ for any positive τ_n and τ_0, and $0 < \varepsilon_0 < 1$. It can also be shown that $\varphi_n(t)$ continuously tends to the solution $\tilde{\chi}_n(t) = \frac{\Phi_n}{\rho c_0}\tau_n(1 - \exp(-t/\tau_n))$ of Equation (8) with $\varepsilon_0 = 0$ and $F(t) = \theta(t)$ as $\varepsilon_0 \to 0$ or $\tau_0 \to 0$.

Let us consider the problem with nonzero ε_0 and τ_0 for the sinusoidal heat source $F_m(t)$. The solution to this problem can be obtained using the Duhamel integral [62,63] (see Appendix A for details). Thus, the solution to Equation (8) for the sinusoidal heat source $F_m(t)$ can be represented by Equation (16):

$$\psi_{n,m}(t) = \int_0^t F_m(\tau)\varphi'_n(t - \tau)d\tau. \quad (16)$$

From Equation (16) we obtain:

$$\psi_{n,m}(t) = \frac{\Phi_n}{\rho c_0}\frac{\tau_n\gamma_n\mu_n}{(\gamma_n - \mu_n)}$$

$$\left[\frac{(\gamma_n\tau_0 - 1)\left[\gamma_n\sin(\pi m t/\tau_p) + \frac{\pi m}{\tau_p}(\exp(-\gamma_n t) - \cos(\pi m t/\tau_p))\right]}{(\gamma_n)^2 + \left(\frac{\pi m}{\tau_p}\right)^2} + \frac{(1 - \mu_n\tau_0)\left[\mu_n\sin(\pi m t/\tau_p) + \frac{\pi m}{\tau_p}(\exp(-\mu_n t) - \cos(\pi m t/\tau_p))\right]}{(\mu_n)^2 + \left(\frac{\pi m}{\tau_p}\right)^2}\right] \quad (17)$$

for $0 \le t \le \tau_p$,

$$\psi_{n,m}(t) = \frac{\Phi_n}{\rho c_0} \frac{\pi m \tau_n}{\tau_p} \frac{\gamma_n \mu_n}{(\gamma_n - \mu_n)}$$

$$\left[\frac{(\gamma_n \tau_0 - 1)\left[exp(-\gamma_n t) - cos(\pi m)exp(\gamma_n(\tau_p - t))\right]}{(\gamma_n)^2 + \left(\frac{\pi m}{\tau_p}\right)^2} + \frac{(1 - \mu_n \tau_0)\left[exp(-\mu_n t) - cos(\pi m)exp(\mu_n(\tau_p - t))\right]}{(\mu_n)^2 + \left(\frac{\pi m}{\tau_p}\right)^2} \right] \qquad (18)$$

for $\tau_p < t$.

It can be shown that $\psi_{n,m}(t)$ continuously tends to the solution $\tilde{\psi}_{n,m}(t)$ as $\varepsilon_0 \to 0$ or $\tau_0 \to 0$. Thus, the solution to Equation (6) with nonzero ε_0 and τ_0 for the sinusoidal heat source $F_m(t)$ is:

$$T_m(t,x) = \sum_{n=1}^{\infty} \psi_{n,m}(t) sin\left(\frac{\pi n x}{d}\right). \qquad (19)$$

Finally, in the case of nonzero ε_0 and τ_0, we obtain the solution to the problem for an arbitrary $F(t)$. Indeed, if $F(t)$ is a continuous piecewise smooth function on the interval $[0, \tau_p]$, then the Fourier series in Equation (20) converges absolutely and uniformly to $F(t)$ [36]. Thus, we represent $F(t)$ on the interval $[0, \tau_p]$ using the Fourier series, as seen in Equation (20):

$$F(t) = \sum_{m=1}^{\infty} C_m sin(\pi m t/\tau_p) \text{ for } 0 \le t \le \tau_p \text{ and } F(t) = 0 \text{ for } \tau_p \le t, \qquad (20)$$

where C_m are the Fourier coefficients $C_m = \frac{2}{\tau_p}\int_0^{\tau_p} F(t)sin(\pi m t/\tau_p)dt$. Since Equation (6) is linear with respect to $T(t,x)$, it follows that the solution $T(t,x)$ of Equation (6) can be represented as a linear combination of solutions $T_m(t,x)$ for sinusoidal heat sources $F_m(t)$ with corresponding coefficients C_m. Therefore, we obtain a solution to Equation (6) for an arbitrary continuous piecewise-smooth heat flux $\Phi(x)F(t)$ distributed on the domain $[0,d]$ and acting during the time interval $[0, \tau_p]$, as seen in Equation (21). The series shown in Equation (21) converges uniformly and absolutely [36]:

$$T(t,x) = \sum_{m=1}^{\infty} \sum_{n=1}^{\infty} C_m \psi_{n,m}(t) sin\left(\frac{\pi n x}{d}\right). \qquad (21)$$

Example 1. *Let the heat flux $\Phi(x)$ be uniformly distributed on the domain $\left[\frac{d-x_0}{2}, \frac{d+x_0}{2}\right]$ with density Φ_0. Then:*

$$\Phi_n = \Phi_0 \frac{4 sin(\pi n/2)}{\pi n} sin\left(\frac{\pi n x_0}{2d}\right). \qquad (22)$$

Denote by $\check{T}_p(t,x)$ the solution to the conventional Equation (6) with $\varepsilon_0 = 0$ for the rectangular heating pulse of duration τ_p, i.e., $F_p(t) = \theta(t)(1 - \theta(t - \tau_p))$. Then,

$$\check{T}_p(t,x) = \sum_{n=1}^{\infty} \left[\tilde{\chi}_n(t) - \tilde{\chi}_n(t - \tau_p)\theta(t - \tau_p)\right] sin\left(\frac{\pi n x}{d}\right), \qquad (23)$$

where $\tilde{\chi}_n(t) = \frac{\Phi_n}{\rho c_0} \tau_n (1 - exp(-t/\tau_n))$ and Φ_n is determined using Equation (22). Note that $\tau_n \sim \frac{1}{n^2}$. Thus, the series in Equation (23) converges as fast as the series $S_N = \sum_{n=1}^{N} 1/n^3$, which converges to the Riemann zeta function $\zeta(3) = 1.20205...$ as $N \to \infty$ [64]. The remainder $(\zeta(3) - S_N)$ is less than 0.1% of $\zeta(3)$ for $N = 20$. Thus, to obtain an accuracy of about 0.1%, it is sufficient to take the series in Equation (23) up to $N = 20$. The examples in this section are calculated with $N = 30$.

The solution of Equation (6) with $\varepsilon_0 = 0$ for the sinusoidal heat source $F_m(t)$ with $m = 1$ is equal to:

$$\check{T}_1(t,x) = \sum_{n=1}^{\infty} \tilde{\psi}_{n,1}(t) sin\left(\frac{\pi n x}{d}\right), \qquad (24)$$

where $\tilde{\psi}_{n,1}(t)$ is determined using Equations (10) and (11) for $m = 1$. Here again, Φ_n is given by Equation (22).

Similarly, solutions to Equation (6) with nonzero ε_0 and τ_0 for the rectangular and sinusoidal heating sources are represented by Equations (25) and (26):

$$T_p(t,x) = \sum_{n=1} \left[\varphi_n(t) - \varphi_n(t-\tau_p)\theta(t-\tau_p)\right] \sin\left(\frac{\pi n x}{d}\right), \quad (25)$$

$$T_1(t,x) = \sum_{n=1} \psi_{n,1}(t) \sin\left(\frac{\pi n x}{d}\right), \quad (26)$$

where $\varphi_n(t)$ and $\psi_{n,1}(t)$ are determined using Equation (14) and Equations (17) and (18), respectively, with Φ_n being the same as in Equations (23) and (24).

For example, let $\varepsilon_0 = 1/3$, $\tau_0 = 30$ ns, and $\tau_p = 10$ ns. Let $d = 100$ nm, $x_0 = 50$ nm, and the thermal parameters of the substance are typical for those of glass-forming polymers. Thus, we take $\rho = 1$ g/cm^3, $\rho c_0 = 2 \times 10^6$ J/m^{3}K, $\lambda = 0.3$ W/mK, $D_0 = 1.5 \times 10^{-7}$ m^2/s, and $\Phi_0 = \rho h_0$, where $h_0 = 200$ J/g is the heat release during crystallization. We focused on the nanometer and nanosecond scales since they are close to real processes during subcritical crystal nucleation [59–61]. First, we verified that Equation (21) gave the correct solution $T(t,x)$ for a rectangular pulsed heat source. Indeed, the solutions $\check{T}_p(t,x)$ and $T_p(t,x)$ represented by Equations (23) and (25) coincided with the results $\check{T}(t,x)$ and $T(t,x)$ calculated using the Fourier series (see Equation (21)) for m up to 45 (see Figure 1b). In addition, Figure 1b shows that the effect of the temporal dispersion of the dynamic heat capacity on the temperature was significant.

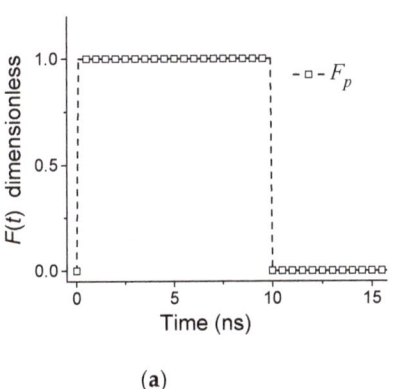 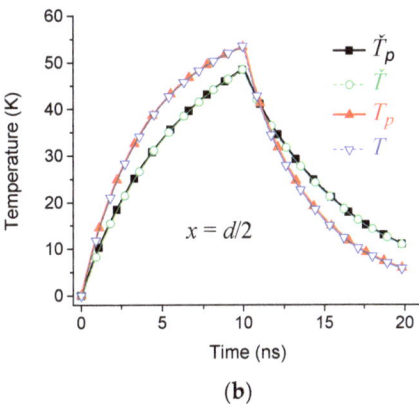

(a) (b)

Figure 1. (a) Rectangular heating pulse $F_p(t)$ with duration $\tau_p = 10$ ns. (b) Time dependences $\check{T}_p\left(t, \frac{d}{2}\right)$ and $\check{T}\left(t, \frac{d}{2}\right)$ (shown using squares and circles), as well as $T_p\left(t, \frac{d}{2}\right)$ and $T\left(t, \frac{d}{2}\right)$ for $\varepsilon_0 = 1/3$ and $\tau_0 = 30$ ns (shown using up and down triangles).

As an example of an arbitrary function $F(t)$, consider the continuous piecewise smooth function $F_A(t)$ shown in Figure 2. Let $\mathcal{F}_A(t)$ denote the Fourier series approximating $F_A(t)$, as seen in Equation (27):

$$\mathcal{F}_A(t) = \sum_{m=1}^{M} C_m^A \sin(\pi m t/\tau_p) \text{ for } 0 \le t \le \tau_p \text{ and } \mathcal{F}_A(t) = 0 \text{ for } \tau_p \le t. \quad (27)$$

For example, let us take $N = 30$ and $M = 30$. Figure 2a shows that $F_A(t)$ was well approximated by $\mathcal{F}_A(t)$ at $M = 30$. The solution $T_A(t,x)$ for the heating source with $F_A(t)$ is represented by Equation (28), where $\psi_{n,m}(t)$ are determined by Equations (17) and (18) with corresponding coefficients C_m^A (see Equation (27)):

$$T_A(t,x) = \sum_{m=1}^{M} \sum_{n=1}^{N} C_m^A \psi_{n,m}(t) \sin\left(\frac{\pi n x}{d}\right). \quad (28)$$

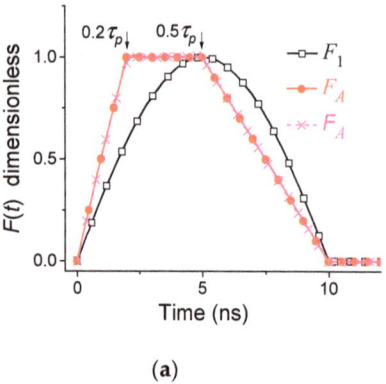
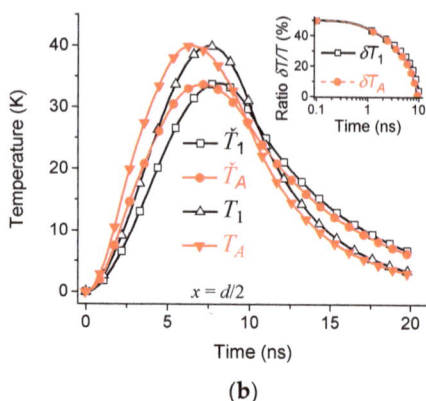

Figure 2. (a) Heating pulses $F_1(t)$, $F_A(t)$, and $\mathcal{F}_A(t)$ (shown using squares, circles, and crosses, respectively). (b) Time dependences of $\check{T}_1\left(t, \frac{d}{2}\right)$ and $\check{T}_A\left(t, \frac{d}{2}\right)$ (shown using squares and circles), as well as $T_1\left(t, \frac{d}{2}\right)$ and $T_A\left(t, \frac{d}{2}\right)$ for $\varepsilon_0 = 1/3$ and $\tau_0 = 30$ ns (shown using up triangles and down triangles). The inset shows the ratios δT_1 and δT_A (shown using squares and circles).

The solutions $T_A(t,x)$ and $T_1(t,x)$ can be compared with $\check{T}_A(t,x)$ and $\check{T}_1(t,x)$, respectively. Let $\delta T(t,x) = (T(t,x) - \check{T}(t,x))/\check{T}(t,x)$ denote the relative contribution to the solution associated with the temporal dispersion of the dynamic heat capacity. This contribution reached about 50% (see Figure 2b). Thus, the effect of the temporal dispersion of the dynamic heat capacity on $T(t,r)$ was significant, especially at the beginning of the heating process. The position of the peak of the time dependence $\check{T}_1\left(t, \frac{d}{2}\right)$ was shifted relative to the peak of the heating pulse $F_1(t)$, as well as $\check{T}_A\left(t, \frac{d}{2}\right)$ relative to $F_A(t)$ (see Figure 2b). Indeed, the thermal response usually lagged behind the heat source. Thus, the peak of the dependences $\check{T}_1\left(t, \frac{d}{2}\right)$ and $\check{T}_A\left(t, \frac{d}{2}\right)$ appeared at around 8 ns and 7 ns, respectively. However, the maxima of $F_1(t)$ and $F_A(t)$ were at 5 ns and about 4 ns, respectively. Interestingly, this shift decreased due to the dynamic heat capacity. Thus, the peaks of the dependences $T_1\left(t, \frac{d}{2}\right)$ and $T_A\left(t, \frac{d}{2}\right)$ appeared around 7.5 ns and 6.5 ns, respectively. In fact, the dynamic heat capacity was less than the equilibrium heat capacity, especially at the beginning of the heating process.

The cylindrically symmetric problem can be solved in the same way as in [29]. Below, we consider only an example of a spherically symmetric problem.

5. Heat Equation with Dynamic Heat Capacity: Spherical Geometry

Consider a spherically symmetric problem with a spherical heat source $\Phi(r)F(t)$ of radius r_0 that is concentrically located in a spherical sample of radius R. The boundary condition is $T(t,R) = 0$ and the initial condition is $T(t,r) = 0$ for $t \leq 0$. From Equation (5), we have:

$$\frac{\partial}{\partial t}T(t,r) - D_0\frac{1}{r}\frac{\partial^2}{\partial r^2}[rT(t,r)] = \frac{\Phi(r)F(t)}{\rho c_0} + \varepsilon_0\frac{\partial}{\partial t}\int_0^t exp\left(-\frac{t-\tau}{\tau_0}\right)T'(\tau,r)d\tau. \quad (29)$$

By replacing $rT(t,r)$ with $U(t,r)$ in Equation (29), we obtain a one-dimensional problem, as seen in Equation (30); this problem is similar to that discussed earlier. The boundary and initial conditions are $U(t,0) = 0$, $U(t,R) = 0$, and $U(t,r) = 0$ at $t \leq 0$.

$$U'(t,r) - D_0\partial^2 U/\partial r^2 = \frac{r\Phi(r)F(t)}{\rho c_0} + \varepsilon_0\frac{\partial}{\partial t}\int_0^t exp\left(-\frac{t-\tau}{\tau_0}\right)U'(\tau,r)d\tau, \quad (30)$$

where the prime means the derivative with respect to the time variable. Thus, the solution $T(t, x)$ is:

$$T(t,r) = \sum_{n=1}^{\infty} \psi_n(t) \frac{\sin(\pi n r/R)}{r}, \tag{31}$$

where the functions $\psi_n(t)$ must satisfy Equation (32):

$$\psi'_n(t) + \frac{\psi_n(t)}{\tau_n} = \frac{r\Phi_n F(t)}{\rho c_0} + \varepsilon_0 \frac{\partial}{\partial t} \int_0^t \exp\left(-\frac{t-\tau}{\tau_0}\right) \psi'_n(\tau) d\tau, \tag{32}$$

where $\Phi_n = \frac{2}{R} \int_0^R r\Phi(r) \sin(\pi n r/R) dr$ and $\tau_n^{-1} = D_0 (\pi n/R)^2$. The functions $\psi_n(t)$ for different $F(t)$ can be obtained in the same way as before (see Example 1).

Example 2. Let the heat flux $\Phi(r)$ be uniformly distributed on the domain $[0, r_0]$ with density Φ_0. Then:

$$\Phi_n = 2R\Phi_0 \frac{\sin(\pi n r_0/R) - (\pi n r_0/R) \cdot \cos(\pi n r_0/R)}{(\pi n)^2}. \tag{33}$$

For heating pulses $F_p(t)$ and $F_A(t)$, similar to those in Figures 1 and 2, consider the solutions $\check{T}_p(t,r)$ and $T_p(t,r)$, as well as $\check{T}_A(t,r)$ and $T_A(t,r)$. In addition, consider the solutions $\check{T}_2(t,r)$ and $T_2(t,r)$ for the sinusoidal heat source $F_m(t)$ with $m = 2$. Let $R = 300$ nm, $r_0 = 30$ nm, $\tau_p = 2$ ns, $\varepsilon_0 = 1/3$, and $\tau_0 = 5$ ns, along with the same thermal parameters as in the above example. The series in Equation (31) converges as fast as the series $S_N = \sum_{n=1}^N 1/n^2$, which converges to $\pi^2/6$ as $N \to \infty$ [64]. The remainder $(\pi^2/6 - S_N)$ is less than 1% of $\pi^2/6$ for $N = 65$. Thus, to obtain an accuracy better than 1%, it is sufficient to take the series in Equation (31) up to $N = 100$. All the examples below were calculated with $N = 100$ and, as before, $M = 30$. Since $R \gg r_0$, it does not matter how large the parameter R is, as long as we consider a sufficiently short time t. This was verified by direct calculation of $\check{T}_A$ and T_A at $R = 300$ nm and 1000 nm (see Figure 3). Moreover, the solutions for different heating pulses practically did not change with the distance at $r > 80$ nm (see Figure 3). As seen in Figure 3, the effect of the temporal dispersion of the dynamic heat capacity was significant. Interestingly, the effect of the temporal dispersion of the dynamic heat capacity could even lead to a change in the sign of the solution $T_2(t,r)$. Indeed, $T_2(t,r)$ had a sign that was opposite to $\check{T}_2(t,r)$ at $r < 21$ nm and $t_0 = 1.5$ ns (see Figure 3b). A similar effect can occur when the temperature changes during the nucleation and decay of subcritical crystals.

The effect of the temporal dispersion of the dynamic heat capacity was most significant in the case of fast and local heating; this effect was most pronounced at the beginning of the heating process. This effect increased with increasing τ_0 and reached saturation at τ_0 in the order of r_0^2/D_0, that is, at about 10 ns at $r_0 = 30$ nm and D_0 in the order of 10^{-7} m^2/s (see Figure 4). Note, in the case of glass-forming substances, the relaxation times τ_0 have a broad distribution. Next, we considered the influence of this distribution on the solution $T(t,r)$ for the spherically symmetric problem.

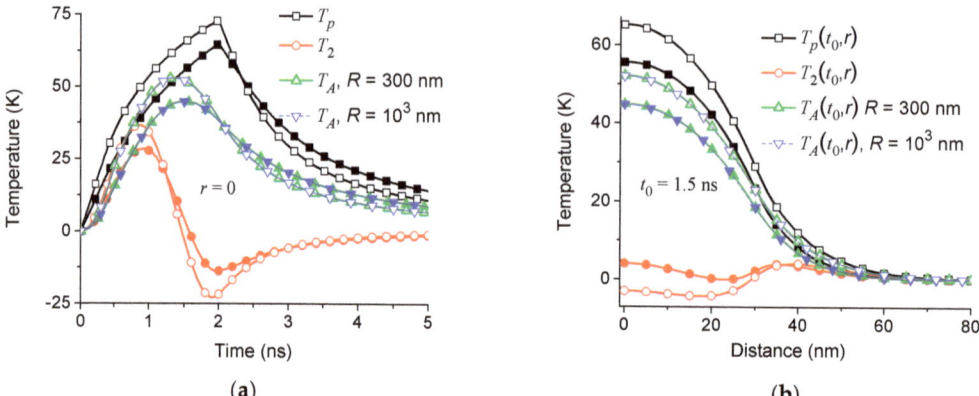

Figure 3. (a) Time dependences of $\check{T}_p(t,0)$, $\check{T}_2(t,0)$, and $\widetilde{T}_A(t,0)$ (shown using filled squares, circles, and up triangles, respectively), as well as $T_p(t,0)$, $T_2(t,0)$, and $T_A(t,0)$ for $\varepsilon_0 = 1/3$ and $\tau_0 = 5$ ns (shown using open squares, circles, and up triangles, respectively) and (b) the corresponding spatial temperature distributions at $t_0 = 1.5$ ns and $R = 300$ nm. Similar dependences for $\check{T}_A$ and T_A were calculated at $R = 1000$ nm (shown using down triangles).

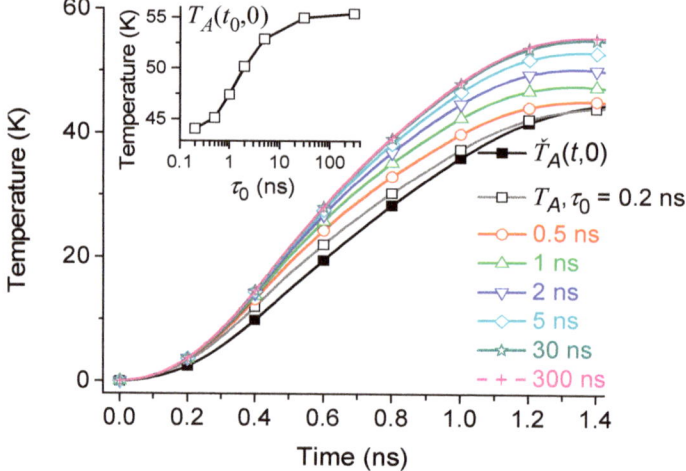

Figure 4. Time dependences of $T_A(t,0)$ at $\tau_0 = 0.2$ ns, 0.5 ns, 1 ns, 2 ns, 5 ns, 30 ns, and 300 ns (shown using squares, circles, up triangles, down triangles, diamonds, stars, and crosses, respectively), as well as $\check{T}_A(t,0)$ (shown using filled squares). The inset shows $T_A(t_0,0)$ as a function of τ_0 at $t_0 = 1.4$ ns.

6. Dependence of the Solution $T(t,r)$ on the Distribution of Relaxation Times τ_0

As before, consider a spherically symmetric problem with a spherical heat source $\Phi(r)F(t)$ of radius r_0 that is concentrically positioned in a spherical sample of radius R. The boundary and initial conditions are the same as in Section 5. Suppose the dynamic heat capacity is represented by Equation (3) or by the same equation in a finite interval $[\tau_{min}, \tau_{max}]$. Then, using the distribution function $H(\tau_0)$, we can obtain the desired solution as a linear combination of solutions for different τ_0. The distribution function $H(\tau_0)$ can be found using broadband heat capacity spectroscopy [19].

Example 3. *Consider the Kohlrausch relaxation law (stretched exponential), which is often used for glass-forming substances. Then, the dynamic heat capacity can be represented using Equation (34):*

$$c_{dyn}(t) = c_0\left[1 - \varepsilon_0 exp\left(-(t/\tau_K)^\beta\right)\right]. \tag{34}$$

For example, let $\beta = 0.5$. Then, the normalized distribution function is $H(\tau_0) = \frac{exp(-\tau_0/4\tau_K)}{\sqrt{4\pi\tau_K\tau_0}}$ [50,51], where the Kohlrausch relaxation time τ_K determines the distribution width, and the average relaxation time is $\langle\tau_0\rangle_{AV} = 2\tau_K$ [64]. In fact, the shape of the distribution function $H(\tau_0)$ is not very significant since the effect of the temporal dispersion of the dynamic heat capacity is saturated with increasing τ_0. For example, consider the uniform distribution $H_u(\tau_0) = \frac{1}{4\tau_K}$ on the interval $[0, 4\tau_K]$ and $H_u(\tau_0) = 0$ outside of this interval. In this case, the average relaxation time is also $2\tau_K$. Comparing the results calculated using the Kohlrausch distribution $H(\tau_0) = \frac{exp(-\tau_0/4\tau_K)}{\sqrt{4\pi\tau_K\tau_0}}$ and the uniform distribution $H_u(\tau_0)$, it can be seen that the results are very close to each other (see Figures 5 and 6). The value of the average relaxation time is a more significant parameter.

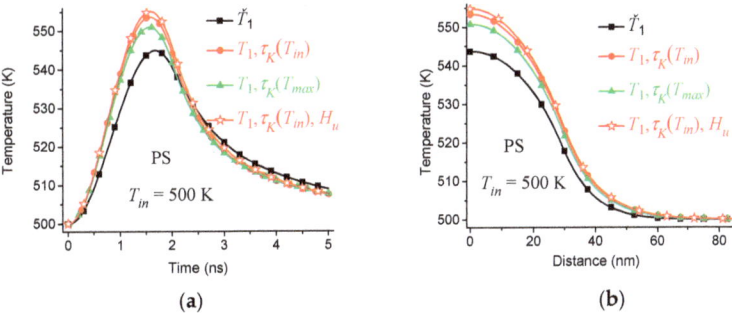

Figure 5. (a) Time dependences of $\check{T}_1(t,0)$, $T_1(t,0,\tau_K(T_{in}))$, and $T_1(t,0,\tau_K(T_{max}))$ (shown using squares, circles, and triangles, respectively) (b) and corresponding spatial temperature distributions at $t_0 = 1.5$ ns for PS at $T_{in} = 500$ K and a sinusoidal heat source $F_1(t)$. Similar dependences were found for $T_1(t,r,\tau_K(T_{in}))$ (shown using stars), which were calculated using the uniform distribution $H_u(\tau_0)$. PS: polystyrene.

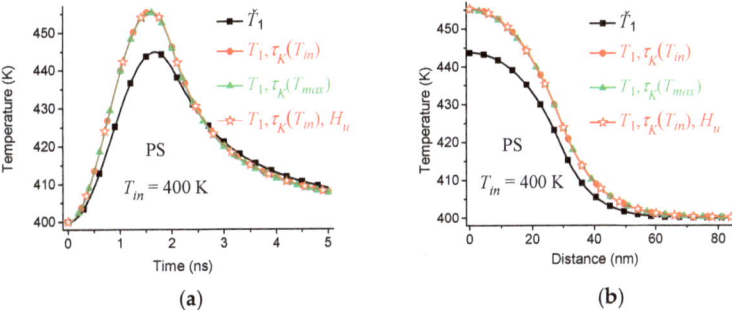

Figure 6. (a) Time dependences of $\check{T}_1(t,0)$, $T_1(t,0,\tau_K(T_{in}))$, and $T_1(t,0,\tau_K(T_{max}))$ (shown using squares, circles, and triangles, respectively) and (b) the corresponding spatial temperature distributions at $t_0 = 1.5$ ns for PS at $T_{in} = 400$ K and a sinusoidal heat source $F_1(t)$. Similar dependences were found for $T_1(t,r,\tau_K(T_{in}))$ (shown using stars), which were calculated using the uniform distribution $H_u(\tau_0)$.

The dynamic heat capacity $c_{dyn}(t)$ can be set on a finite interval $[\tau_{min}, \tau_{max}]$ if this interval is wide enough. It is sufficient to take $[\tau_{min}, \tau_{max}] = [0.001\tau_K, 15\tau_K]$ for an accuracy of about 1%. Thus:

$$c_{dyn}(t, \tau_K) = c_0 \left[1 - \varepsilon_0 \int_{\tau_{min}}^{\tau_{max}} \frac{exp(-\tau_0/4\tau_K)}{\sqrt{4\pi\tau_K\tau_0}} exp(-t/\tau_0) d\tau_0 \right]. \qquad (35)$$

The distribution function in Equation (35) depends on the Kohlrausch relaxation time τ_K; this relaxation time can be found using broadband heat capacity spectroscopy [19], as was done in [28]. Heat capacity spectroscopy allows one to obtain the frequency dependence of the dynamic heat capacity. Let ω_{max} denote the frequency of the maximum of the imaginary part of the dynamic heat capacity. In fact, $\tau_K = 0.74/\omega_{max}$ [28]. ω_{max} can be obtained for different temperatures T from the Vogel–Fulcher–Tammann–Hesse relation $log(\omega_{max}) = A - B/(T - T_0)$, where A, B, and T_0 are measured in the experiment [19]. Thus, one can obtain the Kohlrausch relaxation time $\tau_K(T)$ for different temperatures. Consider the solution $T(t, r, \tau_K(T_{in}))$ for $\tau_K(T_{in})$, where T_{in} is the initial temperature at $t \leq 0$. Compare this solution with $T(t, r, \tau_K(T_{max}))$, where T_{max} is the maximum temperature reached by the sample during the pulsed heating. Since $\tau_K(T)$ changes with heating between $\tau_K(T_{in})$ and $\tau_K(T_{max})$, it follows that the correct solution $T(t, r)$ is between the solutions obtained for $\tau_K(T_{in})$ and $\tau_K(T_{max})$, i.e., $T(t, r, \tau_K(T_{max})) < T(t, r) < T(t, r, \tau_K(T_{in}))$.

For example, let $T_{in} = 500$ K. Consider the solution $T_1(t, r)$ for the sinusoidal heat source $F_m(t)$ with $m = 1$, where $F_m(t) = sin(\pi m t/\tau_p)$ for $t \in [0, \tau_p]$ and $F_m(t) = 0$ outside $[0, \tau_p]$. Let us take the same thermal and size parameters as in the above example. Thus, we obtained $\tau_K(T_{in}) = 13.1$ ns for $T_{in} = 500$ K and $\tau_K(T_{max}) = 3.2$ ns for $T_{max} = 553$ K, using the parameters $A = 10.2$, $B = 388$ K, and $T_0 = 341.5$ K for polystyrene (PS) [19]. The difference between the solutions obtained for $\tau_K(T_{max})$ and $\tau_K(T_{in})$ was less than 5% of the temperature change near the maximum of the curves shown in Figure 5. The influence of the temporal dispersion of the dynamic heat capacity was much greater than this difference (see Figure 5). Moreover, the effect of changing $\tau_K(T)$ upon heating was insignificant at lower T_{in} since $\tau_K(T)$ increased with decreasing T and the influence of the temporal dispersion of the dynamic heat capacity saturated with increasing τ_K (see Figure 6). Therefore, we obtained $\tau_K(T_{in}) = 200$ μs for $T_{in} = 400$ K and $\tau_K(T_{max}) = 120$ ns for $T_{max} = 455$ K using the parameters A, B, and T_0 for PS [19]. Thus, we obtained the solutions $T(t, r, \tau_K(T_{in}))$ and $T(t, r, \tau_K(T_{max}))$, which were practically the same (see Figure 6). In fact, the difference between $T(t, r, \tau_K(T_{in}))$ and $T(t, r, \tau_K(T_{max}))$ was negligible in a wide temperature range from the glass transition temperature to 400 K for PS.

Similar calculations for poly(methyl methacrylate) (PMMA) provided an insignificant difference between the solutions $T(t, r, \tau_K(T_{in}))$ and $T(t, r, \tau_K(T_{max}))$ calculated in a wide temperature range from the glass transition temperature even up to 700 K (see Figure 7). For example, we obtained $\tau_K(T_{in}) = 127$ ns for $T_{in} = 700$ K and $\tau_K(T_{max}) = 107$ ns for $T_{max} = 755$ K using the parameters $A = 7.3$, $B = 185$ K, and $T_0 = 354.3$ K for PMMA [19]. Thus, the calculation of the influence of the temporal dispersion of the dynamic heat capacity for constant $\tau_K(T_{in})$ was quite accurate, even if the temperature change was about 50 K (see Figures 5–7).

Summing up, we concluded that the calculations can be carried out for a constant Kohlrausch relaxation time, determined at T_{in}, and for sufficiently large (about 50 K) temperature changes during heating. This holds for a wide temperature range for T_{in}.

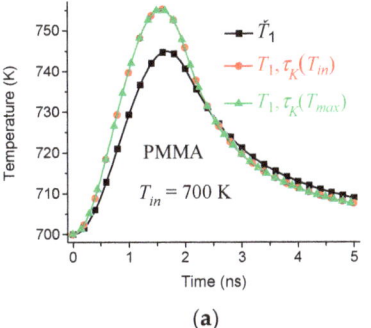 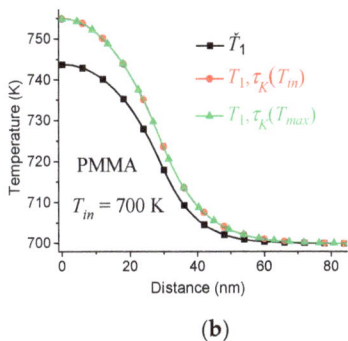

Figure 7. (a) Time dependences of $\breve{T}_1(t,0)$, $T_1(t,0,\tau_K(T_{in}))$, and $T_1(t,0,\tau_K(T_{max}))$ (shown using squares, circles, and triangles, respectively) and (b) the corresponding spatial temperature distributions at $t_0 = 1.5$ ns for PMMA at $T_{in} = 700$ K and a sinusoidal heat source $F_1(t)$. PMMA: poly(methyl methacrylate).

7. Conclusions

An integro-differential equation describes the non-equilibrium thermal response of glass-forming substances with a dynamic (time-dependent) heat capacity to fast thermal perturbations. We found that the corresponding heat transfer problem could be solved analytically for a heat source with an arbitrary time dependence and different geometries. The solutions provide analytical expressions for fast thermal processes on nanosecond and longer timescales. It was shown that the effect of the time dispersion of the dynamic heat capacity on the thermal response $T(t,r)$ upon local heating was significant in glass-forming materials. This effect was enhanced when the relaxation time of the dynamic heat capacity increased. However, this effect reached saturation at tens of nanoseconds. Because of this saturation, in many practical cases, the effect of the time dispersion of the dynamic heat capacity on the thermal response $T(t,r)$ can be calculated using the relaxation time distribution fixed at a constant initial temperature. This effect can be crucial for highly localized and short processes, such as crystal nucleation. The method of analytical calculations described in this work can be applied to an arbitrary time dependence of a local thermal perturbation, for example, the formation of a crystal nucleus in glass-forming materials. The results obtained can be useful for applications and a better understanding of the thermal properties of glass-forming substances, polymers, and nanocomposites.

Author Contributions: Conceptualization, A.A.M.; formal analysis, A.A.M.; funding acquisition, C.S.; methodology, A.A.M.; supervision, C.S.; visualization, A.A.M.; writing—original draft, A.A.M.; writing—review and editing, C.S. All authors have read and agreed to the published version of the manuscript.

Funding: This research was funded by the Ministry of Education and Science of the Russian Federation, grant number 14.Y26.31.0019. The APC was funded by Deutsche Forschungsgemeinschaft and Universität Rostock within the funding program Open Access Publishing.

Institutional Review Board Statement: Not applicable.

Informed Consent Statement: Not applicable.

Data Availability Statement: Not applicable.

Conflicts of Interest: The authors declare no conflict of interest.

Nomenclature
Latin Symbols

$c_{dyn}(t)$	dynamic heat capacity (J·kg^{-1}·K^{-1})
c_{in}	initial part of $c_{dyn}(t)$ (J·kg^{-1}·K^{-1})
c_0	equilibrium heat capacity (J·kg^{-1}·K^{-1})
D_0	thermal diffusivity $D_0 = \lambda/\rho c_0$ (m^2·s^{-1})
d	sample thickness (m)
$F(t)$	heat flux time dependence (dimensionless)
$H(\tau_0)$	distribution function (s^{-1})
h_0	heat release (J·kg^{-1})
l_{ph}	phonon mean-free-path (m)
r, x	space variables (m)
R	radius of spherical sample (m)
r_0	radius of spherical heat source (m)
t	time (s)
$T(t,x)$	solution to non-equilibrium heat equation (K)
$\check{T}(t,x)$	solution to conventional heat equation (K)
$\delta T(t,x)$	non-equilibrium component of the solution $T(t,x)$ (K)
x_0	thickness of the flat heat source (m)

Greek Symbols

β	Kohlrausch coefficient (dimensionless)
γ_n	nth relaxation parameter (s^{-1})
ε_0	$(c_0 - c_{in})/c_0$ (dimensionless)
$\theta(t)$	Heaviside unit step function (dimensionless)
λ	thermal conductivity (W·K^{-1}·m^{-1})
μ_n	nth relaxation parameter (s^{-1})
ρ	density (kg·m^{-3})
τ_K	Kohlrausch relaxation time (s)
τ_n	time constant of nth component (s)
τ_0	Debye relaxation time (s)
τ_p	duration of the heating pulse (s)
$\Phi(\vec{r})F(t)$	volumetric heat flux (W·m^{-3})
$\Phi(\vec{r})$	heat flux space dependence (W·m^{-3})
$\frac{\Phi_n}{\rho c_0}$	nth Fourier components (K/s)
χ_n, φ_n	nth Fourier components (K)
$\psi_{n,m}(t)$	n,mth Fourier component (K)

Appendix A

Consider the solutions $\psi_n(t)$ and $\varphi_n(t)$ of Equations (8) and (13), respectively. These equations are equivalent to Equations (A1) and (A2):

$$\psi'_n(t) + \frac{\psi_n(t)}{\tau_n} = \frac{\Phi_n F(t)}{\rho c_0} + \varepsilon_0 \psi'_n - \frac{\varepsilon_0}{\tau_0} \int_0^t \exp\left(-\frac{t-\tau}{\tau_0}\right) \psi'_n(\tau) d\tau, \qquad (A1)$$

$$\varphi'_n(t) + \frac{\varphi_n(t)}{\tau_n} = \frac{\Phi_n}{\rho c_0} \theta(t) + \varepsilon_0 \varphi'_n - \frac{\varepsilon_0}{\tau_0} \int_0^t \exp\left(-\frac{t-\tau}{\tau_0}\right) \varphi'_n(\tau) d\tau, \qquad (A2)$$

where their Laplace transforms are:

$$\left(p + \tau_n^{-1}\right)\psi_n(p) = \frac{\Phi_n}{\rho c_0} F(p) + p\varepsilon_0 \psi_n(p) - \frac{\varepsilon_0}{\tau_0} \frac{p\psi_n(p)}{\left(p + \tau_0^{-1}\right)}, \qquad (A3)$$

$$\left(p + \tau_n^{-1}\right)\varphi_n(p) = \frac{\Phi_n}{\rho c_0} p^{-1} + p\varepsilon_0 \varphi_n(p) - \frac{\varepsilon_0}{\tau_0} \frac{p\varphi_n(p)}{\left(p + \tau_0^{-1}\right)}, \qquad (A4)$$

where $\psi_n(p)$, $\varphi_n(p)$, and $F(p)$ are the Laplace transforms of $\psi_n(t)$, $\varphi_n(t)$, and $F(t)$, respectively. Let us verify that the solution to Equation (A1) is equal to the following Duhamel integral [50,51]:

$$\psi_n(t) = \int_0^t F(\tau)\varphi'_n(t-\tau)d\tau. \tag{A5}$$

Note that $\varphi_n(0) = 0$ (see Equation (14)). Then, the Laplace transform of Equation (A5) is equal to $\psi_n(p) = F(p)p\varphi_n(p)$ [62]. By assumption, the functions $\varphi_n(p)$ satisfy Equation (A4). Accordingly, substituting $\psi_n(p)/F(p)p$ for $\varphi_n(p)$ into Equation (A4) and multiplying both sides of Equation (A4) by the factor $F(p)p$, we obtain Equation (A3). Then, after the inverse Laplace transform, we have Equation (A1). Therefore, the Duhamel integral (see Equation (A5)) is a solution to Equation (A1). In fact, the existence of the solution $\psi_n(t)$ follows from Equation (A5) since $F(t)$ and $\varphi_n(t)$ are smooth and bounded, and therefore, integrable functions. Moreover, since Equation (A1) is linear with respect to $\psi_n(t)$, the uniqueness of the solution $\psi_n(t)$ follows from the fact that $\psi_n(t) = 0$ if $F(t) = 0$ (see Equation (A5)).

Below, we use some results from our previous article [28]. Equation (A4) is equivalent to Equation (A6):

$$\varphi_n(p) = \frac{\Phi_n}{\rho c_0} \frac{\left(p + \tau_0^{-1}\right)}{p\left[(1-\varepsilon_0)p^2 + p\left(\tau_n^{-1} + \tau_0^{-1}\right) + \tau_n^{-1}\tau_0^{-1}\right]}. \tag{A6}$$

Therefore:

$$\varphi_n(p) = \frac{\Phi_n}{\rho c_0} \frac{\left(p + \tau_0^{-1}\right)}{(1-\varepsilon_0)p(p+\gamma_n)(p+\mu_n)}, \tag{A7}$$

where

$$\gamma_n = \frac{\left(\tau_n^{-1} + \tau_0^{-1}\right) + \sqrt{\left(\tau_n^{-1} - \tau_0^{-1}\right)^2 + 4\varepsilon_0/\tau_n\tau_0}}{2(1-\varepsilon_0)}, \tag{A8}$$

$$\mu_n = \frac{\left(\tau_n^{-1} + \tau_0^{-1}\right) - \sqrt{\left(\tau_n^{-1} - \tau_0^{-1}\right)^2 + 4\varepsilon_0/\tau_n\tau_0}}{2(1-\varepsilon_0)}. \tag{A9}$$

Thus, from Equation (A7) and using the identity $(1-\varepsilon_0)\tau_0\gamma_n\mu_n = \tau_n^{-1}$, we have:

$$\varphi_n(p) = \frac{\Phi_n}{\rho c_0}\tau_n\left[p^{-1} + \frac{\tau_0\gamma_n\mu_n}{(\gamma_n - \mu_n)}\left(\frac{1}{p+\mu_n} - \frac{1}{p+\gamma_n}\right) + \frac{1}{(\gamma_n - \mu_n)}\left(\frac{\mu_n}{p+\gamma_n} - \frac{\gamma_n}{p+\mu_n}\right)\right]. \tag{A10}$$

Note that γ_n and μ_n are real-valued, positive, and $\gamma_n - \mu_n > 0$ for any τ_n, τ_0, and $0 < \varepsilon_0 < 1$. Then, after the inverse Laplace transform of Equation (A10), we obtain the solution $\varphi_n(t)$ of Equation (13) (see Equation (14)).

References

1. Birge, N.O.; Nagel, S.R. Specific-heat spectroscopy of the glass transition. *Phys. Rev. Lett.* **1985**, *54*, 2674–2677. [CrossRef] [PubMed]
2. Herzfeld, K.F.; Litovitz, T.A. *Absorption and Dispersion of Ultrasonic Waves*; Academic Press Inc.: New York, NY, USA; London, UK, 1959.
3. Korus, J.; Beiner, M.; Busse, K.; Kahle, S.; Unger, R.; Donth, E. Heat capacity spectroscopy at the glass transition in polymers. *Thermochim. Acta* **1997**, *305*, 99–110. [CrossRef]
4. Beiner, M.; Korus, J.; Lockwenz, H.; Schröter, K.; Donth, E. Heat Capacity Spectroscopy Compared to Other Linear Response Methods at the Dynamic Glass Transition in Poly(vinyl acetate). *Macromolecules* **1996**, *29*, 5183–5189. [CrossRef]
5. Huth, H.; Beiner, M.; Donth, E. Temperature dependence of glass-transition cooperativity from heat-capacity spectroscopy: Two post-Adam-Gibbs variants. *Phys. Rev. B* **2000**, *61*, 15092–15101. [CrossRef]
6. Donth, E.-J. *The Glass Transition*; Springer Science and Business Media LLC: New York, NY, USA, 2001.
7. Schneider, K.; Donth, E. Unterschiedliche Meßsignale am Glasübergang amorpher Polymere. 1. Die Lage der charakteristischen Frequenzen quer zur Glasübergangszone. *Acta Polym.* **1986**, *37*, 333–335. [CrossRef]
8. Ediger, M.D.; Angell, C.A.; Nagel, S.R. Supercooled Liquids and Glasses. *J. Phys. Chem.* **1996**, *100*, 13200–13212. [CrossRef]

9. Gotze, W.; Sjogren, L.B. Relaxation processes in supercooled liquids. *Rep. Prog. Phys.* **1992**, *55*, 241–376. [CrossRef]
10. Hansen, C.; Stickel, F.; Berger, T.J.; Richert, R.; Fischer, E.W. Dynamics of glass-forming liquids. III. Comparing the dielectric α- and β-relaxation of 1-propanol and o-terphenyl. *J. Chem. Phys.* **1997**, *107*, 1086–1093. [CrossRef]
11. Berthier, L.; Biroli, G. Theoretical perspective on the glass transition and amorphous materials. *Rev. Mod. Phys.* **2011**, *83*, 587–645. [CrossRef]
12. Cangialosi, D. Dynamics and thermodynamics of polymer glasses. *J. Phys. Condens. Matter* **2014**, *26*, 153101. [CrossRef]
13. Angell, C.A.; Ngai, K.L.; McKenna, G.B.; McMillan, P.F.; Martin, S.W. Relaxation in glassforming liquids and amorphous solids. *J. Appl. Phys.* **2000**, *88*, 3113–3157. [CrossRef]
14. Smith, G.D.; Bedrov, D. Relationship between the α- and β-relaxation processes in amorphous polymers: Insight from atomistic molecular dynamics simulations of 1,4-polybutadiene melts and blends. *J. Polym. Sci. Part B Polym. Phys.* **2007**, *45*, 627–643. [CrossRef]
15. Bock, D.; Petzold, N.; Kahlau, R.; Gradmann, S.; Schmidtke, B.; Benoit, N.; Rössler, E. Dynamic heterogeneities in glass-forming systems. *J. Non-Cryst. Solids* **2015**, *407*, 88–97. [CrossRef]
16. Richert, R. Physical Aging and Heterogeneous Dynamics. *Phys. Rev. Lett.* **2010**, *104*. [CrossRef] [PubMed]
17. Saiter-Fourcin, A.; Delbreilh, L.; Couderc, H.; Arabeche, K.; Schönhals, A.; Saiter, J.-M. Temperature dependence of the characteristic length scale for glassy dynamics: Combination of dielectric and specific heat spectroscopy. *Phys. Rev. E* **2010**, *81*, 041805. [CrossRef] [PubMed]
18. Chua, Y.Z.; Young-Gonzales, A.R.; Richert, R.; Ediger, M.D.; Schick, C. Dynamics of supercooled liquid and plastic crystalline ethanol: Dielectric relaxation and AC nanocalorimetry distinguish structural α- and Debye relaxation processes. *J. Chem. Phys.* **2017**, *147*, 014502. [CrossRef]
19. Chua, Y.Z.; Schulz, G.; Shoifet, E.; Huth, H.; Zorn, R.; Scmelzer, J.W.P.; Schick, C. Glass transition cooperativity from broad band heat capacity spectroscopy. *Colloid Polym. Sci.* **2014**, *292*, 1893–1904. [CrossRef]
20. Chua, Y.Z.; Zorn, R.; Holderer, O.; Schmelzer, J.W.P.; Schick, C.; Donth, E. Temperature fluctuations and the thermodynamic determination of the cooperativity length in glass forming liquids. *J. Chem. Phys.* **2017**, *146*, 104501. [CrossRef]
21. Richert, R. Heterogeneous dynamics in liquids: Fluctuations in space and time. *J. Phys. Condens. Matter* **2002**, *14*, R703–R738. [CrossRef]
22. Larini, L.; Ottochian, A.; De Michele, C.; Leporini, D. Universal scaling between structural relaxation and vibrational dynamics in glass-forming liquids and polymers. *Nat. Phys.* **2007**, *4*, 42–45. [CrossRef]
23. Voudouris, P.; Gomopoulos, N.; Le Grand, A.; Hadjichristidis, N.; Floudas, G.; Ediger, M.D.; Fytas, G. Does Brillouin light scattering probe the primary glass transition process at temperatures well above glass transition? *J. Chem. Phys.* **2010**, *132*, 074906. [CrossRef] [PubMed]
24. Lubchenko, V.; Wolynes, P.G. Theory of Structural Glasses and Supercooled Liquids. *Annu. Rev. Phys. Chem.* **2007**, *58*, 235–266. [CrossRef]
25. Pipkin, A.C. *Lectures on Viscoelasticity Theory*; Springer: Berlin/Heidelberg, Germany, 1986; ISBN 978-1-4612-1078-8. Available online: https://www.springer.com/gp/book/9780387963457 (accessed on 2 February 2021).
26. Ikeda, M.; Aniya, M. Bond Strength—Coordination Number Fluctuation Model of Viscosity: An Alternative Model for the Vogel-Fulcher-Tammann Equation and an Application to Bulk Metallic Glass Forming Liquids. *Matererials* **2010**, *3*, 5246–5262. [CrossRef] [PubMed]
27. Minakov, A.; Schick, C. Non-equilibrium fast thermal response of polymers. *Thermochim. Acta* **2018**, *660*, 82–93. [CrossRef]
28. Minakov, A.; Schick, C. Nanometer scale thermal response of polymers to fast thermal perturbations. *J. Chem. Phys.* **2018**, *149*, 074503. [CrossRef]
29. Minakov, A.; Schick, C. Nanoscale Heat Conduction in CNT-POLYMER Nanocomposites at Fast Thermal Perturbations. *Molecules* **2019**, *24*, 2794. [CrossRef]
30. Noeiaghdam, S.; Sidorov, D.; Muftahov, I.R.; Zhukov, A.V. Control of Accuracy on Taylor-Collocation Method for Load Leveling Problem. *Bull. Irkutsk. State Univ. Ser. Math.* **2019**, *30*, 59–72. [CrossRef]
31. Sidorov, D.; Muftahov, I.; Tomin, N.; Karamov, D.N.; Panasetsky, D.A.; Dreglea, A.; Liu, F.; Foley, A. A Dynamic Analysis of Energy Storage With Renewable and Diesel Generation Using Volterra Equations. *IEEE Trans. Ind. Inform.* **2020**, *16*, 3451–3459. [CrossRef]
32. Noeiaghdam, S.; Dreglea, A.; He, J.-H.; Avazzadeh, Z.; Suleman, M.; Araghi, M.A.F.; Sidorov, D.; Sidorov, N.A. Error Estimation of the Homotopy Perturbation Method to Solve Second Kind Volterra Integral Equations with Piecewise Smooth Kernels: Application of the CADNA Library. *Symmetry* **2020**, *12*, 1730. [CrossRef]
33. Wang, Y.; Ezz-Eldien, S.; Aldraiweesh, A.A. A new algorithm for the solution of nonlinear two-dimensional Volterra integro-differential equations of high-order. *J. Comput. Appl. Math.* **2020**, *364*, 112301. [CrossRef]
34. Noeiaghdam, S. Numerical solution of N-th order fredholm integro-differential equations by integral mean value theorem method. *Int. J. Pure Appl. Math.* **2015**, *99*. [CrossRef]
35. Miller, R.K. An integrodifferential equation for rigid heat conductors with memory. *J. Math. Anal. Appl.* **1978**, *66*, 313–332. [CrossRef]
36. Tolstov, G.P. *Fourier Series*; Dover Publications, Inc.: New York, NY, USA, 1962.
37. Koh, Y.K.; Cahill, D.G.; Sun, B. Nonlocal theory for heat transport at high frequencies. *Phys. Rev. B* **2014**, *90*, 205412. [CrossRef]
38. Wingert, M.C.; Zheng, J.; Kwon, S.; Chen, R. Thermal transport in amorphous materials: A review. *Semicond. Sci. Technol.* **2016**, *31*, 113003. [CrossRef]

39. Feng, T.; Ruan, X. Prediction of Spectral Phonon Mean Free Path and Thermal Conductivity with Applications to Thermoelectrics and Thermal Management: A Review. *J. Nanomater.* **2014**, *2014*, 1–25. [CrossRef]
40. Stoner, R.J.; Maris, H.J. Kinetic formula for estimating resistive phonon lifetimes. *Phys. Rev. B* **1993**, *47*, 11826–11829. [CrossRef]
41. He, Y.; Donadio, D.; Galli, G. Heat transport in amorphous silicon: Interplay between morphology and disorder. *Appl. Phys. Lett.* **2011**, *98*, 144101. [CrossRef]
42. Zeller, R.C.; Pohl, R.O. Thermal Conductivity and Specific Heat of Noncrystalline Solids. *Phys. Rev. B* **1971**, *4*, 2029–2041. [CrossRef]
43. Choy, C.L.; Tong, K.W.; Wong, H.K.; Leung, W.P. Thermal conductivity of amorphous alloys above room temperature. *J. Appl. Phys.* **1991**, *70*, 4919–4925. [CrossRef]
44. Sørensen, S.S.; Johra, H.; Mauro, J.C.; Bauchy, M.; Smedskjaer, M.M. Boron anomaly in the thermal conductivity of lithium borate glasses. *Phys. Rev. Mater.* **2019**, *3*, 075601. [CrossRef]
45. Sørensen, S.S.; Pedersen, E.J.; Paulsen, F.K.; Adamsen, I.H.; Laursen, J.L.; Christensen, S.; Johra, H.; Jensen, L.R.; Smedskjaer, M.M. Heat conduction in oxide glasses: Balancing diffusons and propagons by network rigidity. *Appl. Phys. Lett.* **2020**, *117*, 031901. [CrossRef]
46. Choy, C. Thermal conductivity of polymers. *Polymer* **1977**, *18*, 984–1004. [CrossRef]
47. Hartwig, G. *Polymer Properties at Room and Cryogenic Temperatures*; Springer Nature: Berlin/Heidelberg, Germany, 1994.
48. Landau, L.D.; Lifshitz, E.M. *Course of Theoretical Physics 5: Statistical Physics Part 1*, 3rd ed.; Pergamon Press: Oxford, UK, 1980.
49. Landau, L.D.; Lifshitz, E.M. *Course of Theoretical Physics 8: Electrodynamics of Continuous Media*, 2nd ed.; Butterworth-Heinemann: Oxford, UK, 2000.
50. Johnston, D.C. Stretched exponential relaxation arising from a continuous sum of exponential decays. *Phys. Rev. B* **2006**, *74*, 184430. [CrossRef]
51. Berberan-Santos, M.; Bodunov, E.; Valeur, B. Mathematical functions for the analysis of luminescence decays with underlying distributions 1. Kohlrausch decay function (stretched exponential). *Chem. Phys.* **2005**, *315*, 171–182. [CrossRef]
52. Anderssen, R.S.; Loy, R.J. Completely monotone fading memory relaxation modulii. *Bull. Aust. Math. Soc.* **2002**, *65*, 449–460. [CrossRef]
53. Schilling, R.L.; Song, R.; Vondraček, Z. *Bernstein Functions: Theory and Applications*; Hubert: Berlin, Germany, 2010; ISBN 978-3-11-021530-4.
54. Gupta, P.K.; Moynihan, C.T. Prigogine–Defay ratio for systems with more than one order parameter. *J. Chem. Phys.* **1976**, *65*, 4136–4140. [CrossRef]
55. Tournier, R.F. Formation temperature of ultra-stable glasses and application to ethylbenzene. *Chem. Phys. Lett.* **2015**, *641*, 9–13. [CrossRef]
56. You, K.; Yan, G.; Luo, X.; Gilchrist, M.D.; Fang, F. Advances in laser assisted machining of hard and brittle materials. *J. Manuf. Process.* **2020**, *58*, 677–692. [CrossRef]
57. Sahin, A.Z.; Yilbas, B.; Akhtar, S.S. Laser Surface Treatment and Efficiency Analysis. *Compr. Mater. Process.* **2014**, *9*, 307–316. [CrossRef]
58. Dreglea, A.I.; Sidorov, N.A. Integral equations in identification of external force and heat source density dynamics, Buletinul Academiei de Stiinte a Republicii Moldova. *Matematica* **2018**, *88*, 68–77. Available online: http://www.math.md/files/basm/y2018-n3/y2018-n3-(pp68-77).pdf (accessed on 2 February 2021).
59. Luo, C.; Sommer, J.-U. Frozen topology: Entanglements control nucleation and crystallization in polymers. *Phys. Rev. Lett.* **2014**, *112*, 195702. [CrossRef] [PubMed]
60. Wyslouzil, B.E.; Seinfeld, J.H. Nonisothermal homogeneous nucleation. *J. Chem. Phys.* **1992**, *97*, 2661–2670. [CrossRef]
61. Mahata, A.; Zaeem, M.A.; Baskes, M.I. Understanding homogeneous nucleation in solidification of aluminum by molecular dynamics simulations. *Model. Simul. Mater. Sci. Eng.* **2018**, *26*, 025007. [CrossRef]
62. Vladimirov, V.S. *A Collection of Problems on the Equations of Mathematical Physics*; Springer: Berlin/Heidelberg, Germany, 1986; Available online: https://www.springer.com/gp/book/9783662055601 (accessed on 2 February 2021).
63. Polyanin, A.D.; Manzhirov, A.V. *Handbook of mathematics for engineers and scientists*; Chapman & Hall/CRC: Boca Raton, FL, USA; London, UK, 2007.
64. Korn, G.A.; Korn, T.M. *Mathematical Handbook for Scientists and Engineers*, 2nd ed; Dover Publications, Inc.: Mineola, NY, USA, 2000.

Article

A Type of Time-Symmetric Stochastic System and Related Games

Qingfeng Zhu [1,2], Yufeng Shi [2,*], Jiaqiang Wen [3] and Hui Zhang [1]

[1] School of Mathematics and Quantitative Economics, Shandong University of Finance and Economics, and Shandong Key Laboratory of Blockchain Finance, Jinan 250014, China; qfzhu@sdufe.edu.cn (Q.Z.); zhanghui@sdufe.edu.cn (H.Z.)
[2] Institute for Financial Studies and School of Mathematics, Shandong University, Jinan 250100, China
[3] Department of Mathematics, Southern University of Science and Technology, Shenzhen 518055, China; wenjq@sustech.edu.cn
* Correspondence: yfshi@sdu.edu.cn

Abstract: This paper is concerned with a type of time-symmetric stochastic system, namely the so-called forward–backward doubly stochastic differential equations (FBDSDEs), in which the forward equations are delayed doubly stochastic differential equations (SDEs) and the backward equations are anticipated backward doubly SDEs. Under some monotonicity assumptions, the existence and uniqueness of measurable solutions to FBDSDEs are obtained. The future development of many processes depends on both their current state and historical state, and these processes can usually be represented by stochastic differential systems with time delay. Therefore, a class of nonzero sum differential game for doubly stochastic systems with time delay is studied in this paper. A necessary condition for the open-loop Nash equilibrium point of the Pontriagin-type maximum principle are established, and a sufficient condition for the Nash equilibrium point is obtained. Furthermore, the above results are applied to the study of nonzero sum differential games for linear quadratic backward doubly stochastic systems with delay. Based on the solution of FBDSDEs, an explicit expression of Nash equilibrium points for such game problems is established.

Keywords: backward doubly stochastic differential equations; stochastic differential game; maximum principle; Nash equilibrium point; time-delayed generator

1. Introduction

In 1994, Pardoux and Peng [1] put forward the following backward doubly stochastic differential equations (BDSDEs):

$$p(t) = \xi + \int_t^T F(s, p(s), q(s))ds + \int_t^T G(s, p(s), q(s)) \overleftarrow{d} B(s) - \int_t^T q(s) \overrightarrow{d} W(s), \ 0 \leq t \leq T, \quad (1)$$

which can be applied to produce a probabilistic expression of certain quasilinear stochastic partial differential equations (SPDEs). Because of its importance to SPDEs, the interest in BDSDEs has increased considerably (see [2–15]). At the same time, the stochastic control problem of backward doubly stochastic systems has been studied extensively (see [16–21]).

In 2003, Peng and Shi [22] introduced the following time-symmetric fully coupled forward–backward stochastic systems:

$$\begin{cases} y(t) &= x + \int_0^t f(s, p(s), y(s), q(s), z(s))ds - \int_0^t z(s) \overleftarrow{d} B(s) \\ &\quad + \int_0^t g(s, p(s), y(s), q(s), z(s)) \overrightarrow{d} W(s), \\ p(t) &= \Phi(y(T)) + \int_t^T F(s, p(s), y(s), q(s), z(s))ds - \int_t^T q(s) \overrightarrow{d} W(s) \\ &\quad + \int_t^T G(s, p(s), y(s), q(s), z(s)) \overleftarrow{d} B(s), \end{cases} \quad (2)$$

which are the so-called forward–backward doubly stochastic differential equations (FBDS-DEs). The forward and backward equations in Equation (2) are the BDSDE in Equation (1) with stochastic integrals in different directions. Therefore, the FBDSDE in Equation (2) is established to provide a more general framework of fully coupled forward–backward stochastic differential equations. Under some monotone assumptions, Peng and Shi [22] obtained the unique solvability of FBDSDEs (2). Zhu et al. [23,24] have extended the results in [22] to different dimensions and the weaker monotonic assumptions, and gave the probabilistic interpretation for the solutions to SPDEs combined with algebra equations. Zhang and Shi [25] and Shi and Zhu [26] studied the stochastic control problem of FBDSDEs.

Game theory has penetrated into economic theory and attracted more and more research. It was first proposed by Von Neumann and Morgenstern [27]. Nash [28] has done groundbreaking work on non-cooperative games and presents the concept of a classic Nash equilibrium. Zhao et al. [29] studied the optimal investment and reinsurance of insurers in default securities under a mean-variance criterion in the jump-diffusion risk model. Many papers on stochastic differential game problems driven by backward stochastic differential equations have been published (see [30–32]). The differential game problem for forward–backward doubly stochastic differential equations was addressed in [33]. However, the future evolution of a lot of processes depends not only on their current state, but also on their historical state, and these processes can usually be characterized by stochastic differential equations with time delay. The optimal control problem for stochastic differential equations with delay was discussed in [34–39]. The nonzero sum differential game of the stochastic differential delay equation was studied in [40,41]. Shen and Zeng [42] researched the optimal investment and reinsurance with time delay for insurers under a mean-variance criterion.

The extra noise $\{B(t)\}$ in Equation (1) can be regarded as some additional financial information that is not disclosed to the public in practice, such as in the derivative securities market, but is available to some investors. Arriojas et al. [43] and Kazmerchuk et al. [44] obtained the option pricing formula with time delay based on the stock price process with time delay. As far as we know, there is little discussion about differential games of doubly stochastic systems with delay. In this article, we will discuss this direction, that is, the following nonzero sum differential game driven by doubly stochastic systems with time delay. The control system is

$$\begin{cases} dy(t) &= f(t,y(t),z(t),y_\delta(t),z_\delta(t),v_1(t),v_2(t))dt - z(t)\overleftarrow{d}B(t) \\ &\quad + g(t,y(t),z(t),y_\delta(t),z_\delta(t),v_1(t),v_2(t))\overrightarrow{d}W(t), \; t \in [0,T], \\ y(t) &= \phi(t), \; z(t) = \psi(t), \; t \in [-\delta,0], \end{cases}$$

where $(y(\cdot),z(\cdot)) \in \mathbb{R}^n \times \mathbb{R}^{n\times d}$ is the state process pair, $0 < \delta < T$ is a constant time delay parameter, and $y_\delta(t) = y(t-\delta), z_\delta(t) = z(t-\delta)$. We denote $J_1(v(\cdot))$ and $J_2(v(\cdot))$, $v(\cdot) = (v_1(\cdot),v_2(\cdot))$, which are the cost functionals corresponding to the players 1 and 2:

$$J_i(v_1(\cdot),v_2(\cdot)) = \mathbb{E}\left\{\int_0^T l_i(t,y(t),z(t),y_\delta(t),z_\delta(t),v_1(t),v_2(t))dt + \Phi_i(y(T))\right\}, \; i=1,2.$$

Our task is to find $(u_1(\cdot),u_2(\cdot)) \in \mathcal{U}_1 \times \mathcal{U}_2$ such that

$$\begin{cases} J_1(u_1(\cdot),u_2(\cdot)) = \min_{v_1(\cdot)\in\mathcal{U}_1} J_1(v_1(\cdot),u_2(\cdot)), \\ J_2(u_1(\cdot),u_2(\cdot)) = \min_{v_2(\cdot)\in\mathcal{U}_2} J_2(u_1(\cdot),v_2(\cdot)). \end{cases}$$

To figure out the above nonzero sum differential game problem, it is natural to involve the adjoint equation, which is a kind of anticipated BDSDE (see [45,46]). It is therefore necessary to explore the following general FBDSDE with the forward equation being a delayed doubly SDE and the backward equation being the anticipated BDSDE:

$$\begin{cases} dy(t) &= f(t,y(t),p(t),z(t),q(t),y_\delta(t),z_\delta(t))dt - z(t)\overleftarrow{d}B(t) \\ & \quad + g(t,y(t),p(t),z(t),q(t),y_\delta(t),z_\delta(t))\overrightarrow{d}W(t),\, t \in [0,T], \\ -dp(t) &= F(t,y(t),p(t),z(t),q(t),p_{\delta+}(t),q_{\delta+}(t))dt - q(t)\overrightarrow{d}W(t) \\ & \quad + G(t,y(t),p(t),z(t),q(t),p_{\delta+}(t),q_{\delta+}(t))\overleftarrow{d}B(t),\, t \in [0,T], \\ y(t) &= \phi(t),\, t \in [-\delta,0], \\ z(t) &= \psi(t),\, t \in [-\delta,0], \\ p(T) &= \Phi(y(T)),\, p(t) = \xi(t),\, t \in [T,T+\delta], \\ q(t) &= \eta(t),\, t \in [T,T+\delta], \end{cases}$$

where $y_\delta(t) = y(t-\delta), z_\delta = z(t-\delta), p_{\delta+}(t) = p(t+\delta), q_{\delta+} = q(t+\delta)$.

Our work differs from the above in the following distinctions. First of all, we introduce a time-symmetric stochastic system, which generalizes the results in [22] to a more general case: forward doubly stochastic differential equations (SDEs) with delay as forward equations and anticipated backward doubly stochastic differential equations as backward equations. Secondly, we investigate the problem of a nonzero sum differential game driven by doubly stochastic systems with time delay, which enriches the types of stochastic delayed differential game problems. Finally, we explore the linear quadratic (LQ) games for a doubly stochastic system with time delay, and use the solution of the above general FBDSDE to give an explicit expression of the unique equilibrium point.

The structure of this paper is as follows. We give the framework of the doubly stochastic games with delay and a preliminary view on the general FBDSDE in Section 2. We set up a necessary condition for the open-loop Nash equilibrium of such games to form a Pontryagin maximum principle in Section 3. Section 4 is devoted to the verification theorem of a sufficient condition for Nash equilibrium. In order to visually demonstrate the above results, the nonzero sum differential game for LQ double stochastic delay systems is studied in Section 5. By using the results of our FBDSDE, the explicit representation of Nash equilibrium points for LQ game problems is obtained. For the convenience of the reader, we present the skeleton of the proof on uniqueness and existence for the general FBDSDE in Section 6. Finally, we conclude this article with a summary.

2. Formulation of Problems and Preliminaries

2.1. Notations and Formulation of Problems

Suppose $(\Omega,\mathcal{F},P)$ is a probability space, and $[0,T]$ is a fixed arbitrarily large time duration throughout this paper. Let $\{W(t); 0 \leq t \leq T\}$ and $\{B(t); 0 \leq t \leq T\}$ be two mutually independent standard Brownian motions defined on $(\Omega,\mathcal{F},P)$, with values in $\mathbb{R}^d$ and $\mathbb{R}^l$, respectively. Let $\mathcal{N}$ denote the class of P-null elements of $\mathcal{F}$. For each $t \in [0,T]$, we define $\mathcal{F}_t \doteq \mathcal{F}_t^W \vee \mathcal{F}_{t,T}^B$, where $\mathcal{F}_t^W = \mathcal{N} \vee \sigma\{W(r) - W(0); 0 \leq r \leq t\}$, $\mathcal{F}_{t,T}^B = \mathcal{N} \vee \sigma\{B(r) - B(r); t \leq r \leq T\}$. Note that the collection $\{\mathcal{F}_t, t \in [0,T]\}$ is neither increasing nor decreasing, and it does not produce a filtration. $\mathbb{E}$ denotes the expectation on $(\Omega,\mathcal{F},P)$. $\mathbb{E}^{\mathcal{F}_t} := \mathbb{E}[\cdot|\mathcal{F}_t]$ denotes the conditional expectation under $\mathcal{F}_t$. We use the usual inner product $\langle\cdot,\cdot\rangle$ and Euclidean norm $|\cdot|$ in $\mathbb{R}^n, \mathbb{R}^m, \mathbb{R}^{m\times l}$ and $\mathbb{R}^{n\times d}$. The symbol "$\top$" that appears on the superscript indicates the transpose of the matrix. All the equations and inequalities mentioned in this paper are in the sense of $dt \times dP$ almost surely on $[0,T] \times \Omega$. We introduce the following notations:

$$L^2(\mathcal{F}_T;\mathbb{R}^n) = \{\xi : \xi \text{ is an } \mathbb{R}^n\text{-valued, } \mathcal{F}_T\text{-measurable random variable s.t. } \mathbb{E}|\xi|^2 < \infty\},$$

$$L^2_{\mathcal{F}}(0,T;\mathbb{R}^n) = \{v(t), 0 \leq t \leq T : v(t) \text{ is an } \mathbb{R}^n\text{-valued, } \mathcal{F}_t\text{-measurable process}$$

$$\text{s.t. } \mathbb{E}\int_0^T |v(t)|^2 dt < \infty\}.$$

We take into account the following controlled doubly stochastic differential systems with delay:

$$\begin{cases} dy(t) &= f(t,y(t),z(t),y_\delta(t),z_\delta(t),v_1(t),v_2(t))dt - z(t)\overleftarrow{d}B(t) \\ &\quad + g(t,y(t),z(t),y_\delta(t),z_\delta(t),v_1(t),v_2(t))\overrightarrow{d}W(t),\ t \in [0,T], \\ y(t) &= \phi(t),\ z(t) = \psi(t),\ t \in [-\delta,0], \end{cases} \quad (3)$$

where $(y(\cdot),z(\cdot)) \in \mathbb{R}^n \times \mathbb{R}^{n\times d}$ is the state process pair, $0 < \delta < T$ is a constant time delay parameter, and $y_\delta(t) = y(t-\delta), z_\delta(t) = z(t-\delta)$. Here, $f : [0,T] \times \mathbb{R}^n \times \mathbb{R}^{n\times d} \times \mathbb{R}^n \times \mathbb{R}^{n\times d} \times \mathbb{R}^{k_1} \times \mathbb{R}^{k_2} \to \mathbb{R}^n$, $g : [0,T] \times \mathbb{R}^n \times \mathbb{R}^{n\times d} \times \mathbb{R}^n \times \mathbb{R}^{n\times d} \times \mathbb{R}^{k_1} \times \mathbb{R}^{k_2} \to \mathbb{R}^{n\times l}$ are given functions, and $\phi(\cdot), \psi(\cdot) \in L^2_{\mathcal{F}}(-\delta,0;\mathbb{R}^n)$ are the initial paths of y, z, respectively.

Let U_i be a nonempty convex subset of $\mathbb{R}^i$ and $v_i(\cdot)$ be the control process of player $i, i = 1,2$. We denote by $\mathcal{U}_i$ the set of U_i-valued control processes $v_i \in L^2_{\mathcal{F}}(0,T;\mathbb{R}^{k_i})$ and it is called the admissible control set for player $i, i = 1,2$. Each element of $\mathcal{U}_i$ is called an (open-loop) admissible control for player $i, i = 1,2$. In addition, $\mathcal{U} = \mathcal{U}_1 \times \mathcal{U}_2$ is called the set of admissible controls for the two players.

We assume that

Hypothesis 1 (H1). *f and g are continuously differentiable with respect to $(y, y_\delta, z, z_\delta, v_1, v_2)$, and their partial derivatives are bounded.*

Now, if both $v_1(\cdot)$ and $v_2(\cdot)$ are admissible controls, and assumption (H1) holds, then doubly stochastic differential equation with delay (3) admits a unique solution $(y(\cdot), z(\cdot)) \in L^2_{\mathcal{F}}(-\delta,T;\mathbb{R}^n) \times L^2_{\mathcal{F}}(-\delta,T;\mathbb{R}^{n\times l})$ (see [20]). The two players have their own benefits, which are described by the cost functional

$$J_i(v_1(\cdot),v_2(\cdot)) = \mathbb{E}\left\{ \int_0^T l_i(t,y(t),z(t),y_\delta(t),z_\delta(t),v_1(t),v_2(t))dt + \Phi_i(y(T)) \right\},$$

where $l_i : [0,T] \times \mathbb{R}^n \times \mathbb{R}^{n\times d} \times \mathbb{R}^n \times \mathbb{R}^{n\times d} \times \mathbb{R}^{k_1} \times \mathbb{R}^{k_2} \to \mathbb{R}$ and $\Phi_i : \mathbb{R}^n \to \mathbb{R}$ are given functions, $i = 1,2$.

We also assume

Hypothesis 2 (H2). *l_i is continuously differentiable in $(y, z, y_\delta, z_\delta, v_1, v_2)$, its partial derivatives are continuous in $(y, z, y_\delta, z_\delta, v_1, v_2)$ and bounded by $c(1 + |y| + |z| + |y_\delta| + |z_\delta| + |v_1| + |v_2|)$. Moreover, $\Phi_i(y)$ is continuously differentiable in y and $\Phi_{iy}(y)$ is bounded by $c(1 + |y|)$.*

Assume that each participant wants to minimize her/his cost functional $J_i(v_1(\cdot),v_2(\cdot))$ by selecting an appropriate admissible control $v_i(\cdot)(i = 1,2)$. Then the problem is to find a pair of admissible controls $(u_1(\cdot), u_2(\cdot)) \in \mathcal{U}_1 \times \mathcal{U}_2$ such that

$$\begin{cases} J_1(u_1(\cdot),u_2(\cdot)) &= \min_{v_1(\cdot) \in \mathcal{U}_1} J_1(v_1(\cdot),u_2(\cdot)), \\ J_2(u_1(\cdot),u_2(\cdot)) &= \min_{v_2(\cdot) \in \mathcal{U}_2} J_2(u_1(\cdot),v_2(\cdot)). \end{cases} \quad (4)$$

We call the above problem a doubly stochastic differential game with time delay. For simplicity's sake, let us write it as Problem (A). If we can find an admissible control $u(\cdot) = (u_1(\cdot), u_2(\cdot))$ satisfying Equation (4), then we call it an equilibrium point of Problem (A) and denote the corresponding state trajectory by $(y(\cdot), z(\cdot)) = (y^u(\cdot), z^u(\cdot))$.

2.2. The General FBDSDE

We deal with the following general FBDSDE, in which the forward equation is a delayed doubly SDE, and the backward equation is the anticipated BDSDE:

$$\begin{cases} dy(t) &= f(t,y(t),p(t),z(t),q(t),y_\delta(t),z_\delta(t))dt - z(t)\overleftarrow{d}B(t) \\ & \quad + g(t,y(t),p(t),z(t),q(t),y_\delta(t),z_\delta(t))\overrightarrow{d}W(t),\ t \in [0,T], \\ -dp(t) &= F(t,y(t),p(t),z(t),q(t),p_{\delta+}(t),q_{\delta+}(t))dt - q(t)\overrightarrow{d}W(t) \\ & \quad + G(t,y(t),p(t),z(t),q(t),p_{\delta+}(t),q_{\delta+}(t))\overleftarrow{d}B(t),\ t \in [0,T], \\ y(t) &= \phi(t),\ t \in [-\delta,0], \\ z(t) &= \psi(t),\ t \in [-\delta,0], \\ p(T) &= \Phi(y(T)),\ p(t) = \xi(t),\ t \in [T,T+\delta], \\ q(t) &= \eta(t),\ t \in [T,T+\delta], \end{cases} \quad (5)$$

where $y_\delta(t) = y(t-\delta), z_\delta = z(t-\delta), p_{\delta+}(t) = p(t+\delta), q_{\delta+} = q(t+\delta)$, and

$$\begin{aligned} F &: \Omega \times [0,T] \times \mathbb{R}^n \times \mathbb{R}^m \times \mathbb{R}^{n\times l} \times \mathbb{R}^{m\times d} \times \mathbb{R}^m \times \mathbb{R}^{m\times d} \to \mathbb{R}^m, \\ f &: \Omega \times [0,T] \times \mathbb{R}^n \times \mathbb{R}^m \times \mathbb{R}^{n\times l} \times \mathbb{R}^{m\times d} \times \mathbb{R}^n \times \mathbb{R}^{n\times l} \to \mathbb{R}^n, \\ G &: \Omega \times [0,T] \times \mathbb{R}^n \times \mathbb{R}^m \times \mathbb{R}^{n\times l} \times \mathbb{R}^{m\times d} \times \mathbb{R}^m \times \mathbb{R}^{m\times d} \to \mathbb{R}^{m\times l}, \\ g &: \Omega \times [0,T] \times \mathbb{R}^n \times \mathbb{R}^m \times \mathbb{R}^{n\times l} \times \mathbb{R}^{m\times d} \times \mathbb{R}^n \times \mathbb{R}^{n\times l} \to \mathbb{R}^{n\times d}, \\ \Phi &: \Omega \times \mathbb{R}^n \to \mathbb{R}^m. \end{aligned}$$

Given an $m \times n$ full-rank matrix H. Let us introduce some notation:

$$u = \begin{pmatrix} y \\ p \\ z \\ q \end{pmatrix},\ \begin{pmatrix} y_\delta(\cdot) \\ p_{\delta+}(\cdot) \\ z_\delta(\cdot) \\ q_{\delta+}(\cdot) \end{pmatrix} = \begin{pmatrix} \alpha \\ \mu \\ \beta \\ \nu \end{pmatrix},\ A(t,u,\alpha,\mu,\beta,\nu) = \begin{pmatrix} -H^\top F(t,u,\mu,\nu) \\ Hf(t,u,\alpha,\beta) \\ -H^\top G(t,u,\mu,\nu) \\ Hg(t,u,,\alpha,\beta) \end{pmatrix}.$$

where $H^\top G = (H^\top G_1 \cdots H^\top G_l)$ and $Hg = (Hg_1 \cdots Hg_d)$.

Similar to [23,35,47], we present the definition of solution to FBDSDEs (5) as follows:

Definition 1. *A quadruple of $\mathcal{F}_t$-measurable stochastic processes $(y,p,z,q) : \Omega \times [-\delta,T] \times [0,T+\delta] \times [-\delta,T] \times [0,T+\delta] \to \mathbb{R}^n \times \mathbb{R}^m \times \mathbb{R}^{n\times l} \times \mathbb{R}^{m\times d}$ is called a solution of FBDSDE (5), if $(y,p,z,q) \in L^2_\mathcal{F}(-\delta,T;\mathbb{R}^n) \times L^2_\mathcal{F}(0,T+\delta;\mathbb{R}^m) \times L^2_\mathcal{F}\left(-\delta,T;\mathbb{R}^{n\times l}\right) \times L^2_\mathcal{F}\left(0,T+\delta;\mathbb{R}^{m\times d}\right)$ and satisfies FBDSDE (5).*

We suppose the following Assumption (H3) holds:

Hypothesis 3 (H3).

(i) $\int_0^T \langle A(t,u,\alpha,\mu,\beta,\nu) - A(t,\bar{u},\bar{\alpha},\bar{\mu},\bar{\beta},\bar{\nu}), u - \bar{u}\rangle dt$

$\leq \int_0^T [-\mu_1 \left(|H(y-\bar{y})|^2 + |H(z-\bar{z})|^2\right) - \mu_2 \left(\left|H^\top(p-\bar{p})\right|^2 + \left|H^\top(q-\bar{q})\right|^2\right)]dt,$

$\forall u = (y,p,z,q),\ \bar{u} = (\bar{y},\bar{p},\bar{z},\bar{q}) \in \mathbb{R}^n \times \mathbb{R}^m \times \mathbb{R}^{n\times l} \times \mathbb{R}^{m\times d};$

(ii) $\langle \Phi(y) - \Phi(\bar{y}), H(y-\bar{y})\rangle \geq \beta_1 |H(y-\bar{y})|^2,\ \forall y,\bar{y} \in \mathbb{R}^n;$

where μ_1, μ_2 and β_1 are given non-negative constants with $\mu_1 + \mu_2 > 0$, and $\mu_2 + \beta_1 > 0$. Moreover we have $\mu_1 > 0, \beta_1 > 0$ (resp., $\mu_2 > 0$) when $m > n$ (resp., $m < n$);

(iii) for each $u, \alpha, \mu, \beta, \nu,\ A(\cdot, u, \alpha, \mu, \beta, \nu)$ is an $\mathcal{F}_t$-measurable vector process defined on $[0,T]$ with $A(\cdot, 0) \in L^2_\mathcal{F}(0,T)$, and for each $y \in \mathbb{R}^n,\ \Phi(y)$ is an $\mathcal{F}_T$-measurable random vector with $\Phi(0) \in L^2(\mathcal{F}_T;\mathbb{R}^m);$

(iv) $A(t,u,\alpha,\mu,\beta,\nu)$ and Φ satisfy the Lipschitz conditions: there exist constants $k > 0$ and $0 < \lambda < 1$ such that $\forall u, \bar{u}, \alpha, \bar{\alpha}, \mu, \bar{\mu}, \beta, \bar{\beta}, \nu, \bar{\nu}, \forall t \in [0,T]$,

$\left|f(t,u,\alpha,\beta) - f(t,\bar{u},\bar{\alpha},\bar{\beta})\right| \leq k(|u-\bar{u}| + |\alpha - \bar{\alpha}| + |\beta - \bar{\beta}|),$

$\left|F(t,u,\mu,\nu) - F(t,\bar{u},\bar{\mu},\bar{\nu})\right| \leq k(|u-\bar{u}| + \mathbb{E}^{\mathcal{F}_t}[|\mu - \bar{\mu}| + |\nu - \bar{\nu}|]),$

$$|g(t,u,\alpha,\beta) - g(t,\bar{u},\bar{\alpha},\bar{\beta})|$$
$$\leq k(|y-\bar{y}| + |p-\bar{p}| + |q-\bar{q}| + |\alpha-\bar{\alpha}|) + \lambda(|z-\bar{z}| + |\beta-\bar{\beta}|),$$
$$|G(t,u,\mu,v) - G(t,\bar{u},\bar{\mu},\bar{v})|$$
$$\leq k(|y-\bar{y}| + |p-\bar{p}| + |z-\bar{z}| + \mathbb{E}^{\mathcal{F}_t}[|\mu-\bar{\mu}|]) + \lambda(|q-\bar{q}| + \mathbb{E}^{\mathcal{F}_t}[|v-\bar{v}|]),$$
$$|\Phi(y) - \Phi(\bar{y})| \leq k|y-\bar{y}|.$$

By the similar method of [23,35,47], we can prove the following Theorem 1. For the convenience of the reader, we present the skeleton of the proof in Section 6.

Theorem 1. *Under assumption (H3), then FBDSDE (5) has a unique solution* $(y,p,z,q) \in L^2_{\mathcal{F}}(-\delta,T;\mathbb{R}^n) \times L^2_{\mathcal{F}}(0,T+\delta;\mathbb{R}^m) \times L^2_{\mathcal{F}}\left(-\delta,T;\mathbb{R}^{n\times l}\right) \times L^2_{\mathcal{F}}\left(0,T+\delta;\mathbb{R}^{m\times d}\right).$

3. Necessary Maximum Principle

For convex admissible control sets, the classical method to obtain the necessary optimality condition is the convex perturbation method. Let $u(\cdot) = (u_1(\cdot), u_2(\cdot))$ be an equilibrium point of Problem (A) and $(y(\cdot), z(\cdot))$ be the corresponding optimal trajectory. Let $(v_1(\cdot), v_2(\cdot))$ be such that $(u_1(\cdot) + v_1(\cdot), u_2(\cdot) + v_2(\cdot)) \in \mathcal{U}_1 \times \mathcal{U}_2$. Since $\mathcal{U}_1$ and $\mathcal{U}_2$ are convex, for any $0 \leq \rho \leq 1$, $(u_1^\rho(\cdot), u_2^\rho(\cdot)) = (u_1(\cdot) + \rho v_1(\cdot), u_1(\cdot) + \rho v_1(\cdot))$ is also in $\mathcal{U}_1 \times \mathcal{U}_2$. As illustrated before, we denote by $(y^{u_1^\rho}(\cdot), z^{u_1^\rho}(\cdot))$ and $(y^{u_2^\rho}(\cdot), z^{u_2^\rho}(\cdot))$ the corresponding state trajectories of the game system in Equation (3) along with the controls $(u_1^\rho(\cdot), u_2(\cdot))$ and $(u_1(\cdot), u_2^\rho(\cdot))$.

For convenience, we use the following notations throughout this paper:
$$\varphi(t) = \varphi(t, y(t), z(t), y_\delta(t), z_\delta(t), u_1(t), u_2(t)),$$
$$\varphi^{u_1^\rho}(t) = \varphi(t, y^\rho(t), z^\rho(t), y_\delta^\rho(t), z_\delta^\rho(t), u_1^\rho(t), u_2(t)),$$
$$\varphi^{u_2^\rho}(t) = \varphi(t, y^\rho(t), z^\rho(t), y_\delta^\rho(t), z_\delta^\rho(t), u_1(t), u_2^\rho(t)),$$

where φ means one of $f, g, l_i, i = 1, 2$.

We bring in the following variational equation:
$$\begin{cases} dy_i^1(t) = [f_y(t)y_i^1(t) + f_z(t)z_i^1(t) + f_{y_\delta}(t)y_{i\delta}^1(t) + f_{z_\delta}(t)z_{i\delta}^1(t) + f_{v_i}(t)v_i(t)]dt \\ \qquad + [g_y(t)y_i^1(t) + g_z(t)z_i^1(t) + g_{y_\delta}(t)y_{i\delta}^1(t) + g_{z_\delta}(t)z_{i\delta}^1(t) + g_{v_i}(t)v_i(t)]\overrightarrow{d}W(t) \\ \qquad - z_i^1(t)\overleftarrow{d}B(t), \\ y_i^1(t) = 0, \ z_i^1(t) = 0, \ t \in [-\delta,0], \ (i=1,2). \end{cases} \quad (6)$$

By (H1) and Theorem 3.1.1 in [20], it is easy to see that there is a unique adapted solution to Equation (6).

For $t \in [0,T], \rho > 0$, we set
$$\tilde{y}_i^\rho(t) = \frac{y^{u_i^\rho}(t) - y(t)}{\rho} - y_i^1(t),$$
$$\tilde{z}_i^\rho(t) = \frac{z^{u_i^\rho}(t) - z(t)}{\rho} - z_i^1(t), \ (i=1,2).$$

We have the following:

Lemma 1. *Let the hypotheses (H1) and (H2) be true. Then, for $i=1,2$,*

$$\lim_{\rho \to 0} \sup_{0 \leq t \leq T} \mathbb{E}|\tilde{y}_i^\rho(t)|^2 = 0, \quad (7)$$

$$\lim_{\rho \to 0} \mathbb{E}\int_0^T |\tilde{z}_i^\rho(t)|^2 dt = 0. \quad (8)$$

Proof of Lemma 1. For $i = 1$, we have

$$\begin{cases} d\tilde{y}_1^\rho(t) = [\frac{1}{\rho}\left(f^{u_1^\rho}(t) - f(t)\right) - f_y(t)y_1^1(t) - f_z(t)z_1^1(t) - f_{y_\delta}(t)y_{1\delta}^1(t) \\ \qquad\qquad - f_{z_\delta}(t)z_{1\delta}^1(t) - f_{v_1}(t)v_1(t)]dt \\ \qquad\qquad + [\frac{1}{\rho}\left(g^{u_1^\rho}(t) - g(t)\right) - g_y(t)y_1^1(t) - g_z(t)z_1^1(t) - g_{y_\delta}(t)y_{1\delta}^1(t) \\ \qquad\qquad - g_{z_\delta}(t)z_{1\delta}^1(t) - g_{v_1}(t)v_1(t)]\overrightarrow{d}W(t) - \tilde{z}_1^\rho(t)\overleftarrow{d}B(t), \\ \tilde{y}_1^\rho(t) = 0,\ \tilde{z}_1^\rho(t) = 0,\ t \in [-\delta, 0], \end{cases}$$

or

$$\begin{cases} d\tilde{y}_1^\rho(t) = [A_1^\rho(t,\cdot)\tilde{y}_1^\rho(t) + \bar{A}_1^\rho(t,\cdot)\tilde{y}_{1\delta}^\rho(t) + B_1^\rho(t,\cdot)\tilde{z}_1^\rho(t) \\ \qquad\qquad + \bar{B}_1^\rho(t,\cdot)\tilde{z}_{1\delta}^\rho(t) + G_1^\rho(t)]dt \\ \qquad\qquad + [C_1^\rho(t,\cdot)\tilde{y}_1^\rho(t) + \bar{C}_1^\rho(t,\cdot)\tilde{y}_{1\delta}^\rho(t) + D_1^\rho(t,\cdot)\tilde{z}_1^\rho(t) \\ \qquad\qquad + \bar{D}_1^\rho(t,\cdot)\tilde{z}_{1\delta}^\rho(t) + G_2^\rho(t)]\overrightarrow{d}W(t) - \tilde{z}_1^\rho(t)\overleftarrow{d}B(t), \\ \tilde{y}_1^\rho(t) = 0,\ \tilde{z}_1^\rho(t) = 0,\ t \in [-\delta, 0], \end{cases}$$

where we denote

$$A_1^\rho(t) = \int_0^1 f_y(t, y_1(t) + \lambda(y^{u_1^\rho}(t) - y(t)), z^{u_1^\rho}(t), y_\delta^{u_1^\rho}(t), z_\delta^{u_1^\rho}(t), u_1^\rho(t), u_2(t))d\lambda,$$

$$B_1^\rho(t) = \int_0^1 f_z(t, y(t), z_1(t) + \lambda(z^{u_1^\rho}(t) - z(t)), y_\delta^{u_1^\rho}(t), z_\delta^{u_1^\rho}(t), u_1^\rho(t), u_2(t))d\lambda,$$

$$\bar{A}_1^\rho(t) = \int_0^1 f_{y_\delta}(t, y(t), z(t), y_{1\delta}(t) + \lambda\rho(y_{1\delta}^1(t) + \tilde{y}_{1\delta}^\rho(t)), z_\delta^{u_1^\rho}(t), u_1^\rho(t), u_2(t))d\lambda,$$

$$\bar{B}_1^\rho(t) = \int_0^1 f_{z_\delta}(t, y(t), z(t), y_\delta(t), z_{1\delta}(t) + \lambda\rho(z_{1\delta}^1(t) + \tilde{z}_{1\delta}^\rho(t)), u_1^\rho(t), u_2(t))d\lambda,$$

$$C_1^\rho(t) = \int_0^1 g_y(t, y_1(t) + \lambda(y^{u_1^\rho}(t) - y(t)), z^{u_1^\rho}(t), y_\delta^{u_1^\rho}(t), z_\delta^{u_1^\rho}(t), u_1^\rho(t), u_2(t))d\lambda,$$

$$D_1^\rho(t) = \int_0^1 g_z(t, y(t), z_1(t) + \lambda(z^{u_1^\rho}(t) - z(t)), y_\delta^{u_1^\rho}(t), z_\delta^{u_1^\rho}(t), u_1^\rho(t), u_2(t))d\lambda,$$

$$\bar{C}_1^\rho(t) = \int_0^1 g_{y_\delta}(t, y(t), z(t), y_{1\delta}(t) + \lambda\rho(y_{1\delta}^1(t) + \tilde{y}_{1\delta}^\rho(t)), z_\delta^{u_1^\rho}(t), u_1^\rho(t), u_2(t))d\lambda,$$

$$\bar{D}_1^\rho(t) = \int_0^1 g_{z_\delta}(t, y(t), z(t), y_\delta(t), z_{1\delta}(t) + \lambda\rho(z_{1\delta}^1(t) + \tilde{z}_{1\delta}^\rho(t)), u_1^\rho(t), u_2(t))d\lambda,$$

$$G_1^\rho(t) = \int_0^1 (f_{v_1}(t, y(t), z(t), y_\delta(t), z_\delta(t), u_1(t) + \rho\lambda v_1(t), u_2(t)) - f_{v_1}(t))v_1(t)d\lambda$$
$$\qquad + [A_1^\rho(t) - f_y(t)]y_1^1(t) + [B_1^\rho(t) - f_z(t)]z_1^1(t)$$
$$\qquad + [\bar{A}_1^\rho(t) - f_{y_\delta}(t)]y_{1\delta}^1(t) + [\bar{B}_1^\rho(t) - f_{z_\delta}(t)]z_{1\delta}^1(t),$$

$$G_2^\rho(t) = \int_0^1 (g_{v_1}(t, y(t), z(t), y_\delta(t), z_\delta(t), u_1(t) + \rho\lambda v_1(t), u_2(t)) - g_{v_1}(t))v_1(t)d\lambda$$
$$\qquad + [C_1^\rho(t) - g_y(t)]y_1^1(t) + [D_1^\rho(t) - g_z(t)]z_1^1(t)$$
$$\qquad + [\bar{C}_1^\rho(t) - g_{y_\delta}(t)]y_{1\delta}^1(t) + [\bar{D}_1^\rho(t) - g_{z_\delta}(t)]z_{1\delta}^1(t).$$

Using Itô's formula to $\left|\tilde{y}_1^\rho(t)\right|^2$ on $[0, t]$, through (H1), we get

$$\mathbb{E}\left|\tilde{y}_1^\rho(t)\right|^2 + \mathbb{E}\int_0^t |\tilde{z}_1^\rho(s)|^2 ds$$
$$\leq C_0\mathbb{E}\int_0^t \left|\tilde{y}_1^\rho(s)\right|^2 ds + \frac{1}{2}\mathbb{E}\int_0^t \left|\tilde{z}_1^\rho(s)\right|^2 ds + C_1(\mathbb{E}\int_0^t (|G_1^\rho(s)|^2 + |G_2^\rho(s)|^2)ds.$$

Applying Grownwall's inequalities, we can easily get the desired result. Again, we can prove that for $i = 2$. The proof is complete. □

Based on $(u_1(\cdot), u_2(\cdot))$ being an equilibrium point of Problem (A), then

$$\rho^{-1}[J_1(u_1^\rho(\cdot), u_2(\cdot)) - J_1(u_1(\cdot), u_2(\cdot))] \geq 0, \tag{9}$$

$$\rho^{-1}[J_2(u_1(\cdot), u_2^\rho(\cdot)) - J_2(u_1(\cdot), u_2(\cdot))] \geq 0. \tag{10}$$

From Equations (9) and (10) and Lemma 1, we obtain the following variational inequality.

Lemma 2. *Let assumptions (H1) and (H2) hold. Then*

$$\mathbb{E}\int_0^T \left[l_{iy}(t)y_i^1(t) + l_{iz}(t)z_i^1(t) + l_{iy_\delta}(t)y_{i\delta}^1(t) + l_{iz_\delta}(t)z_{i\delta}^1(t) + l_{iv_i}(t)v_i(t)\right]dt$$
$$+\mathbb{E}[\Phi_{iy}(y(T))y_i^1(T)] \geq 0, \ (i = 1, 2). \tag{11}$$

Proof of Lemma 2. For $i = 1$, from Equation (7), we derive

$$\rho^{-1}\mathbb{E}[\Phi_1(y^{u_1^\rho}(T)) - \Phi_1(y(T))]$$
$$= \rho^{-1}\mathbb{E}\int_0^1 \Phi_{1y}(y(T) + \lambda(y^{u_1^\rho}(T) - y(T)))(y^{u_1^\rho}(T) - y(T))d\lambda$$
$$\to \mathbb{E}[\Phi_{1y}(y(T))y_1^1(T)], \ \rho \to 0.$$

Similarly, we have

$$\rho^{-1}\left\{\mathbb{E}\int_0^T [l_1^{u_1^\rho}(t) - l_1(t)]dt\right\}$$
$$\to \mathbb{E}\int_0^T \left[l_{1y}(t)y_1^1(t) + l_{1z}(t)z_1^1(t) + l_{1y_\delta}(t)y_{1\delta}^1(t) + l_{1z_\delta}(t)z_{1\delta}^1(t) + l_{1v_1}(t)v_1(t)\right]dt, \ \rho \to 0.$$

Let $\rho \to 0$ in Equation (9), so, for $i = 1$, Equation (11) is established. Similarly, we can prove that the conclusion holds for $i = 2$. The proof is complete. □

Let us define the Hamiltonian function $H_i : [0, T] \times \mathbb{R}^n \times \mathbb{R}^{n \times d} \times \mathbb{R}^n \times \mathbb{R}^{n \times d} \times \mathbb{R}^{k_1} \times \mathbb{R}^{k_2} \times \mathbb{R}^n \times \mathbb{R}^{n \times l} \to \mathbb{R}$, $i = 1, 2$ as follows:

$$H_i(t, y, z, y_\delta, z_\delta, v_1, v_2, p_i, q_i)$$
$$= \langle f(t, y, z, y_\delta, z_\delta, v_1, v_2), p_i \rangle + \langle g(t, y, z, y_\delta, z_\delta, v_1, v_2), q_i \rangle + l_i(t, y, z, y_\delta, z_\delta, v_1, v_2), \ i = 1, 2.$$

We introduce the following adjoint equation

$$\begin{cases} -dp_i(t) &= \{H_{iy}(t, \Theta(t), u_1(t), u_2(t), p_i(t), q_i(t)) \\ & \quad + \mathbb{E}^{\mathcal{F}_t}[H_{iy_\delta}(t+\delta, \Theta(t+\delta), u_1(t+\delta), u_2(t+\delta), p_i(t+\delta), q_i(t+\delta))]\}dt \\ & \quad + \{H_{iz}(t, \Theta(t), u_1(t), u_2(t), p_i(t), q_i(t)) \\ & \quad + \mathbb{E}^{\mathcal{F}_t}[H_{iz_\delta}(t+\delta, \Theta(t+\delta), u_1(t+\delta), u_2(t+\delta), p_i(t+\delta), q_i(t+\delta))]\}\overleftarrow{d}B(t) \\ & \quad - q_i(t)\overrightarrow{d}W(t), \\ p_i(T) &= \Phi_{iy}(y(T)), \ p_i(t) = 0, \ q_i(t) = 0, \ t \in [T, T+\delta], \ (i = 1, 2). \end{cases} \tag{12}$$

where $\Theta(t) := (y(t), z(t), y_\delta(t), z_\delta(t))$.

Remark 1. *It is easy to see that the adjoint Equation (12) above is a linear anticipated BDSDE, then the unique solvability of Equation (12) can be guaranteed by theorem 3.2 in [45] and theorem 2.4 in [46].*

Theorem 2 (Necessary maximum principle). *Suppose (H1) and (H2) hold, and $(u_1(\cdot), u_2(\cdot))$ is an equilibrium point of Problem (A) and $(y(\cdot), z(\cdot))$ is the corresponding state trajectory. Then we have*

$$\langle H_{1v_1}(t,\Theta(t),u_1(t),u_2(t),p_1(t),q_1(t)),v_1-u_1(t)\rangle \geq 0,$$
$$\langle H_{2v_2}(t,\Theta(t),u_1(t),u_2(t),p_2(t),q_2(t)),v_2-u_2(t)\rangle \geq 0,$$

hold for any $(v_1,v_2) \in U_1 \times U_2$, a.e., a.s., where $(p_i(\cdot),q_i(\cdot))(i=1,2)$ is the solution of the adjoint Equation (12).

Proof of Theorem 2. For $i=1$. Using Itô's formula to $\langle p_1(t), y_1^1(t)\rangle$, we obtain

$$\mathbb{E}\langle \Phi_{1y}(y(T)), y_1^1(T)\rangle$$
$$= \mathbb{E}\int_0^T \langle p_1(t), f_y(t)y_1^1(t) + f_z(t)z_1^1(t) + f_{y_\delta}(t)y_{1\delta}^1(t) + f_{z_\delta}(t)z_{1\delta}^1(t) + f_{v_1}(t)v_1(t)\rangle dt$$
$$+ \mathbb{E}\int_0^T \langle -f_y^\top(t)p_1(t) - g_y^\top(t)q_1(t) - l_{1y}(t)$$
$$\qquad + \mathbb{E}^{\mathcal{F}_t}[-f_{y_\delta}^\top(t+\delta)p_1(t+\delta) - g_{y_\delta}^\top(t+\delta)q_1(t+\delta) - l_{1y_\delta}(t+\delta)], y_1^1(t)\rangle dt$$
$$+ \mathbb{E}\int_0^T \langle q_1(t), g_y(t)y_1^1(t) + g_z(t)z_1^1(t) + g_{y_\delta}(t)y_{1\delta}^1(t) + g_{z_\delta}(t)z_{1\delta}^1(t) + g_{v_1}(t)v_1(t)\rangle dt$$
$$+ \mathbb{E}\int_0^T \langle -f_z^\top(t)p_1(t) - g_z^\top(t)q_1(t) - l_{1z}(t)$$
$$\qquad + \mathbb{E}^{\mathcal{F}_t}[-f_{z_\delta}^\top(t+\delta)p_1(t+\delta) - g_{z_\delta}^\top(t+\delta)q_1(t+\delta) - l_{1z_\delta}(t+\delta)], z_1^1(t)\rangle dt.$$

Combining the initial conditions and the termination conditions, we get

$$\mathbb{E}\int_0^T [\langle p_1(t), f_{y_\delta}(t)y_{1\delta}^1(t)\rangle - \langle \mathbb{E}^{\mathcal{F}_t}[f_{y_\delta}^\top(t+\delta)p_1(t+\delta)], y_1^1(t)\rangle]dt$$
$$= \mathbb{E}\int_0^T \langle p_1(t), f_{y_\delta}(t)y_{1\delta}^1(t)\rangle dt - \mathbb{E}\int_\delta^{T+\delta} \langle f_{y_\delta}^\top(t)p_1(t), y_{1\delta}^1(t)\rangle dt$$
$$= \mathbb{E}\int_0^\delta \langle p_1(t), f_{y_\delta}(t)y_{1\delta}^1(t)\rangle dt - \mathbb{E}\int_T^{T+\delta} \langle f_{y_\delta}^\top(t)p_1(t), y_{1\delta}^1(t)\rangle dt$$
$$= 0.$$

Similarly, we have

$$\mathbb{E}\int_0^T [\langle p_1(t), f_{z_\delta}(t)z_{1\delta}^1(t)\rangle - \langle \mathbb{E}^{\mathcal{F}_t}[f_{z_\delta}^\top(t+\delta)p_1(t+\delta)], z_1^1(t)\rangle]dt = 0,$$
$$\mathbb{E}\int_0^T [\langle q_1(t), g_{y_\delta}(t)y_{1\delta}^1(t)\rangle - \langle \mathbb{E}^{\mathcal{F}_t}[g_{y_\delta}^\top(t+\delta)q_1(t+\delta)], y_1^1(t)\rangle]dt = 0,$$
$$\mathbb{E}\int_0^T [\langle q_1(t), g_{z_\delta}(t)z_{1\delta}^1(t)\rangle - \langle \mathbb{E}^{\mathcal{F}_t}[g_{z_\delta}^\top(t+\delta)q_1(t+\delta)], z_1^1(t)\rangle]dt = 0.$$

Then, we get

$$\mathbb{E}\int_0^T \left[l_{1y}(t)y_1^1(t) + l_{1z}(t)z_1^1(t) + l_{1y_\delta}(t)y_{1\delta}^1(t) + l_{1z_\delta}(t)z_{1\delta}^1(t) + l_{1v_1}(t)v_1(t) \right] dt$$
$$+ \mathbb{E}\langle \Phi_{1y}(y(T)), y_1^1(T)\rangle$$
$$= \mathbb{E}\int_0^T \langle f_{v_1}^\top(t)p_1(t) + g_{v_1}^\top(t)q_1(t) + l_{1v_1}(t), v_1(t)\rangle dt.$$

According to Lemma 2, we have

$$\mathbb{E}\int_0^T \langle H_{1v_1}(t,\Theta(t),u_1(t),u_2(t),p_1(t),q_1(t)), v_1(t)\rangle dt \geq 0.$$

Because $v_1(t)$ satisfies $u_1(t) + v_1(t) \in \mathcal{U}_1$, we have

$$\mathbb{E}\int_0^T \langle H_{1v_1}(t,\Theta(t),u_1(t),u_2(t),p_1(t),q_1(t)), v_1 - u_1(t)\rangle dt \geq 0, \forall v_1 \in \mathcal{U}_1,$$

which means that

$$\mathbb{E}\langle H_{1v_1}(t,\Theta(t),u_1(t),u_2(t),p_1(t),q_1(t)),v_1-u_1(t)\rangle \geq 0, \text{ a.s.} \quad (13)$$

At present, take an arbitrary element F of σ-algebra $\mathcal{F}_t$, and set

$$w(t) = v\mathbf{1}_F + u(t)\mathbf{1}_{\Omega-F}.$$

Obviously, $w(t)$ is an admissible control.
We apply the inequality in Equation (13) to $w(t)$, and get

$$\mathbb{E}[\mathbf{1}_F\langle H_{1v_1}(t,\Theta(t),u_1(t),u_2(t),p_1(t),q_1(t)),v_1-u_1(t)\rangle] \geq 0, \forall F \in \mathcal{F}_t,$$

which contains that

$$\mathbb{E}[\langle H_{1v_1}(t,\Theta(t),u_1(t),u_2(t),p_1(t),q_1(t)),v_1-u_1(t)\rangle|\mathcal{F}_t] \geq 0.$$

The expression within the conditional expectation is $\mathcal{F}_t$-measurable, so the result follows. Following the above proof, we can prove that the other inequality is true for any $v_2 \in \mathcal{U}_2$. The proof is completed. □

4. Sufficient Maximum Principle

In this section, the sufficient maximum principle for Problem (A) is investigated. Let $(y(t), z(t), u_1(t), u_2(t))$ be a quintuple satisfying Equation (3), and suppose there exists a solution $(p_i(t), q_i(t))$ of the corresponding adjoint forward SDE (12). We assume that:

Hypothesis 4 (H4). *For $i = 1, 2$, for all $t \in [0, T]$, $H_i(t, y, z, y_\delta, z_\delta, v_1, v_2, p_i, q_i)$ is convex in $(y, z, y_\delta, z_\delta, v_1, v_2)$, and $\Phi_i(y)$ is convex in y.*

Theorem 3 (Sufficient maximum principle). *Suppose (H1), (H2) and (H4) are true. In addition, the following conditions hold*

$$H_1(t,\Theta(t),u_1(t),u_2(t),p_1(t),q_1(t)) = \min_{v_1 \in \mathcal{U}_1} H_1(t,\Theta(t),v_1(t),u_2(t),p_1(t),q_1(t)), \quad (14)$$

$$H_2(t,\Theta(t),u_1(t),u_2(t),p_2(t),q_2(t)) = \min_{v_2 \in \mathcal{U}_2} H_2(t,\Theta(t),u_1(t),v_2(t),p_2(t),q_2(t)). \quad (15)$$

Then $(u_1(\cdot), u_2(\cdot))$ is an equilibrium point of Problem (A).

Proof of Theorem 3. For any $v_1(\cdot) \in \mathcal{U}_1$, we consider

$$\begin{aligned}&J_1(v_1(\cdot),u_2(\cdot)) - J_1(u_1(\cdot),u_2(\cdot))\\ &= \mathbb{E}\int_0^T [l_1(t,\Theta(t),v_1(t),u_2(t)) - l_1(t,\Theta(t),u_1(t),u_2(t))]dt\\ &\quad + \mathbb{E}[\Phi_1(y^{v_1}(T)) - \Phi_1(y(T))].\end{aligned}$$

Now we put into use Itô's formula to $\langle p_1(t), y^{v_1}(t) - y(t)\rangle$ on $[0, T]$, and get

$$\begin{aligned}&\mathbb{E}\langle \Phi_{1y}(y(T)), y^{v_1}(T) - y(T)\rangle\\ &= -\mathbb{E}\int_0^T \langle y^{v_1}(t) - y(t), H_{1y}(t,\Theta(t),u_1(t),u_2(t),p_1(t),q_1(t))\\ &\qquad + \mathbb{E}^{\mathcal{F}_t}[H_{1y_\delta}(t+\delta,\Theta(t+\delta),u_1(t+\delta),u_2(t+\delta),p_1(t+\delta),q_1(t+\delta))]\rangle dt\\ &\quad + \mathbb{E}\int_0^T \langle p_1(t), f(t,\Theta(t),v_1(t),u_2(t)) - f(t,\Theta(t),u_1(t),u_2(t))\rangle dt\\ &\quad - \mathbb{E}\int_0^T \langle z^{v_1}(t) - z(t), H_{1z}(t,\Theta(t),u_1(t),u_2(t),p_1(t),q_1(t))\\ &\qquad + \mathbb{E}^{\mathcal{F}_t}[H_{1z_\delta}(t+\delta,\Theta(t+\delta),u_1(t+\delta),u_2(t+\delta),p_1(t+\delta),q_1(t+\delta))]\rangle dt\\ &\quad + \mathbb{E}\int_0^T \langle q_1(t), g(t,\Theta(t),v_1(t),u_2(t)) - g(t,\Theta(t),u_1(t),u_2(t))\rangle dt.\end{aligned}$$

Since Φ_1 is convex, we have

$$\Phi_1(y^{v_1}(T)) - \Phi_1(y(T)) \geq \langle \Phi_{1y}(y(T)), y^{v_1}(T) - y(T) \rangle.$$

Then, we have

$$\begin{aligned}
& J_1(v_1(\cdot), u_2(\cdot)) - J_1(u_1(\cdot), u_2(\cdot)) \\
\geq\ & \mathbb{E} \int_0^T [H_1(t, \Theta(t), v_1(t), u_2(t), p_1(t), q_1(t)) - H_1(t, \Theta(t), u_1(t), u_2(t), p_1(t), q_1(t))] dt \\
& - \mathbb{E} \int_0^T \langle y^{v_1}(t) - y(t), H_{1y}(t, \Theta(t), u_1(t), u_2(t), p_1(t), q_1(t)) \\
& \quad + \mathbb{E}^{\mathcal{F}_t}[H_{1y_\delta}(t+\delta, \Theta(t+\delta), u_1(t+\delta), u_2(t+\delta), p_1(t+\delta), q_1(t+\delta))] \rangle dt \\
& - \mathbb{E} \int_0^T \langle z^{v_1}(t) - z(t), H_{1z}(t, \Theta(t), u_1(t), u_2(t), p_1(t), q_1(t)) \\
& \quad + \mathbb{E}^{\mathcal{F}_t}[H_{1z_\delta}(t+\delta, \Theta(t+\delta), u_1(t+\delta), u_2(t+\delta), p_1(t+\delta), q_1(t+\delta))] \rangle dt.
\end{aligned}$$

Based on the convexity of H_1 with respect to $(y, z, y_\delta, z_\delta, v_1, v_2)$, we achieve

$$\begin{aligned}
& H_1(t, \Theta(t), v_1(t), u_2(t), p_1(t), q_1(t)) - H_1(t, \Theta(t), u_1(t), u_2(t), p_1(t), q_1(t)) \\
\geq\ & \langle y^{v_1}(t) - y(t), H_{1y}(t, \Theta(t), u_1(t), u_2(t), p_1(t), q_1(t)) \rangle \\
& + \langle z^{v_1}(t) - z(t), H_{1z}(t, \Theta(t), u_1(t), u_2(t), p_1(t), q_1(t)) \rangle \\
& + \langle y_\delta^{v_1}(t) - y_\delta(t), H_{1y_\delta}(t, \Theta(t), u_1(t), u_2(t), p_1(t), q_1(t)) \rangle \\
& + \langle z_\delta^{v_1}(t) - z_\delta(t), H_{1z_\delta}(t, \Theta(t), u_1(t), u_2(t), p_1(t), q_1(t)) \rangle \\
& + \langle v_1(t) - u_1(t), H_{1v_1}(t, \Theta(t), u_1(t), u_2(t), p_1(t), q_1(t)) \rangle.
\end{aligned}$$

Noticing the fact that

$$\begin{aligned}
& \mathbb{E} \int_0^T \langle y_\delta^{v_1}(t) - y_\delta(t), H_{1y_\delta}(t, \Theta(t), u_1(t), u_2(t), p_1(t), q_1(t)) \rangle dt \\
& - \mathbb{E} \int_0^T \langle y^{v_1}(t) - y(t), \mathbb{E}^{\mathcal{F}_t}[H_{1y_\delta}(t+\delta, \Theta(t+\delta), u_1(t+\delta), u_2(t+\delta), p_1(t+\delta), q_1(t+\delta))] \rangle dt \\
=\ & \mathbb{E} \int_0^T \langle y_\delta^{v_1}(t) - y_\delta(t), H_{1y_\delta}(t, \Theta(t), u_1(t), u_2(t), p_1(t), q_1(t)) \rangle dt \\
& - \mathbb{E} \int_\delta^{T+\delta} \langle y_\delta^{v_1}(t) - y_\delta(t), H_{1y_\delta}(t, \Theta(t), u_1(t), u_2(t), p_1(t), q_1(t)) \rangle dt \\
=\ & \mathbb{E} \int_0^\delta \langle y_\delta^{v_1}(t) - y_\delta(t), H_{1y_\delta}(t, \Theta(t), u_1(t), u_2(t), p_1(t), q_1(t)) \rangle dt \\
& - \mathbb{E} \int_T^{T+\delta} \langle y_\delta^{v_1}(t) - y_\delta(t), H_{1y_\delta}(t, \Theta(t), u_1(t), u_2(t), p_1(t), q_1(t)) \rangle dt \\
=\ & 0.
\end{aligned}$$

Similarly, we have

$$\begin{aligned}
& \mathbb{E} \int_0^T \langle z_\delta^{v_1}(t) - z_\delta(t), H_{1z_\delta}(t, \Theta(t), u_1(t), u_2(t), p_1(t), q_1(t)) \rangle dt \\
& - \mathbb{E} \int_0^T \langle z^{v_1}(t) - z(t), \mathbb{E}^{\mathcal{F}_t}[H_{1z_\delta}(t+\delta, \Theta(t+\delta), u_1(t+\delta), u_2(t+\delta), p_1(t+\delta), q_1(t+\delta))] \rangle dt \\
=\ & 0.
\end{aligned}$$

Then, we get

$$\begin{aligned}
& J_1(v_1(\cdot), u_2(\cdot)) - J_1(u_1(\cdot), u_2(\cdot)) \\
\geq\ & \mathbb{E} \int_0^T \langle H_{1v_1}(t, \Theta(t), u_1(t), u_2(t), p_1(t), q_1(t)), v_1(t) - u_1(t) \rangle dt.
\end{aligned}$$

Finally, by the necessary optimality conditions in Equation (14), we obtain

$$J_1(v_1(\cdot), u_2(\cdot)) - J_1(u_1(\cdot), u_2(\cdot)) \geq 0.$$

This implies that

$$J_1(u_1(\cdot), u_2(\cdot)) = \min_{v_1(\cdot) \in \mathcal{U}_1} J_1(v_1(\cdot), u_2(\cdot)).$$

In the same way

$$J_2(u_1(\cdot), u_2(\cdot)) = \min_{v_2(\cdot) \in \mathcal{U}_2} J_2(u_1(\cdot), v_2(\cdot)).$$

Hence, the desired conclusion is drawn. The proof is completed. □

5. Applications in LQ Doubly Stochastic Games with Delay

In this section, our maximal principle is used for the nonzero sum differential game problem of LQ doubly stochastic systems with delay. To simplify the notation, let us assume that $d = l = 1$. The control system is

$$\begin{cases} dy(t) &= [A(t)y(t) + B(t)z(t) + \bar{A}(t)y_\delta(t) + \bar{B}(t)z_\delta(t) + E_1(t)v_1(t) \\ &\quad + E_2(t)v_2(t)]dt - z(t)\overleftarrow{d}B(t) \\ &\quad + [C(t)y(t) + D(t)z(t) + \bar{C}(t)y_\delta(t) + \bar{D}(t)z_\delta(t)]\overrightarrow{d}W(t), \ t \in [0,T], \\ y(t) &= \xi(t), \ t \in [-\delta, 0], \\ z(t) &= \eta(t), \ t \in [-\delta, 0], \end{cases} \quad (16)$$

where $(\xi(\cdot), \eta(\cdot)) \in L^2_{\mathcal{F}}(-\delta, T; \mathbb{R}^n)$ is the initial path of (y, z). $A, \bar{A}, B, \bar{B}, C, \bar{C}, D, \bar{D}$ are $n \times n$ bounded matrices, $v_1(t)$ and $v_2(t), t \in [0, T]$ are two admissible control processes, i.e., $\mathcal{F}_t$-measurable square-integrable processes taking values in $\mathbb{R}^k$. E_1 and E_2 are $n \times k$ bounded matrices. We denote $J_1(v_1(\cdot), v_2(\cdot))$ and $J_2(v_1(\cdot), v_2(\cdot))$, which are the cost functionals corresponding to the players 1 and 2:

$$J_i(v_1(\cdot), v_2(\cdot)) = \frac{1}{2}\mathbb{E}\{\int_0^T [\langle M_i(t)y(t), y(t)\rangle + \langle R_i(t)z(t), z(t)\rangle + \langle N_i(t)v_i(t), v_i(t)\rangle]dt \\ + \langle Q_i y(T), y(T)\rangle\}, \ i = 1, 2, \quad (17)$$

where $M_i(t), R_i(t), Q_i, i = 1, 2$ are $n \times n$ non-negative symmetric bounded matrices, and $N_i(t), i = 1, 2$ are $k \times k$ positive symmetric bounded matrices and the inverse $N_i^{-1}(t), i = 1, 2$ are also bounded. Our task is to find $(u_1(\cdot), u_2(\cdot)) \in \mathbb{R}^k \times \mathbb{R}^k$ such that

$$\begin{cases} J_1(u_1(\cdot), u_2(\cdot)) &= \min_{v_1(\cdot) \in \mathcal{U}_1} J_1(v_1(\cdot), u_2(\cdot)), \\ J_2(u_1(\cdot), u_2(\cdot)) &= \min_{v_2(\cdot) \in \mathcal{U}_2} J_2(u_1(\cdot), v_2(\cdot)). \end{cases} \quad (18)$$

We need the following assumption:

Hypothesis 5 (H5). $E_i(N_i)^{-1}E_i^\top S = SE_i(N_i)^{-1}E_i^\top$

where $S = A, \bar{A}, B, \bar{B}, C, \bar{C}, D, \bar{D}$, and $i = 1, 2$. Now, with the help of the above general FBDSDE, the explicit expression for the Nash equilibrium point of the above game problem can be obtained.

Theorem 4. *The mapping*

$$(u_1(t), u_2(t)) = (N_1^{-1}(t)E_1^\top(t)p_1(t), N_2^{-1}(t)E_2^\top(t)p_2(t)), \ t \in [0, T], \quad (19)$$

is one Nash equilibrium point for the above game problems in Equations (16)–(18), where $(y(t), z(t), p_1(t), p_2(t), q_1(t), q_2(t))$ is the solution of the following general FBDSDE:

$$\begin{cases} dy(t) &= [A(t)y(t) + B(t)z(t) + \bar{A}(t)y_\delta(t) + \bar{B}(t)z_\delta(t) \\ & \quad -E_1(t)N_1^{-1}(t)E_1^\top(t)p_1(t) - E_2(t)N_2^{-1}(t)E_2^\top(t)p_2(t)]dt - z(t)\overleftarrow{d}B(t) \\ & \quad +[C(t)y(t) + D(t)z(t) + \bar{C}(t)y_\delta(t) + \bar{D}(t)z_\delta(t)]\overrightarrow{d}W(t),\ t \in [0,T], \\ -dp_i(t) &= \{A^\top(t)p_i(t) + C^\top(t)q_i(t) + \mathbb{E}^{\mathcal{F}_t}[\bar{A}_{\delta+}^\top(t)p_{i\delta+}(t)] \\ & \quad + \mathbb{E}^{\mathcal{F}_t}[\bar{C}_{\delta+}^\top(t)q_{i\delta+}(t)] + M_i(t)y(t)\}dt - q_i(t)\overrightarrow{d}W(t) \\ & \quad + \{B^\top(t)p_i(t) + D^\top(t)q_i(t) + \mathbb{E}^{\mathcal{F}_t}[\bar{B}_{\delta+}^\top(t)p_{i\delta+}(t)] \\ & \quad + \mathbb{E}^{\mathcal{F}_t}[\bar{D}_{\delta+}^\top(t)q_{i\delta+}(t)] + R_i(t)z(t)\}\overleftarrow{d}B(t),\ t \in [0,T], \\ y(t) &= \xi(t),\ t \in [-\delta,0], \\ z(t) &= \eta(t),\ t \in [-\delta,0], \\ p_i(T) &= Q_i y(T),\ p_i(t) = 0,\ t \in [T, T+\delta],\ i = 1,2, \\ q_i(t) &= 0,\ t \in [T, T+\delta],\ i = 1,2. \end{cases} \qquad (20)$$

Similar to [31,48], the proof of Theorem 4 is easy to give, and we have therefore excluded it.

For sake of clarity, we give the following Problem (S), which is a special case of Problem (A). To simplify the notation, let us assume that $n = d = l = k = 1$. The control system is

$$\begin{cases} dy(t) &= [y_\delta(t) + z_\delta(t) + v_1(t) + v_2(t)]dt - z(t)\overleftarrow{d}B(t) + y_\delta(t)\overrightarrow{d}W(t),\ t \in [0,T], \\ y(t) &= \xi(t),\ t \in [-\delta,0], \\ z(t) &= \eta(t),\ t \in [-\delta,0], \end{cases}$$

where $(\xi(\cdot), \eta(\cdot)) \in L^2_\mathcal{F}(-\delta, T; \mathbb{R})$ is the initial path of (y, z). $v_1(t)$ and $v_2(t)$, $t \in [0, T]$ are two admissible control processes, i.e., $\mathcal{F}_t$-measurable square-integrable processes taking values in $\mathbb{R}$. We denote $J_1(v_1(\cdot), v_2(\cdot))$ and $J_2(v_1(\cdot), v_2(\cdot))$, which are the cost functionals corresponding to the players 1 and 2:

$$J_i(v_1(\cdot), v_2(\cdot)) = \mathbb{E}\left\{\frac{1}{2}\int_0^T v_i^2(t)dt + y(T)\right\},\ i = 1, 2.$$

Our task is to find $(u_1(\cdot), u_2(\cdot)) \in \mathbb{R}^k \times \mathbb{R}^k$ such that

$$\begin{cases} J_1(u_1(\cdot), u_2(\cdot)) &= \min_{v_1(\cdot) \in \mathcal{U}_1} J_1(v_1(\cdot), u_2(\cdot)), \\ J_2(u_1(\cdot), u_2(\cdot)) &= \min_{v_2(\cdot) \in \mathcal{U}_2} J_2(u_1(\cdot), v_2(\cdot)). \end{cases}$$

Then the Hamiltonian functions are

$$H_i(t, y_\delta, z_\delta, v_1, v_2, p_i, q_i) = [y_\delta + z_\delta + v_1 + v_2]p_i + y_\delta q_i + \frac{1}{2},\ i = 1, 2,$$

where $(p_1(t), p_2(t), q_1(t), q_2(t))$ is the solution of the following adjoint equations:

$$\begin{cases} -dp_i(t) &= \{\mathbb{E}^{\mathcal{F}_t}[p_{i\delta+}(t)] + \mathbb{E}^{\mathcal{F}_t}[q_{i\delta+}(t)]\}dt - q_i(t)\overrightarrow{d}W(t) \\ & \quad + \mathbb{E}^{\mathcal{F}_t}[p_{i\delta+}(t)]\overleftarrow{d}B(t),\ t \in [0, T], \\ p_i(T) &= 1,\ p_i(t) = 0,\ t \in [T, T+\delta],\ i = 1,2, \\ q_i(t) &= 0,\ t \in [T, T+\delta],\ i = 1,2. \end{cases}$$

It is easy to see that the above equation is the anticipated BDSDE, which is solvable theorem 3.2 in [45] and theorem 2.4 in [46]. From the maximum principle, we get that $(u_1(t), u_2(t)) = (p_1(t), p_2(t))$, $t \in [0, T]$ is one Nash equilibrium point for the above game in Equations (16)–(18).

6. The Proof of Theorem 1

Proof of Theorem 1. Since the initial path of (y, z) in $[-\delta, 0]$ and the terminal conditions and trajectories of (p, q) in $[T, T+\delta]$ are given in advance, we only need to consider $(y_t, p_t, z_t, q_t), 0 \leq t \leq T$.

Uniqueness Let $U = (y, p, z, q)$ and $\bar{U} = (\bar{y}, \bar{p}, \bar{z}, \bar{q})$ be two solutions of Equation (3). We set $\hat{U} = U - \bar{U} = (\hat{y}, \hat{p}, \hat{z}, \hat{q}) = (y - \bar{y}, p - \bar{p}, z - \bar{z}, q - \bar{q})$.

Applying Itô's formula to $\langle H\hat{y}, \hat{p}\rangle$ on $[0, T]$, we have

$$\mathbb{E}\langle H\hat{y}(T), \Phi(y(T)) - \Phi(\bar{y}(T))\rangle$$
$$= \mathbb{E}\int_0^T \langle A(t, U(t), \alpha(t), \mu(t), \beta(t), \nu(t)) - A(t, \bar{U}(t), \bar{\alpha}(t), \bar{\mu}(t), \bar{\beta}(t), \bar{\nu}(t)), \widehat{U}(t)\rangle dt$$
$$\leq -\mu_1 \mathbb{E}\int_0^T \left(|H(y(t) - \bar{y}(t))|^2 + |H(z(t) - \bar{z}(t))|^2\right) dt$$
$$-\mu_2 \mathbb{E}\int_0^T \left(\left|H^\top(p(t) - \bar{p}(t))\right|^2 + \left|H^\top(q(t) - \bar{q}(t))\right|^2\right) dt.$$

By virtue of (H3), it follows that

$$\mu_1 \mathbb{E}\int_0^T \left(|H(y(t) - \bar{y}(t))|^2 + |H(z(t) - \bar{z}(t))|^2\right) dt$$
$$+\mu_2 \mathbb{E}\int_0^T \left(\left|H^\top(p(t) - \bar{p}(t))\right|^2 + \left|H^\top(q(t) - \bar{q}(t))\right|^2\right) dt \leq 0.$$

If $m > n$, $\mu_1 > 0$, then we have $|H(y(t) - \bar{y}(t))|^2 \equiv 0$ and $|H(z(t) - \bar{z}(t))|^2 \equiv 0$. Thus $y(t) \equiv \bar{y}(t)$ and $z(t) \equiv \bar{z}(t)$. In particular, $\Phi(y(T)) = \Phi(\bar{y}(T))$. Consequently, from the uniqueness result of the anticipated BDSDE (see [45,46]), it follows that $p(t) \equiv \bar{p}(t)$ and $q(t) \equiv \bar{q}(t)$.

If $m < n$, $\mu_2 > 0$, then we have $\left|H^\top(p(t) - \bar{p}(t))\right|^2 \equiv 0$ and $\left|H^\top(q(t) - \bar{q}(t))\right|^2 \equiv 0$. Thus $p(t) \equiv \bar{p}(t)$ and $q(t) \equiv \bar{q}(t)$. From the uniqueness result of the delayed doubly SDE (see [20]), it follows that $y(t) \equiv \bar{y}(t)$ and $z(t) \equiv \bar{z}(t)$.

Similarly to the above arguments, the desired result can be obtained easily in the case $n = m$. The uniqueness is proved. □

The proof of the existence is a combination of the above technique and a priori estimate technique introduced by Peng [49]. We divide the proof of existence into three cases: $m > n$, $m < n$ and $m = n$.

Case 1 If $m > n$, then $\mu_1 > 0$, $\beta_1 > 0$. We consider the following family of FBDSDEs parametrized by $\alpha \in [0, 1]$

$$\begin{cases} dy(t) &= [\alpha f(t, U(t), y_\delta(t), z_\delta(t)) + f_0(t)]dt - z(t)\overleftarrow{d}B(t) \\ &\quad + [\alpha g(t, U(t), y_\delta(t), z_\delta(t)) + g_0(t)]\overrightarrow{d}W(t),\ t \in [0, T], \\ -dp(t) &= [\alpha F(t, U(t), p_{\delta+}(t), q_{\delta+}(t)) + (1-\alpha)\mu_1 Hy(t) + F_0(t)]dt \\ &\quad + [\alpha \overrightarrow{G}(t, U(t), p_{\delta+}(t), q_{\delta+}(t)) + (1-\alpha)\mu_1 Hz(t) + G_0(t)]\overleftarrow{d}B(t) \\ &\quad - q(t)\overrightarrow{d}W(t),\ t \in [0, T], \\ y(t) &= \phi(t),\ t \in [-\delta, 0], \\ z(t) &= \psi(t),\ t \in [-\delta, 0], \\ p(T) &= \alpha\Phi(y(T)) + (1-\alpha)Hy(T) + \varphi, \\ p(t) &= \xi(t),\ t \in [T, T+\delta], \\ q(t) &= \eta(t),\ t \in [T, T+\delta], \end{cases} \quad (21)$$

where $U = (y, p, z, q)$ and $(F_0, f_0, G_0, g_0) \in L_\mathcal{F}^2\left(0, T; \mathbb{R}^{m+n+m\times l+n\times d}\right)$ and $\varphi \in L^2(\mathcal{F}_T; \mathbb{R}^m)$ are arbitrarily given vector-valued random variables. When $\alpha = 1$ the existence of the solution of Equation (21) implies clearly that of Equation (5). Due to the existence and uniqueness of the delayed doubly SDE (see [20]), when $\alpha = 0$, the Equation (21) is uniquely solvable. The following a priori lemma is a key step in the proof of the method of continuation. It shows that for a fixed $\alpha = \alpha_0 \in [0, 1)$, if Equation (21) is uniquely solvable, then it is also uniquely solvable for any $\alpha \in [\alpha_0, \alpha_0 + \gamma_0]$, for some positive constant γ_0 independent of α_0.

Lemma 3. *We assume $m > n$. Under assumptions (H3), there exists a positive constant γ_0 such that if a priori, for each $\varphi \in L^2(\mathcal{F}_T; \mathbb{R}^m)$, $(F_0, f_0, G_0, g_0) \in L_\mathcal{F}^2\left(0, T; \mathbb{R}^{m+n+m\times l+n\times d}\right)$, Equation (16) is uniquely solvable for some $\alpha_0 \in [0, 1)$, then for each $\alpha \in [\alpha_0, \alpha_0 + \gamma_0]$, and $\varphi \in L^2(\mathcal{F}_T; \mathbb{R}^m)$, $(F_0, f_0, G_0, g_0) \in L_\mathcal{F}^2\left(0, T; \mathbb{R}^{m+n+m\times l+n\times d}\right)$, Equation (16) is also uniquely solvable in $L_\mathcal{F}^2(-\delta, T; \mathbb{R}^n) \times L_\mathcal{F}^2(0, T+\delta; \mathbb{R}^m) \times L_\mathcal{F}^2(-\delta, T; \mathbb{R}^{n\times l}) \times L_\mathcal{F}^2\left(0, T+\delta; \mathbb{R}^{m\times d}\right)$.*

Proof of Lemma 3. Since for $\varphi \in L^2(\mathcal{F}_T; \mathbb{R}^m)$, $(F_0, f_0, G_0, g_0) \in L^2_{\mathcal{F}}(0, T; \mathbb{R}^{m+n+m\times l+n\times d})$, there exists a unique solution to Equation (16) for $\alpha = \alpha_0$. Thus for each $\bar{U} = (\bar{y}, \bar{p}, \bar{z}, \bar{q}) \in L^2_{\mathcal{F}}(-\delta, T; \mathbb{R}^n) \times L^2_{\mathcal{F}}(0, T + \delta; \mathbb{R}^m) \times L^2_{\mathcal{F}}(-\delta, T; \mathbb{R}^{n\times l}) \times L^2_{\mathcal{F}}(0, T + \delta; \mathbb{R}^{m\times d})$, there exists a unique quadruple $U = (y, p, z, q) \in L^2_{\mathcal{F}}(-\delta, T; \mathbb{R}^n) \times L^2_{\mathcal{F}}(0, T + \delta; \mathbb{R}^m) \times L^2_{\mathcal{F}}(-\delta, T; \mathbb{R}^{n\times l}) \times L^2_{\mathcal{F}}(0, T + \delta; \mathbb{R}^{m\times d})$ satisfying the following equations

$$\begin{cases} dy(t) &= [\alpha_0 f(t, U(t), y_\delta(t), z_\delta(t)) + \gamma f(t, \bar{U}(t), \bar{y}_\delta(t), \bar{z}_\delta(t)) + f_0(t)]dt \\ &\quad + [\alpha_0 g(t, U(t), y_\delta(t), z_\delta(t)) + \gamma g(t, \bar{U}(t), \bar{y}_\delta(t), \bar{z}_\delta(t)) \\ &\quad + g_0(t)]\overrightarrow{d}W(t) - z(t)\overleftarrow{d}B(t),\ t \in [0, T], \\ -dp(t) &= [\alpha_0 F(t, U(t), p_{\delta+}(t), q_{\delta+}(t)) + (1-\alpha_0)\mu_1 Hy(t) \\ &\quad + \gamma(F(t, \bar{U}(t), \bar{p}_{\delta+}(t), \bar{q}_{\delta+}(t)) - \mu_1 H\bar{y}(t)) + F_0(t)]dt \\ &\quad + [\alpha_0 G(t, U(t), p_{\delta+}(t), q_{\delta+}(t)) + (1-\alpha_0)\mu_1 Hz(t) \\ &\quad + \gamma(G(t, \bar{U}(t), \bar{p}_{\delta+}(t), \bar{q}_{\delta+}(t)) - \mu_1 H\bar{y}(t)) + G_0(t)]\overleftarrow{d}B(t) \\ &\quad - q(t)\overrightarrow{d}W(t),\ t \in [0, T], \\ y(t) &= \phi(t),\ t \in [-\delta, 0], \\ z(t) &= \psi(t),\ t \in [-\delta, 0], \\ p(T) &= \alpha_0 \Phi(y(T)) + (1-\alpha_0)Hy(T) + \gamma(\Phi(\bar{y}(T)) - H\bar{y}(T)) + \varphi, \\ p(t) &= \xi(t),\ t \in [T, T+\delta], \\ q(t) &= \eta(t),\ t \in [T, T+\delta], \end{cases}$$

where $\gamma \in (0, 1)$ is independent of α_0. We will prove that the mapping defined by

$$U = I_{\alpha_0+\gamma}(\bar{U}):$$
$$L^2_{\mathcal{F}}(-\delta, T; \mathbb{R}^n) \times L^2_{\mathcal{F}}(0, T+\delta; \mathbb{R}^m) \times L^2_{\mathcal{F}}(-\delta, T; \mathbb{R}^{n\times l}) \times L^2_{\mathcal{F}}(0, T+\delta; \mathbb{R}^{m\times d})$$
$$\to L^2_{\mathcal{F}}(-\delta, T; \mathbb{R}^n) \times L^2_{\mathcal{F}}(0, T+\delta; \mathbb{R}^m) \times L^2_{\mathcal{F}}(-\delta, T; \mathbb{R}^{n\times l}) \times L^2_{\mathcal{F}}(0, T+\delta; \mathbb{R}^{m\times d})$$

is contractive for a small enough $\gamma > 0$.

Let $\bar{U}' = (\bar{y}', \bar{p}', \bar{z}', \bar{q}') \in L^2_{\mathcal{F}}(-\delta, T; \mathbb{R}^n) \times L^2_{\mathcal{F}}(0, T+\delta; \mathbb{R}^m) \times L^2_{\mathcal{F}}(-\delta, T; \mathbb{R}^{n\times l}) \times L^2_{\mathcal{F}}(0, T+\delta; \mathbb{R}^{m\times d})$ and $(y', p', z', q') = U' = I_{\alpha_0+\delta}(\bar{U}')$.

$$\hat{\bar{U}} = \bar{U} - \bar{U}' = (\hat{\bar{y}}, \hat{\bar{p}}, \hat{\bar{z}}, \hat{\bar{q}}) = (\bar{y} - \bar{y}', \bar{p} - \bar{p}', \bar{z} - \bar{z}', \bar{q} - \bar{q}'),$$
$$\hat{U} = U - U' = (\hat{y}, \hat{p}, \hat{z}, \hat{q}) = (y - y', p - p', z - z', q - q').$$

Applying Itô's formula to $\langle H\hat{y}, \hat{p}\rangle$ on $[0, T]$, it follows that

$$(1 - \alpha_0 + \alpha_0\beta_1)\mathbb{E}|H\hat{y}(T)|^2 + \mu_1 \mathbb{E}\int_0^T \left(|H\hat{y}(t)|^2 + |H\hat{z}(t)|^2\right)dt$$
$$\leq \gamma C\mathbb{E}\int_0^T \left(|\hat{\bar{U}}(t)|^2 + |\hat{\bar{y}}_\delta(t)|^2 + |\hat{\bar{z}}_\delta(t)|^2 + |\hat{\bar{p}}_{\delta+}(t)|^2 + |\hat{\bar{q}}_{\delta+}(t)|^2\right)dt$$
$$\quad + \gamma C\mathbb{E}\int_0^T |\hat{U}(t)|^2 dt + \gamma C\left(\mathbb{E}|\hat{y}(T)|^2 + \mathbb{E}|\hat{\bar{y}}(T)|^2\right)$$
$$\leq \gamma C\left[\mathbb{E}\int_0^T \left(|\hat{\bar{y}}(t)|^2 + |\hat{\bar{z}}(t)|^2\right)dt + \mathbb{E}\int_0^{T+\delta}\left(|\hat{\bar{p}}(t)|^2 + |\hat{\bar{q}}(t)|^2\right)dt\right]$$
$$\quad + \gamma C\mathbb{E}\int_0^T |\hat{U}(t)|^2 dt + \gamma C\left(\mathbb{E}|\hat{y}(T)|^2 + \mathbb{E}|\hat{\bar{y}}(T)|^2\right), \tag{22}$$

with some constant $C > 0$. Hereafter, C will be some generic constant, which can be different from line to line and depends only on the Lipschitz constants k, λ, μ_1, β_1, H and T. It is obvious that $1 - \alpha_0 + \alpha_0\beta_1 \geq \beta$, $\beta = \min(1, \beta_1) > 0$.

On the other hand, for the difference of the solutions $(\hat{p}, \hat{q}) = (p - p', q - q')$, we apply a standard method of estimation. Applying Itô's formula to $|\hat{p}(t)|^2$ on $[t, T]$, we have

$$\mathbb{E}\int_0^T \left(|\widehat{p}(t)|^2 + |\widehat{q}(t)|^2\right) dt$$
$$\leq \gamma C\left[\mathbb{E}\int_0^T \left(|\widehat{y}(t)|^2 + |\widehat{z}(t)|^2\right) dt + \mathbb{E}\int_0^{T+\delta}\left(|\widehat{p}(t)|^2 + |\widehat{q}(t)|^2\right) dt\right]$$
$$+ C\left(\mathbb{E}|\widehat{y}(T)|^2 + \delta\mathbb{E}|\widehat{y}(T)|^2\right) + C\mathbb{E}\int_0^T \left(|\widehat{y}(t)|^2 + |\widehat{z}(t)|^2\right) dt. \qquad (23)$$

Combining the estimates in Equations (22) and (23), for a sufficiently large constant $C > 0$, we have

$$\mathbb{E}\int_{-\delta}^T \left(|\widehat{y}(t)|^2 + |\widehat{z}(t)|^2\right) dt + \mathbb{E}\int_0^{T+\delta}\left(|\widehat{p}(t)|^2 + |\widehat{q}(t)|^2\right) dt + \mathbb{E}|\widehat{y}(T)|^2$$
$$\leq \gamma C\mathbb{E}\int_{-\delta}^T \left(|\widehat{y}(t)|^2 + |\widehat{z}(t)|^2\right) dt + \mathbb{E}\int_0^{T+\delta}\left(|\widehat{p}(t)|^2 + |\widehat{q}(t)|^2\right) dt + \mathbb{E}|\widehat{y}(T)|^2.$$

We now choose $\gamma_0 = \dfrac{1}{2C}$. It is clear that, for each fixed $\gamma \in [0, \gamma_0]$, the mapping $I_{\alpha_0 + \gamma}$ is contractive in the sense that

$$\mathbb{E}\int_{-\delta}^T \left(|\widehat{y}(t)|^2 + |\widehat{z}(t)|^2\right) dt + \mathbb{E}\int_0^{T+\delta}\left(|\widehat{p}(t)|^2 + |\widehat{q}(t)|^2\right) dt + \mathbb{E}|\widehat{y}(T)|^2$$
$$\leq \frac{1}{2}\mathbb{E}\int_{-\delta}^T \left(|\widehat{y}(t)|^2 + |\widehat{z}(t)|^2\right) dt + \mathbb{E}\int_0^{T+\delta}\left(|\widehat{p}(t)|^2 + |\widehat{q}(t)|^2\right) dt + \mathbb{E}|\widehat{y}(T)|^2.$$

Thus this mapping has a unique fixed point $U = (y, p, z, q) \in L^2_{\mathcal{F}}(-\delta, T; \mathbb{R}^n) \times L^2_{\mathcal{F}}(0, T + \delta; \mathbb{R}^m) \times L^2_{\mathcal{F}}(-\delta, T; \mathbb{R}^{n \times l}) \times L^2_{\mathcal{F}}(0, T + \delta; \mathbb{R}^{m \times d})$, which is the solution of Equation (16) for $\alpha = \alpha_0 + \gamma$, as $\gamma \in [0, \gamma_0]$. The proof is complete. □

Case 2 If $m < n$, then $\mu_2 > 0$. We consider the following equations

$$\begin{cases} dy(t) &= \left[\alpha f(t, U(t), y_\delta(t), z_\delta(t)) + (1-\alpha)\mu_1 H^\top p(t) + f_0(t)\right] dt \\ &\quad + \left[\alpha g(t, U(t), y_\delta(t), z_\delta(t)) + (1-\alpha)\mu_1 H^\top q(t) + g_0(t)\right] \overrightarrow{d}W(t) \\ &\quad - z(t) \overleftarrow{d}B(t), \ t \in [0, T], \\ -dp(t) &= [\alpha F(t, U(t), p_{\delta+}(t), q_{\delta+}(t)) + F_0(t)] dt - q(t) \overrightarrow{d}W(t) \\ &\quad + [\alpha G(t, U(t), p_{\delta+}(t), q_{\delta+}(t)) + G_0(t)] \overleftarrow{d}B(t), \ t \in [0, T], \\ y(t) &= \phi(t), \ t \in [-\delta, 0], \\ z(t) &= \psi(t), \ t \in [-\delta, 0], \\ p(T) &= \alpha \Phi(y(T)) + \varphi, \\ p(t) &= \xi(t), \ t \in [T, T + \delta], \\ q(t) &= \eta(t), \ t \in [T, T + \delta]. \end{cases} \qquad (24)$$

When $\alpha = 1$, the existence of the solution of Equation (24) implies clearly that of Equation (16). Due to the existence and uniqueness of the anticipated BDSDE (see [45,46]), when $\alpha = 0$, we know that Equation (24) is uniquely solvable. By the same techniques, we can also prove the following lemma similar to Lemma 3.

Lemma 4. *Assume $m < n$. Under assumption (H3), there exists a positive constant γ_0 such that if a priori, for each $\varphi \in L^2(\mathcal{F}_T; \mathbb{R}^m)$, and $(F_0, f_0, G_0, g_0) \in L^2_{\mathcal{F}}\left(0, T; \mathbb{R}^{m+n+m \times l+n \times d}\right)$, Equation (24) is uniquely solvable for some $\alpha_0 \in [0, 1)$, then for each $\alpha \in [\alpha_0, \alpha_0 + \gamma_0]$, and $\varphi \in L^2(\mathcal{F}_T; \mathbb{R}^m)$, $(F_0, f_0, G_0, g_0) \in L^2_{\mathcal{F}}\left(0, T; \mathbb{R}^{m+n+m \times l+n \times d}\right)$, Equation (24) is also uniquely solvable in $L^2_{\mathcal{F}}(-\delta, T; \mathbb{R}^n) \times L^2_{\mathcal{F}}(0, T + \delta; \mathbb{R}^m) \times L^2_{\mathcal{F}}(-\delta, T; \mathbb{R}^{n \times l}) \times L^2_{\mathcal{F}}\left(0, T + \delta; \mathbb{R}^{m \times d}\right)$.*

Case 3 $m = n$. From (H3), we only need to consider two cases as follows:

(1) If $\mu_1 > 0$, $\mu_2 \geq 0$, $\beta_1 > 0$, we can have the same result as Lemma 3.
(2) If $\mu_1 \geq 0$, $\mu_2 > 0$, $\beta_1 \geq 0$, the same result as Lemma 4 holds.

Now we give the proof of the existence of Theorem 1.

Proof of the Existence of Theorem 1. For the first case where $m > n$, we know that for each $\varphi \in L^2(\mathcal{F}_T; \mathbb{R}^m)$, and $(F_0, f_0, G_0, g_0) \in L^2_{\mathcal{F}}(0, T; \mathbb{R}^{m+n+m \times l+n \times d})$, Equation (21) has a unique solution as $\alpha = 0$. It follows from Lemma 3 that there exists a positive constant $\gamma_0 = \gamma_0(k, \lambda, \beta_1, \mu_1, H, T)$ such that for any $\gamma \in [0, \gamma_0]$ and $\varphi \in L^2(\mathcal{F}_T; \mathbb{R}^m)$, and $(F_0, f_0, G_0, g_0) \in L^2_{\mathcal{F}}(0, T; \mathbb{R}^{m+n+m \times l+n \times d})$, Equation (21) has a unique solution for $\alpha = \gamma$. Since γ_0 depends only on $k, \lambda, \beta_1, \mu_1, H, T$, we can repeat this process for N times with $1 \leq N\gamma_0 < 1 + \gamma_0$. In particular, for $\alpha = 1$ with $(F_0, f_0, G_0, g_0) \equiv 0$, and $\varphi \equiv 0$, $\psi \equiv 0$, Equation (21) has a unique solution in $L^2_{\mathcal{F}}(-\delta, T; \mathbb{R}^n) \times L^2_{\mathcal{F}}(0, T + \delta; \mathbb{R}^m) \times L^2_{\mathcal{F}}(-\delta, T; \mathbb{R}^{n \times l}) \times L^2_{\mathcal{F}}(0, T + \delta; \mathbb{R}^{m \times d})$.

In the case where $m < n$ and $m = n$, our desired result can be obtained similarly. The proof of the existence of Theorem 1 is complete. □

Remark 2. *In the proof of the Existence of Theorem 1, (i) and (ii) in (H3) can be replaced by*

(i)' $\int_0^T \langle A(t, u, \alpha, \mu, \beta, \nu) - A(t, \bar{u}, \bar{\alpha}, \bar{\mu}, \bar{\beta}, \bar{\nu}), u - \bar{u} \rangle dt$
$\geq \int_0^T [\mu_1(|H(y - \bar{y})|^2 + |H(z - \bar{z})|^2) + \mu_2(|H^\top(p - \bar{p})|^2 + |H^\top(q - \bar{q})|^2)]dt,$
$\forall u = (y, p, z, q), \bar{u} = (\bar{y}, \bar{p}, \bar{z}, \bar{q}) \in \mathbb{R}^n \times \mathbb{R}^m \times \mathbb{R}^{n \times l} \times \mathbb{R}^{m \times d}, \forall t \in [0, T].$

(ii)' $\langle \Phi(y) - \Phi(\bar{y}), H(y - \bar{y}) \rangle \leq -\beta_1 |H(y - \bar{y})|^2, \forall y, \bar{y} \in \mathbb{R}^n.$

where μ_1, μ_2 and β_1 are given non-negative constants with $\mu_1 + \mu_2 > 0$ and $\mu_2 + \beta_1 > 0$. Moreover we have $\mu_1 > 0, \beta_1 > 0$ (resp., $\mu_2 > 0$) when $m > n$ (resp., $m < n$).

7. Conclusions

The future evolution of a lot of processes depends not only on their current state, but also on their historical state, and these processes can usually be characterized by stochastic differential equations with time delay. In this article, we have discussed a class of differential games driven by doubly stochastic systems with time delay. To deal with the above nonzero sum differential game problem, it is natural to involve the adjoint equation, which is a kind of anticipated BDSDE. It is therefore necessary to explore a kind of general FBDSDE with the forward equation being a delayed doubly SDE and the backward equation being an anticipated BDSDE, which are so-called time-symmetry stochastic systems. This kind of FBDSDE covers a lot of the previous results, which promotes the results in [35] to doubly stochastic integrals, and extends the results in [23] to the case that involves the time delay and anticipation. We have adopted the convex variational method, and established a necessary condition and a sufficient condition for the equilibrium point of the game. In the LQ game problem, the state equation and the adjoint equation are completely coupled, then a class of linear FBDSDE is constructed, in which the forward equation is an anticipated forward doubly SDE and the backward equation is a delayed backward doubly SDE. By means of the unique solvability of the FBDSDE, the explicit expression for the Nash equilibrium point of the LQ game is obtained. Many financial and economic phenomena can be modeled by the LQ model, and we expect that the LQ game driven by doubly stochastic systems with time delay can be widely applied in these fields.

Notwithstanding that we are committed to the above game problem, we are also able to progress some consequences of optimal control for BDSDEs with time delay, for example Xu and Han [19,20].

Author Contributions: Writing—original draft preparation, writing—review and editing, Q.Z. and H.Z.; supervision, Y.S.; Conceptualization, J.W. All authors have read and agreed to the published version of the manuscript.

Funding: This research was funded by the National Key R&D Program of China (2018YFA0703900), National Natural Science Foundation of China (11871309, 11671229, 71871129, 11371226, 11301298), Southern University of Science and Technology Start up fund (Y01286233), Natural Science Foundation of Shandong Province of China (ZR2020MA032, ZR2019MA013), Special Funds of Taishan Scholar Project (tsqn20161041), and Fostering Project of Dominant Discipline and Talent Team of Shandong Province Higher Education Institutions.

Acknowledgments: The authors express their sincerest thanks to the reviewers for their valuable comments, which further improve the conclusion and proof process of the article.

Conflicts of Interest: The authors declare no conflict of interest.

References

1. Pardoux, P.; Peng, S.G. Backward doubly stochastic differential equations and systems of quasilinear parabolic SPDEs. *Probab. Theory Relat. Fields* **1994**, *98*, 209–227.
2. Bahlali, K.; Gatt, R.; Mansouri, B.; Mtiraoui, A. Backward doubly SDEs and SPDEs with superlinear growth generators. *Stoch. Dynam.* **2017**, *17*, 1–31.
3. Bally, V.; Matoussi, A. Weak solutions for SPDEs and backward doubly stochastic differential equations. *J. Theoret. Probab.* **2001**, *14*, 125–164.
4. Hu, L.Y.; Ren, Y. Stochastic PDIEs with nonlinear Neumann boundary conditions and generalized backward doubly stochastic differential equations driven by Lévy processes. *J. Comput. Appl. Math.* **2009**, *229*, 230–239.
5. Ma, N.; Wu, Z. Backward doubly stochastic differential equations with Markov chains and a comparison theorem. *Symmetry* **2020**, *12*, 1953.
6. Matoussi, A.; Piozin, L.; Popier, A. Stochastic partial differential equations with singular terminal condition. *Stoch. Proc. Appl.* **2017**, *127*, 831–876.
7. Ren, Y.; Lin, A.H.; Hu, L.Y. Stochastic PDIEs and backward doubly stochastic differential equations driven by Lévy processes. *J. Comput. Appl. Math.* **2009**, *223*, 901–907.
8. Shi, Y.F.; Gu, Y.L.; Liu, K. Comparison theorems of backward doubly stochastic differential equations and applications. *Stoch. Anal. Appl.* **2005**, *23*, 97–110.
9. Shi, Y.F.; Wen, J.Q.; Xiong, J. Backward doubly stochastic Volterra integral equations and their applications. *J. Differ. Equ.* **2020**, *269*, 6492–6528.
10. Wen, J.Q.; Shi, Y.F. Backward doubly stochastic differential equations with random coefficients and quasilinear stochastic PDEs. *J. Math. Anal. Appl.* **2019**, *476*, 86–100.
11. Wu, Z.; Zhang, F. BDSDEs with locally monotone coefficients and Sobolev solutions for SPDEs. *J. Differ. Equ.* **2011**, *251*, 759–784.
12. Zhang, Q.; Zhao, H.Z. Stationary solutions of SPDEs and infinite horizon BDSDEs. *J. Funct. Anal.* **2007**, *252*, 171–219.
13. Zhang, Q.; Zhao, H.Z. Stationary solutions of SPDEs and infinite horizon BDSDEs under non-Lipschitz coefficients. *J. Differ. Equ.* **2010**, *248*, 953–991.
14. Zhang, Q.; Zhao, H.Z. SPDEs with polynomial growth coefficients and the Malliavin calculus method. *Stoch. Proc. Appl.* **2013**, *123*, 2228–2271.
15. Zhang, Q.; Zhao, H.Z. Backward doubly stochastic differential equations with polynomial growth coefficients. *Discrete Con. Dyn. A* **2015**, *35*, 5285–5315.
16. Bahlali, S.; Gherbal, B. Optimality conditions of controlled backward doubly stochastic differential equations. *Random Oper. Stoch. Equ.* **2010**, *18*, 247–265.
17. Han, Y.C.; Peng, S.G.; Wu, Z. Maximum principle for backward doubly stochastic control systems with applications. *SIAM J. Control Optim.* **2010**, *48*, 4224–4241.
18. Wu, J.B.; Liu, Z.M. Optimal control of mean-field backward doubly stochastic systems driven by Itô-Lévy processes. *Int. J. Control* **2020**, *93*, 953–970.
19. Xu, J.; Han, Y.C. Stochastic maximum principle for delayed backward doubly stochastic control systems. *J. Nonlinear Sci. Appl.* **2017**, *10*, 215–226.
20. Xu, J. Stochastic maximum principle for delayed doubly stochastic control systems and their applications. *Int. J. Control* **2020**, *93*, 1371–1380.
21. Zhu, Q.F.; Shi, Y.F. Optimal control of backward doubly stochastic systems with partial information. *IEEE Trans. Autom. Control* **2015**, *60*, 173–178.
22. Peng, S.G.; Shi, Y.F. A type of time-symmetric forward-backward stochastic differential equations. *C. R. Acad. Sci. Paris Ser. I* **2003**, *336*, 773–778.
23. Zhu, Q.F.; Shi, Y.F.; Gong, X.J. Solutions to general forward-backward doubly stochastic differential equations. *Appl. Math. Mech.* **2009**, *30*, 517–526.
24. Zhu, Q.F.; Shi, Y.F. Forward-backward doubly stochastic differential equations and related stochastic partial differential equations. *Sci. China Math.* **2012**, *55*, 2517–2534.
25. Zhang, L.Q.; Shi, Y.F. Maximum principle for forward-backward doubly stochastic control systems and applications. *ESAIM Control Optim. Calc. Var.* **2011**, *17*, 1174–1197.
26. Shi, Y.F.; Zhu, Q.F. Partially observed optimal control of forward-backward doubly stochastic systems. *ESAIM Control Optim. Calc. Var.* **2013**, *19*, 828–843.
27. Von Neumann, J.; Morgenstern, O. *The Theory of Games and Economic Behavior*; Princeton University Press: Princeton, NJ, USA, 1944.
28. Nash, J. Non-cooperative games. *Ann. Math.* **1951**, *54*, 286–295.
29. Zhao, H.; Shen, Y.; Zeng, Y. Time-consistent investment-reinsurance strategy for mean-variance insurers with a defaultable security. *J. Math. Anal. Appl.* **2016**, *437*, 1036–1057.
30. Wang, G.C.; Yu, Z.Y. A Pontryagin's maximum principle for nonzero sum differential games of BSDEs with applications. *IEEE Trans. Autom. Control* **2010**, *55*, 1742–1747.
31. Wang, G.C.; Yu, Z.Y. A partial information non-zero sum differential game of backward stochastic differential equations with applications. *Automatica* **2012**, *48*, 342–352.

32. Yu, Z.Y.; Ji, S.L. Linear-quadratic non-zero sum differential game of backward stochstic differential equations. In Proceedings of the 27th Chinese Control Conference, Kunming, China, 16–18 July 2008; pp. 562–566.
33. Hui, E.; Xiao, H. Differential games of partial information forward-backward doubly SDE and applications. *ESAIM Control Optim. Calc. Var.* **2014**, *20*, 78–94.
34. Chen, L.; Wu, Z. Maximum principle for the stochastic optimal control problem with delay and application. *Automatica* **2010**, *46*, 1074–1080.
35. Chen, L.; Wu, Z. A type of generalized forward-backward stochastic differential equations and applications. *Chin. Ann. Math. Ser. B* **2011**, *32*, 279–292.
36. Chen, L.; Huang, J.H. Stochastic maximum principle for controlled backward delayed system via advanced stochastic differential equation. *J. Optim. Theory Appl.* **2015**, *167*, 1112–1135.
37. Meng, Q.X.; Shen, Y. Optimal control of mean-field jump-diffusion systems with delay: A stochastic maximum principle approach. *J. Comput. Appl. Math.* **2015**, *279*, 13–30.
38. Meng, Q.X.; Shen, Y. Optimal control for stochastic delay evolution equations. *Appl. Math. Optim.* **2016**, *74*, 53–89.
39. Shen, Y.; Meng, Q.X.; Shi, P. Maximum principle for jump-diffusion mean-field stochastic delay differential equations and its application to finance. *Automatica* **2014**, *50*, 1565–1579.
40. Chen, L.; Yu, Z.Y. Maximum principle for nonzero-sum stochastic differential game with delays. *IEEE Trans. Autom. Control* **2015**, *60*, 1422–1426.
41. Shi, J.T.; Wang, G.C. A nonzero sum differential game of BSDE with time-delayed generator and applications. *IEEE Trans. Autom. Control* **2016**, *61*, 1959–1964.
42. Shen, Y.; Zeng Y. Optimal investment-reinsurance with delay for mean-variance insurers: A maximum principle approach. *Insur. Math. Econ.* **2014**, *57*, 1–12.
43. Arriojas, M.; Hu, Y.Z.; Mohammed, S.E.; Pap, G. A delayed black and scholes formula. *Stoch. Anal. Appl.* **2007**, *25*, 471–492.
44. Kazmerchuk, Y.; Swishchuk, A.; Wu, J.H. The pricing of option for securities markets with delayed response. *Math. Comput. Simul.* **2007**, *75*, 69–79.
45. Xu, X.M. Anticipated backward doubly stochastic differential equations. *Appl. Math. Comput.* **2013**, *220*, 53–62.
46. Zhang, F. Anticipated backward doubly stochastic differential equations. *Sci. Sin. Math.* **2013**, *43*, 1223–1236. (In Chinese)
47. Peng, S.G.; Wu, Z. Fully coupled forward-backward stochastic differential equations and applications to optimal control. *SIAM J. Control Optim.* **1999**, *37*, 825–843.
48. Zhu, Q.F.; Shi, Y.F. Nonzero-sum differential game of backward doubly stochastic systems with delay and applications. *Math. Control Relat. F* **2021**, *11*, 73–94.
49. Peng, S.G. Probabilistic interpretation for systems of quasilinear parabolic partial differential equations. *Stochastics* **1991**, *37*, 61–74.

Article

Effects of Second-Order Velocity Slip and the Different Spherical Nanoparticles on Nanofluid Flow

Jing Zhu *, Ye Liu and Jiahui Cao

School of Mathematics and Physics, University of Science and Technology Beijing, Beijing 100083, China; s20200705@xs.ustb.edu.cn (Y.L.); s20190728@xs.ustb.edu.cn (J.C.)
* Correspondence: zhujing@ustb.edu.cn; Tel.: +86-1368-121-2703

Abstract: The paper theoretically investigates the heat transfer of nanofluids with different nanoparticles inside a parallel-plate channel. Second-order slip condition is adopted due to the microscopic roughness in the microchannels. After proper transformation, nonlinear partial differential systems are converted to ordinary differential equations with unknown constants, and then solved by homotopy analysis method. The residual plot is drawn to verify the convergence of the solution. The semi-analytical expressions between Nu_B and N_{BT} are acquired. The results show that both first-order slip parameter and second-order slip parameter have positive effects on Nu_B of the MHD flow. The effect of second-order velocity slip on Nu_B is obvious, and Nu_B in the alumina–water nanofluid is higher than that in the titania–water nanofluid. The positive correlation between slip parameters and N_{dp} is significant for the titania–water nanofluid.

Keywords: nanofluid; second-order slip velocity; nanoparticles migration; homotopy analysis method

Citation: Zhu, J.; Liu, Y.; Cao, J. Effects of Second-Order Velocity Slip and the Different Spherical Nanoparticles on Nanofluid Flow. *Symmetry* 2021, *13*, 64. https://doi.org/10.3390/sym13010064

Received: 12 November 2020
Accepted: 28 December 2020
Published: 31 December 2020

Publisher's Note: MDPI stays neutral with regard to jurisdictional claims in published maps and institutional affiliations.

Copyright: © 2020 by the authors. Licensee MDPI, Basel, Switzerland. This article is an open access article distributed under the terms and conditions of the Creative Commons Attribution (CC BY) license (https://creativecommons.org/licenses/by/4.0/).

1. Introduction

Modern industrial applications are expected to achieve higher heat transfer rates, so how to improve the heat transfer performance of heat exchanger becomes the main problem concerned by researchers. Meanwhile, microchannels have many applications such as automobile cooling systems and electronic devices in micro-sized cooling systems. Li et al. [1] and Duan et al. [2] studied the heat transfer rates of nanofluid in microchannels.

To study the flow of nanofluid, homogeneous flow models and dispersion models have been proposed. In 2006, Buongiorno [3] showed that the dispersed effects can be completely ignored due to the size of nanoparticles, and Brownian diffusion and thermophoresis are important in nanofluids. Based on the above analysis, he proposed that the homogeneous models are more appropriate for predicting the heat transfer coefficient. By using this model, Yang et al. [4] studied the variation of forced convection transport with temperature jump in continuous flow and slip flow regimes. F. Hedayati et al. [5] studied the variation of $TiO_2 - H_2O$ nanofluid mixing convection within vertical microchannel of nanoparticle migration and asymmetric heating. R.S.Andhare et al. [6] studied pressure drop characteristics of a flat plate manifold microchannel heat exchanger. O.D. Makinde et al. [7] studied MHD variable viscosity reacting flow with thermophoresis and radiative heat transfer. A.Malvandi et al. [8] discussed effects of nanoparticle migration on alumina–water nanofluid.

Boundary conditions are critical to the model; initially, the common velocity slip is the Maxwell [9] slip condition. Kou et al. [10] studied the effects of wall slip and temperature jump on heat and mass transfer characteristics of evaporative films. A.A. Avramenko et al. [11] investigated mixed convection in a circular microchannel with the slip boundary conditions. As micro/nanotechnology develops, the size of micro/nanodevices are getting smaller and smaller. The Navier slip condition will break down at higher shear rates. In 1997, Thompson [12] developed a nonlinear slip model based on the first-order slip model proposed by Maxwell. However, many researchers found that the model could not predict the flow at a high Kn number. The values calculated by the second-order slip boundary condition are closer to the experimental data. Beskok and Karniadakis [13] improved a second-order slip conditions. Based on Beskok and Karniadakis, Wu [14] improved the slip condition. Zhu et al. [15] and Almutairi et al. [16] described the effects of second-order velocity slip.

However, as a result of the migration of nanoparticles under second-order slip condition and the influence of different nanoparticles, the heat transfer of nanofluids is limited. Besides, there is little attention paid to the analytic solution [17]. In this paper, the overall goal is to study the fully developed convection of nanofluids in a parallel plate channel theoretically. Two water-based nanofluids, containing alumina and titania nanoparticles, respectively, are considered. The governing partial differential equations are transformed into ordinary differential equations with an unknown constant by using similar variables, which are solved by the homotopy analysis method (HAM).

2. Mathematical Analysis

Considering a stable, incompressible, laminar flow in a parallel-plate channel with a uniform magnetic field, the upper wall of the parallel plate channel remains insulated, while the lower wall receives a constant cooling heat flow. Taking parallel to the wall as the x-axis and perpendicular to the wall as the y-axis, a two-dimensional coordinate frame is established. Nanofluids have been studied using an improved two-component heterogeneous model. Hence, the mass, momentum, thermal energy, and nanoparticle fraction equations of the flow system can be expressed as follows:

$$\partial_i(\rho U_i) = 0 \tag{1}$$

$$\partial_t(\rho U_i) + \partial_j(\rho U_i U_j) = -\partial_i P + \partial_j \mu(\partial_i U_j + \partial_j U_i) - \sigma_0 B_0^2 U_i \tag{2}$$

$$\partial_t(\rho c T) + \partial_i(\rho c U_i T) = \partial_i(k \partial_i T) + \rho c(D_B \partial_i \phi + \frac{D_T}{T_C} \partial_i T) \partial_i T + Q_0(T - T_w) - \frac{\partial q_r}{\partial y} \tag{3}$$

$$\partial_t(\phi) + \partial_i(U_i \phi) = \partial_i(D_B \partial_i \phi + \frac{D_T}{T_C} \partial_i T) \tag{4}$$

when the nanoparticle volume fractions are different, ρ, μ, k, and c also change. The expressions are as follows:

$$\mu(\phi) = \begin{cases} \mu_{bf}(1 + 39.11\phi + 533.9\phi^2), & Alumina--water \\ \mu_{bf}(1 + 5.45\phi + 108.2\phi^2), & Titania--water \end{cases} \tag{5}$$

$$k(\phi) = \begin{cases} k_{bf}(1 + 7.47\phi), & Alumina--water \\ k_{bf}(1 + 2.92\phi - 11.99\phi^2), & Titania--water \end{cases} \tag{6}$$

$$\rho = \phi \rho_p + (1-\phi)\rho_{bf}, \quad c = \frac{\phi \rho_p c_p + (1-\phi)\rho_{bf} c_{bf}}{\rho} \tag{7}$$

where p stands for particle and bf stands for base fluid. Moreover, the thermal physical properties of Al_2O_3 nanoparticles, TiO_2 nanoparticles, and the base fluid (water) are also analyzed as follows:

$c_{p_{bf}} = 4182 \text{ J/kgK}$, $k_{bf} = 0.597 \text{ W/mK}$, $\rho_{bf} = 998.2 \text{ kg/m}^3$, $\mu_{bf} = 9.93 * 10^{-4} \text{ kg/ms}$
$c_{p_{Al_2O_3}} = 773 \text{ J/kgK}$, $k_{Al_2O_3} = 36 \text{ W/mK}$, $\rho_{Al_2O_3} = 3380 \text{ kg/m}^3$
$c_{p_{TiO_2}} = 385 \text{ J/kgK}$, $k_{TiO_2} = 8.4 \text{ W/mK}$, $\rho_{TiO_2} = 4175 \text{ kg/m}^3$

Based on material performance of a typical water-based nanofluid with alumina (titania/water) nanoparticles, the coefficients of Equation (3) can be calculated [3] by scale analysis. Scale analysis indicates that the heat conduction term is about 1000 times more than virtue of nanoparticle diffusion. Actually, heat transfer in connection with the diffusion of nanoparticles $\rho c(D_B \partial_i \phi + \frac{D_T}{T_C}\partial_i T)\partial_i T$ can be neglected in comparison with heat conduction and convection. When the flow velocity is very low, the Re is very small. Therefore, compared with viscous resistance [8], inertia effect can be ignored. Assuming hydrodynamically and thermally fully developed conditions, Equations (1)–(4) can be simplified as follows [18]:

$$-\frac{dP}{dx} + \frac{d}{dy}\left(\mu(\phi)\frac{dU}{dy}\right) - \sigma_0 B_0^2 U = 0 \tag{8}$$

$$\frac{\partial}{\partial y}\left(k(\phi)\frac{\partial T}{\partial y}\right) + Q_0(T - T_w) - \frac{\partial q_r}{\partial y} = 0 \tag{9}$$

$$\frac{\partial}{\partial y}\left(D_B \frac{\partial \phi}{\partial y} + \frac{D_T}{T_C}\frac{\partial T}{\partial y}\right) = 0 \tag{10}$$

Radiant heat flux q_r is described by Rosseland approximation [17] as follows:

$$q_r = -\frac{4\sigma^*}{3\delta}\frac{\partial T^4}{\partial y} \tag{11}$$

Assuming that temperature difference is small enough in the flow, using Taylor series to expand T^4, and ignoring the higher-order terms, T^4 can be expressed as a linear function [19]. The approximate expression is as follows:

$$T^4 \cong 4T_\infty^3 T - 3T_\infty^4 \tag{12}$$

The following appropriate transformations are:

$$\eta = \frac{y}{H}, \quad u = \frac{U}{U_m}, \quad Ha^2 = \frac{\sigma B_0^2 H^2}{\mu_w}, \quad \theta = \frac{k_w(T - T_w)}{q_w H}$$
$$\sigma = \frac{(dp/dx)}{U_m/H^2}, \quad N_{BT} = \frac{D_B}{D_T}\frac{k_w T_C}{q_w H}, \quad \gamma = \frac{Q_0 q_w H^3}{k_w}, \quad \alpha = \frac{dP/dx}{(\mu_{bf} u_B)/H^2} \tag{13}$$

Equations (8)–(10) can be reduced to:

$$\mu(\phi)\frac{d^2 u}{d\eta^2} + \frac{d\mu(\phi)}{d\eta}\frac{du}{d\eta} - Ha^2 u - \alpha = 0 \tag{14}$$

$$k(\phi)\frac{d^2\theta}{d\eta^2} + \frac{dk(\phi)}{d\eta}\frac{d\theta}{d\eta} + \gamma\theta + \frac{16\sigma^*}{3k^*}\frac{d^2\theta}{d\eta^2} = 0 \tag{15}$$

$$N_{BT}(1+\gamma\theta)^2 \frac{\partial \phi}{\partial \eta} - \phi\frac{\partial \theta}{\partial \eta} = 0 \tag{16}$$

3. Boundary Conditions

At micro- or nanoscale, the slip boundary condition can be used to predict accurately. In current investigations, the most common velocity slip is the Maxwell [9] slip condition. The Maxwell expression is:

$$\overrightarrow{u_{slip}} = -\frac{2-\beta}{\beta\mu_w}\xi\overrightarrow{\tau} - \frac{3}{4}\frac{N_{Pr}(\delta-1)}{\delta p}\overrightarrow{q} \tag{17}$$

where $\overrightarrow{\tau} = S \cdot (n \cdot \Pi)$, $\overrightarrow{q} = \overrightarrow{Q} \cdot S$.

Beskok and Karniadakis [13] improved the second-order slip conditions:

$$u_s - u_w = \frac{2 - \sigma_v}{\sigma_v}[Kn(\frac{\partial u}{\partial n})_s + \frac{Kn^2}{2}(\frac{\partial^2 u}{\partial n^2})_s] \tag{18}$$

where $(\frac{\partial}{\partial n})$ shows gradients normal to the wall surface. Based on Beskok and Karniadakis, Wu [14] improved the slip condition in detail:

$$u_{slip} = \frac{2}{3}(\frac{3 - \omega l^3}{\omega} - \frac{3}{2}\frac{1 - l^2}{Kn})\lambda\frac{\partial u}{\partial y} - \frac{1}{4}[l^4 + \frac{2}{Kn^2}(1 - l^2)]\lambda^2\frac{\partial^2 u}{\partial y^2} = A\frac{\partial u}{\partial y} + B\frac{\partial^2 u}{\partial y^2} \tag{19}$$

where $l = min[\frac{1}{Kn}, 1]$. The expression of velocity boundary condition is as follows:

$$y = 0 : U = N_1\frac{\partial U}{\partial y} + N_2\frac{\partial^2 U}{\partial y^2} \tag{20}$$

$$y = H : U = -N_1\frac{\partial U}{\partial y} - N_2\frac{\partial^2 U}{\partial y^2} \tag{21}$$

The other boundary conditions are as follows:

$$y = 0 : -k_w\frac{\partial T}{\partial y} = q_w, \frac{\partial \phi}{\partial y} = -\frac{D_T}{D_B}\frac{1}{T_C}\frac{\partial T}{\partial y} \tag{22}$$

$$y = H : \frac{\partial T}{\partial y} = 0, \frac{\partial \phi}{\partial y} = -\frac{D_T}{D_B}\frac{1}{T_C}\frac{\partial T}{\partial y} \tag{23}$$

Substituting Equation (13) into Equations (20)–(23), the boundary conditions are as follows:

$$\eta = 0 : u = \lambda_1\frac{\partial u}{\partial \eta} + \lambda_2\frac{\partial^2 u}{\partial \eta^2}, \frac{\partial \theta}{\partial \eta} = 1, \theta = 0, \phi = \phi_w \tag{24}$$

$$\eta = 1 : u = -\lambda_1\frac{\partial u}{\partial \eta} - \lambda_2\frac{\partial^2 u}{\partial \eta^2} \tag{25}$$

In actual applications, the mass flow rate is specified through the channels. Therefore, the average fluid velocity is introduced:

$$U_m = \frac{\int_0^H U dy}{\int_0^H dy}$$

Dimensionless variables can be obtained as follows:

$$\int_0^1 u d\eta = 1 \tag{26}$$

The average of the parameters on the cross section can be calculated using the following formula [20]:

$$\langle \Gamma \rangle = \frac{1}{A}\int_0^1 dA = \int_0^1 \Gamma d\eta$$

Further, θ_B and ϕ_B can be worked out as follows:

$$\theta_B = \frac{<\rho c u \theta>}{<\rho c u>}, \phi_B = \frac{<u\phi>}{<u>} \tag{27}$$

According to the bulk properties and hydraulic diameter of nanofluids, the Nusselt number can be assessed as [21]:

$$Nu_B = \frac{hH}{k_B} = \frac{1}{\theta_B}\frac{k_w}{k_B} \tag{28}$$

The non-dimensional pressure drop can be defined as:

$$N_{dp} = \frac{-(dp/dx)}{(\mu_{bf}u_B)/H^2} = -\alpha \tag{29}$$

In addition, the semi-analytical relationship between Nu_B and N_{BT} in the alumina–water nanofluid can be obtained as:

$$a_1 = 2.1494(-248469 - 551.028 N_{BT} + 72.1567 N_{BT}^2 - 0.0184761 N_{BT}^3) \tag{30}$$

$$b_1 = -68012.1 + 2498.11 N_{BT} + 45.6722 N_{BT}^2 - 1.80232 N_{BT}^3 + 0.000305721 N_{BT}^4 \tag{31}$$

$$c_1 = \frac{(-0.00184879 + 0.000522536 N_{BT} - 0.0000613978 N_{BT}^2 + 1.57212 \times 10^{-8} N_{BT}^3)}{-0.0600415 + 0.0000151093 N_{BT}} \tag{32}$$

$$\times 7.47 + 1$$

$$Nu_{B_{Al_2O_3}} = \frac{a_1}{b_1 c_1} \tag{33}$$

The semi-analytical relation between Nu_B and N_{BT} in the titania–water nanofluid can be obtained as:

$$a_2 = 2.1494(-202053 - 426.941 N_{BT} + 46.4139 N_{BT}^2 - 0.00159493 N_{BT}^3) \tag{34}$$

$$b_2 = -20516.7 + 623.982 N_{BT} + 7.42706 N_{BT}^2 - 0.334142 N_{BT}^3 + 7.1142 \times 10^{-6} N_{BT}^4 \tag{35}$$

$$c_2 = \frac{(-0.00148734 + 0.000369449 N_{BT} - 0.0000394933 N_{BT}^2 + 1.35712 \times 10^{-9} N_{BT}^3)}{-0.0488208 + 1.73602 \times 10^{-6} N_{BT}} \tag{36}$$

$$\times 7.47 + 1$$

$$Nu_{B_{TiO_2}} = \frac{a_2}{b_2 c_2} \tag{37}$$

4. Application of HAM

In this article, to obtain the series solutions, we adopt homotopy analysis method (HAM). HAM is one of the well-known semi-analytical methods for solving various types of linear and nonlinear differential equations (ordinary as well as partial). This method is based on coupling of the traditional perturbation method and homotopy in topology. By this method, one may obtain an exact solution or a power series solution which converges in general to the exact solution. HAM consists of the convergence control parameter, which controls the convergent region and rate of convergence of the series solution. We select the initial guess solutions:

$$u_0(\eta) = -0.1 + \eta - \eta^2$$
$$\theta_0(\eta) = \eta - 2\eta^2 \tag{38}$$
$$\phi_0(\eta) = \phi_B$$

What calls for special attention is that the boundary condition (26) is not yet used, which can be used to determine the unknown parameter α_{k-1}. For example, when $k = 1$, we are able to obtain $u_1(\eta)$ and its integration with η in the range $[0,1]$, which is the function of α_0. Using the boundary condition (26), we obtain:

$$\alpha_0(\eta) = 1 \tag{39}$$

In this way, $u_k(\eta), \theta_k(\eta), \phi_k(\eta), \alpha_k(\eta)$ can be successively worked out one after another according to the order $k = 0, 1, 2, \ldots$. At mth-order, we obtain:

$$u(\eta) = u_0(\eta) + \sum_{k=1}^{m} u_k(\eta)$$

$$\theta(\eta) = \theta_0(\eta) + \sum_{k=1}^{m} \theta_k(\eta) \tag{40}$$

$$\phi(\eta) = \phi_0(\eta) + \sum_{k=1}^{m} \phi_k(\eta)$$

$$\alpha = \alpha_0 + \sum_{k=1}^{m} \alpha_k$$

The auxiliary linear operators are:

$$L_u = \frac{d^2 u}{d\eta^2}, \quad L_\theta = \frac{d^2 \theta}{d\eta^2}, \quad L_\phi = \frac{d\phi}{d\eta} \tag{41}$$

The properties of the auxiliary linear operator are as follows:

$$L_u[C_1 + C_2\eta + C_3\eta^2] = 0, \quad L_\theta[C_4 + C_5\eta + C_6\eta^2] = 0, \quad L_\phi[C_7 + C_8\eta] = 0 \tag{42}$$

where $C_i, i = 1, \ldots, 8$ are constants.

Next, construct the mth-order deformation equation as follows:

$$L_u[u_m(\eta) - \chi_m u_{m-1}(\eta)] = qh_u R_m(\eta)$$
$$L_\theta[\theta_m(\eta) - \chi_m \theta_{m-1}(\eta)] = qh_\theta R_m(\eta) \tag{43}$$
$$L_\phi[\phi_m(\eta) - \chi_m \phi_{m-1}(\eta)] = qh_\phi(\eta) R_m(\eta)$$

5. Convergence of the HAM Solutions

Liao [22] showed that the values of auxiliary parameters h_u, h_θ, and h_ϕ can adjust and control the convergence of the series solutions. Directly selecting the appropriate values of h_u, h_θ, and h_ϕ ensures the convergence of the series solutions. Figures 1 and 2 give the respective valid ranges of h_θ, h_ϕ. The valid ranges are as follows:

$$\begin{cases} 0 \leq h_u \leq 0.4 \\ -0.8 \leq h_\theta \leq -0.35 \\ -3 \leq h_\phi \leq 0.1 \end{cases}$$

In addition, one way to find the appropriate h_u, h_θ, and h_ϕ is to utilize the residual error. In this article, the residual error $E_{m,t}$ [23] is defined as follows:

$$E_{m,\theta} = \int_0^1 k\theta'' + k'\theta' + \gamma\theta + \frac{16\sigma^*}{3k^*}\theta'' d\eta \tag{44}$$

Using the square residual error function, it is found that the residual error becomes more and more accurate as the order of HAM approximation increases (Figure 3). Finally, α of the HAM solution agrees well with the BVPh 2.0 solution (Table 1). BVPh 2.0 is a free software package for nonlinear boundary-value and eigenvalue problems based on HAM.

In addition, it serves to show that the current results are in accordance with the results given by Yang et al. [24] (Table 2) greatly.

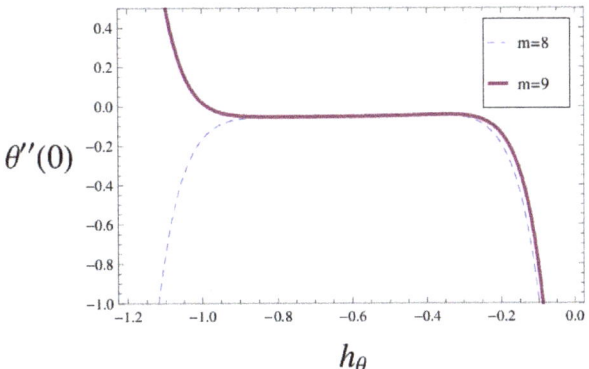

Figure 1. h_θ-curve of $\theta''(0)$.

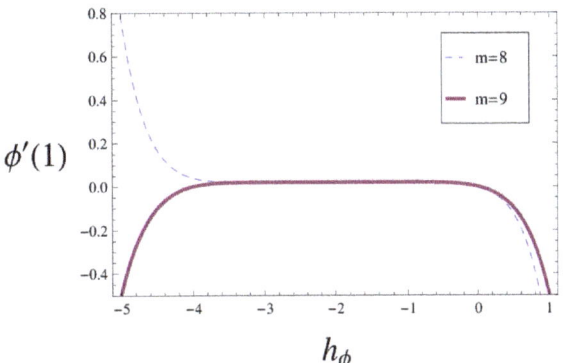

Figure 2. h_ϕ-curve of $\phi'(1)$.

Table 1. Comparison of HAM results with BVPh2.0 results.

ϕ_B	α		
	BVPh2.0	HAM	Relative Error(%)
0.01	−0.00286834	−0.00286673	0.05624769
0.02	−0.00396349	−0.00394892	0.36760532
0.03	−0.00527046	−0.00529243	0.33803511
0.04	−0.00678941	−0.00671253	1.13235171

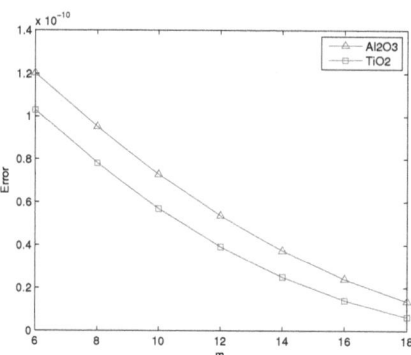

Figure 3. The residual errors with HAM approximations order m in different nanofluids.

Table 2. Comparison of HAM results with those of C. Yang et al. [24].

N_{BT}	Nu_B		
	Yang et al. [24]	HAM	Relative Error(%)
0.1	7.26823	7.26679	0.01981
0.2	7.55883	7.55889	0.00079
0.3	7.69768	7.69418	0.04547
0.4	7.79492	7.79163	0.04225
0.5	7.85227	7.85200	0.00344
0.6	7.90000	7.90338	0.04278
0.7	7.94920	7.94526	0.04956
0.8	7.95957	7.95947	0.00126
0.9	7.97313	7.97791	0.05995
1	8.04496	8.04478	0.00224
2	8.12940	8.12983	0.00529
10	8.21841	8.21630	0.02567

6. Results and Discussion

The effects of N_{BT}, λ_2 and λ_1 on the nanoparticle velocity u/u_B, the nanoparticle volume fraction ϕ/ϕ_B, temperature profiles θ/θ_B, and Nusselt number Nu_B are shown in Figures 4–14. In these figures, $\eta = 1$ corresponds to the adiabatic wall, whereas $\eta = 0$ corresponds to the cooled wall.

The slip parameter characterizes slip resistance at the surface. The first-order velocity slip parameters λ_1 and second-order velocity slip parameters λ_2 affect the flow and heat. Figures 4–6 depict the effects of second-order velocity slip λ_2 on u/u_B, ϕ/ϕ_B, and θ/θ_B. Figures 7–9 illustrate the effects of first order velocity slip λ_1 on u/u_B, ϕ/ϕ_B, and θ/θ_B. Figures 4 and 7 show that u/u_B is lower near the walls and peaks near the middle of the microchannel. As Figure 4 reveals, the increase in λ_2 causes momentum to build up in the core area, with the velocity profile becoming more uniform as the slip parameters decrease. Figure 7 shows that an increase in λ_1 results in the momentum accumulation at the core region. The second-order slip condition shows a prominent effect on the velocity profile u/u_B in Figures 4 and 7. Assuming that mass flows are constant, in order to satisfy continuity, they must increase in the core region as the magnitude of the velocities at the boundary decreases. Meanwhile, in Figures 5 and 8, the temperature profile θ/θ_B decreases and then increases toward the upper wall. The titania–water nanofluid temperature changes more gently than that of the alumina–water nanofluid. The minimum of the temperature profile is increasing and shifts toward the upper wall with increasing λ_2. Figure 8 shows no significant variation in the dimensionless temperature for λ_1. In addition, with increasing λ_2 or λ_1, the volume fraction ϕ/ϕ_B of nanoparticles shows an increasing trend in Figures 6 and 9. Hence, s more uniform distribution of the volume fraction emerges.

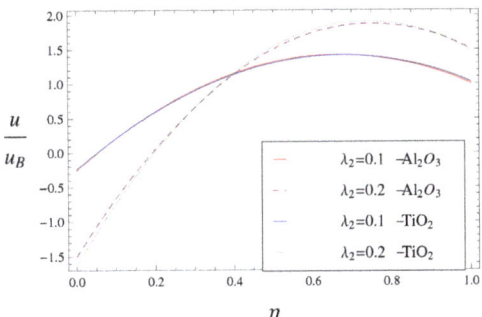

Figure 4. The effects of λ_2 on the nanoparticle velocity u/u_B.

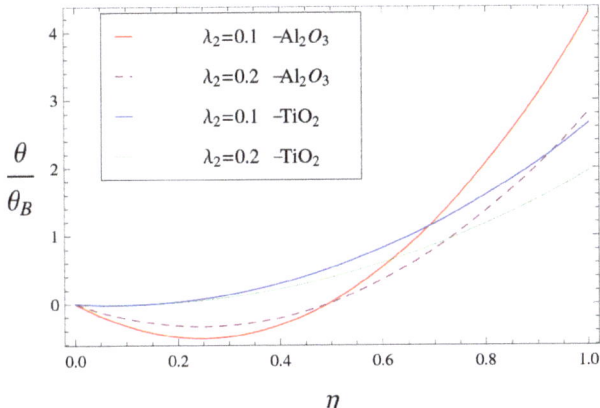

Figure 5. The effects of λ_2 on the temperature profiles θ/θ_B.

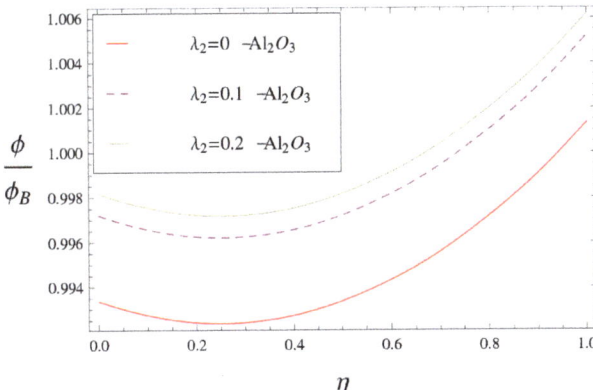

Figure 6. The effects of λ_2 on the nanoparticle volume fraction ϕ/ϕ_B.

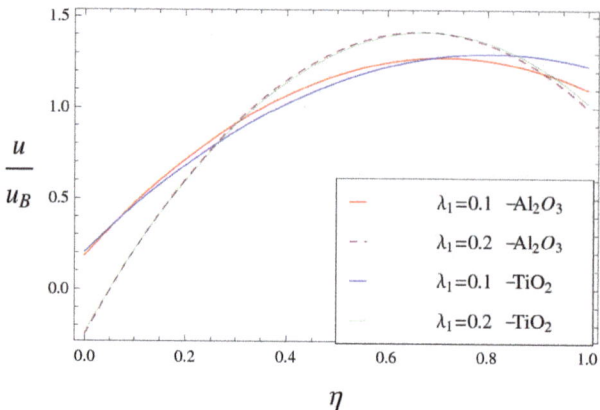

Figure 7. The effects of first-order velocity slip parameters λ_1 on the nanoparticle velocity u/u_B.

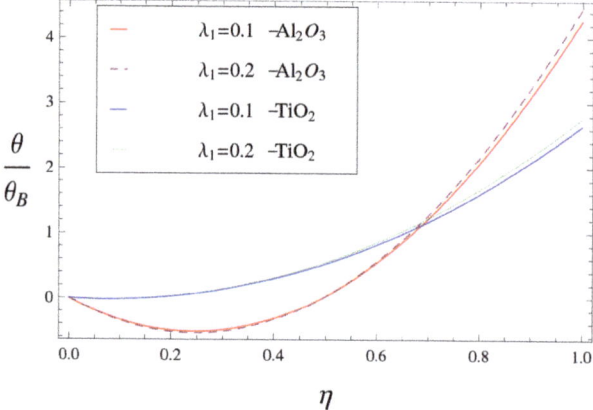

Figure 8. The effects of first-order velocity slip parameters λ_1 on θ/θ_B.

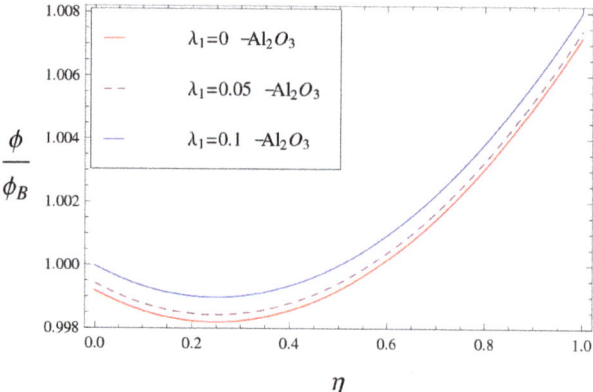

Figure 9. The effects of first-order velocity slip parameters λ_1 on ϕ/ϕ_B.

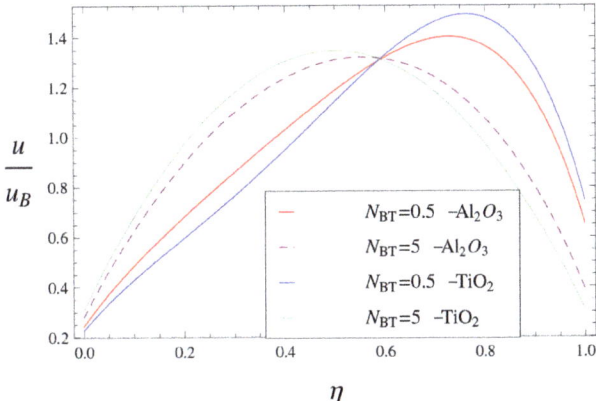

Figure 10. The effects of N_{BT} on the nanoparticle velocity u/u_B.

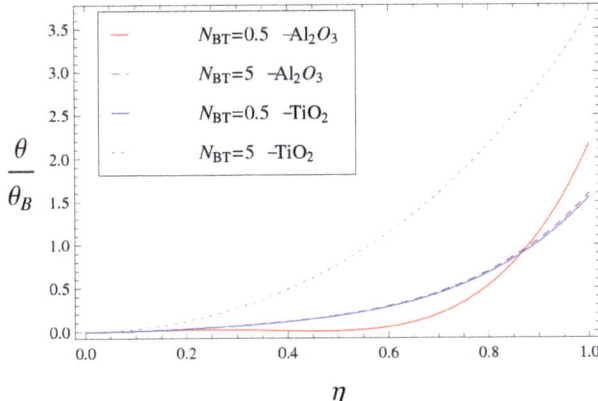

Figure 11. The effects of N_{BT} on the temperature profiles θ/θ_B.

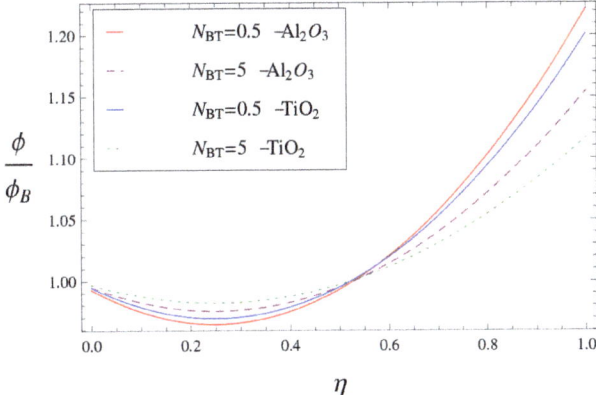

Figure 12. The effects of N_{BT} on the nanoparticle volume fraction ϕ/ϕ_B.

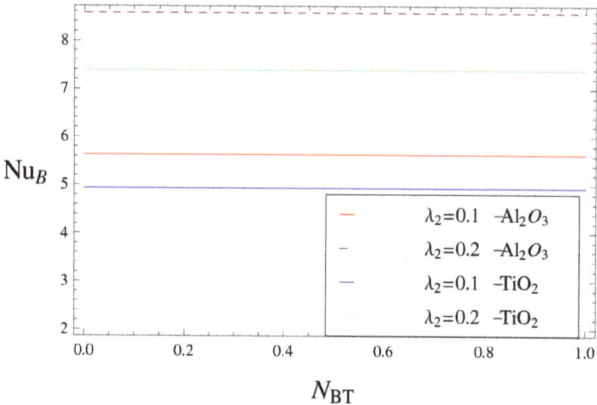

Figure 13. The effects of second-order velocity slip parameters λ_2 on Nu_B.

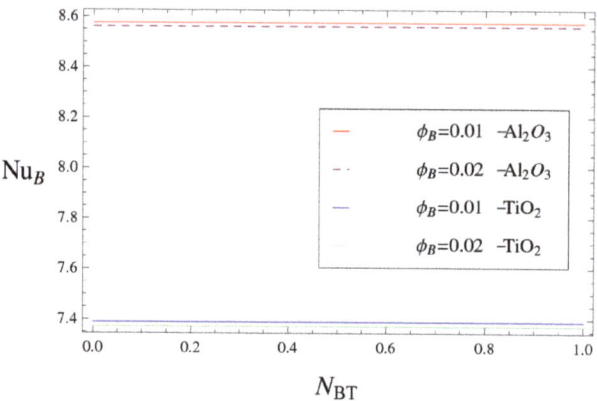

Figure 14. The effects of ϕ_B on Nu_B.

Figures 10–12 plot the effects of N_{BT} on u/u_B, ϕ/ϕ_B, and θ/θ_B. Apparently, on the cooling wall, the concentration of nanoparticles is higher; at the adiabatic wall, the nanoparticle concentration is lower. Hence, the trend of the nanoparticle motion is moving from the adiabatic wall toward the cooled wall; accordingly, an uneven distribution of nanoparticles is constructed. This motion makes the viscosity near the cold wall much greater than that near the adiabatic wall, thus increasing the velocity near the adiabatic wall and decreasing the velocity near the cold wall. Therefore, the velocity profile deforms and its peak moves toward the adiabatic wall. As a result, at higher values of N_{BT}, ϕ/ϕ_B becomes more uniform, which can be observed in Figure 12. At higher value of N_{BT}, with momentum enhanced, the heat transfer rate of the cooling wall also increases. Therefore, the increase of N_{BT} gives rise to an increase in the temperature gradient of the cooling wall, as shown in Figure 11.

Figure 13 depicts the effect of λ_2 on the Nusselt number Nu_B. One thing to note is that the growth trend for Nu_B with the increase in λ_2 comes from the momentum accumulation near the wall. Therefore, the second-order slip parameter plays a positive role in Nu_B of the MHD flow. As a result, compared with Navier's condition, under second-order slip conditions, nanofluids transfer heat more efficiently. It also must be stated that the effect of the slip parameters on Nu_B is quite protensive; it rests with the type of nanoparticle. In alumina–water nanofluid, the sensitivity on Nu_B is much higher than that of titania–water nanofluid, since the corresponding Nu_B of alumina–water nanofluid is higher than that of titania–water nanofluid. Figure 14 depicts the ϕ_B on the Nu_B. It can be obtained that increasing ϕ_B leads to a decrease Nu_B because the increasing ϕ_B can increase the thermal conductivity and viscosity near walls.

Table 3 gives N_{dp} with different λ_2, λ_1, N_{BT}, and ϕ_B when $Ha = 0$, respectively. It can be concluded that both one-slip parameter and second-order slip parameter have positive correlation with the pressure drop ratio of the nanofluid to base fluid N_{dp}. As λ_2 or λ_1 decreases, the frictional forces on the walls diminish because of the velocity jumps at the walls. However, the positive correlation is significant to titania–water nanofluid. Because of a slight increase in viscosity at the wall, N_{BT} has a minor positive effect on N_{dp} and ϕ_B has a minor negative effect on N_{dp}.

Table 3. N_{dp} with different λ_2, λ_1, N_{BT}, and ϕ_B when $Ha = 0$.

λ_2	λ_1	N_{BT}	ϕ_B	Types of Fluids	
				Al_2O_3-Water	TiO_2-Water
0.1	0.1	0.5	0.01	0.000127706	0.000129222
0.2				0.000131924	0.014154800
	0.2			0.000179880	0.020517000
		10		0.000127722	0.000127916
			0.04	0.000119804	0.000127917

Table 4 gives Nu_B with different λ_2, λ_1, N_{BT}, and ϕ_B when $Ha = 0$, respectively. It can be deduced that the velocity gradient at the wall of the microchannel increases. The slip velocity increases with the increase of velocity gradient. Thus, momentum closer to the wall increases and causes convective heat transfer to rise.

Table 4. Nu_B with different λ_2, λ_1, N_{BT}, and ϕ_B when $Ha = 0$.

λ_2	λ_1	N_{BT}	ϕ_B	Types of Fluids	
				Al_2O_3-Water	TiO_2-Water
0.1	0.1	0.5	0.01	5.62714	4.93726
0.2				8.57429	7.38838
	0.2			8.95671	7.71938
		10		8.69342	7.40811
			0.04	8.56113	7.37139

7. Conclusions

In this paper, we conduct a theoretical study on the heat transfer of alumina/water and titania/water nanofluids in a parallel-plate channel. We discuss the effects of Brownian motion and thermophoresis. Their effects are characterized by the ratio of the Brownian to thermophoretic diffusion coefficients N_{BT}. Moreover, The second-order velocity slip condition is considered. Analytic solutions are obtained by HAM. The main conclusions of this paper can be drawn as follows:

a The semi-analytical relation between Nu_B and N_{BT} is obtained.
b Both first-order slip parameter and second-order slip parameter have positive effects on Nu_B of the MHD flow, but nanofluids can transfer heat more efficiently with a second-order slip condition than with a Navier's condition.
c In the alumina–water nanofluid, Nu_B is higher than that of titania–water nanofluid.
d The positive correlation between slip parameters and N_{dp} is significant for the titania-water nanofluid.

Author Contributions: J.Z. conducted the original research, modified the model, and contributed analysis tools. J.C. analyzed the data, simulated the modified model, and prepared original draft. Y.L. revised the manuscript. All authors have read and agreed to the published version of the manuscript.

Funding: This research received no external funding

Institutional Review Board Statement: Not applicable.

Informed Consent Statement: Not applicable.

Data Availability Statement: Not applicable.

Acknowledgments: The work is supported by the Fundamental Research Funds for the Central Universities (FRF-BR-18-008B).

Conflicts of Interest: The authors declare that there was no conflict of interest regarding the publication of this paper.

Symbol	Description
B_0	magnetic field strength
C_p	specific heat (m^2/s^2K)
D_B	Brownian motion constant
D_T	thermophoresis diffusion coefficient
H	radius (m)
h	heat transfer coefficient (W/m^2K)
Ha	Hartmann number
HTC	dimensionless heat transfer coefficient
k	thermal conductivity (W/mK)
T_∞	free stream temperature
N_{BT}	ratio of the Brownian to thermophoretic diffusivities
N_p	non-dimensional pressure drop
Nu	Nusselt number
p	pressure (Pa)
q_w	surface heat flux
q_r	radiative heat flux
ϕ	nanoparticle volume fraction
ρ	density
η	transverse direction
λ_1, λ_2	slip parameters of velocity
B	bulk mean
U	axial velocity (m/s)
T	temperature (K)
k	thermal conductivity
μ	dynamic viscosity (kg/m s)
σ^*	Stefan–Boltzman constant
γ	ratio of wall and fluid temperature difference to absolute temperature
Subscripts	
x, y	coordinate system
p	nanoparticle
bf	base fluid
i	velocity components

References

1. Li, T.; Liu, B.; Zhou, J.Z.; Xi, W.X.; Huai, X.L.; Zhang, H. A Comparative Study of Cavitation Characteristics of Nano-Fluid and Deionized Water in Micro-Channels. *Mathematics* **2020**, *11*, 310. [CrossRef] [PubMed]
2. Duan, Z.P.; Lv, X.; Ma, H.H.; Su, L.B.; Zhang, M.Q. Analysis of Flow Characteristics and Pressure Drop for an Impinging Plate Fin Heat Sink with Elliptic Bottom Profiles. *Appl. Sci.* **2020**, *10*, 225. [CrossRef]
3. Buongiorno, J. Convective transport in nanofluids. *J. Heat Transf.* **2006**, *128*, 240–250. [CrossRef]
4. Yang, C.; Wang, Q.L.; Nakayama, A.; Qiu, T. Effect of temperature jump on forced convective transport of nanofluids in the continuum flow and slip flow regimes. *Chem. Eng. Sci.* **2015**, *137*, 730–739. [CrossRef]

5. Hedayati, F.; Domairry, G. Effects of nanoparticle migration and asymmetric heating on mixed convection of $TiO_2 - H_2O$ nanofluid inside a vertical microchannel. *Powder Technol.* **2015**, *272*, 250–259. [CrossRef]
6. Andhare, R.S.; Shooshtari, A.; Dessiatoun, S.V.; Ohadi, M.M. Heat transfer and pressure drop characteristics of a flat plate manifold microchannel heat exchanger in counter flow configuration. *Appl. Thermal Eng.* **2016**, *96*, 178–189. [CrossRef]
7. Ooi, E.H.; Popov, V. Numerical study of on the natural convection in Cu-water nanofluid. *Int. J. Thermal Sci.* **2013**, *65*, 178–188. [CrossRef]
8. Ravnik, J.; Šušnjara, A.; Tibaut, J.; Poljak, D.; Cvetkovi, M. Stochastic modelling of nanofluids using the fast Boundary-Domain Integral Method. *Eng. Anal. Boundary Elem.* **2019**, *107*, 185–197. [CrossRef]
9. Maxwell, J.C. Temperature. On Stresses in Rarefied Gases Arising from Inequalities of Temperature. *Philos. Trans. R. Soc.* **1879**, *170*, 231–256.
10. Kou, Z.H.; Bai, M.L. Effects of wall slip and temperature jump on heat and mass transfer characteristics of an evaporating thin film. *Int. Commun. Heat Mass Transf.* **2011**, *38*, 874–878. [CrossRef]
11. Avramenko, A.A.; Tyrinov, A.I.; Shevchuk, I.V.; Dmitrenko, N.P.; Kravchuk, AV.; Shevchuk, V.I. Mixed convection in a vertical circular microchannel. *Int. J. Therm. Sci.* **2017**, *121*, 1–12 . [CrossRef]
12. Thompson, P.A.; Troian, S.M. A general boundary condition for liquid flowat solid surfaces. *Nature* **1997**, *389*, 360–362 . [CrossRef]
13. Beskok, A.; Karniadakis, G.E. A model for flows in channels, pipes, and ducts at micro and nano scales. *Microsc. Thermophys. Eng.* **1999**, *3*, 43–77 .
14. Wu, L.A. A slip model for rarefied gas flows at arbitrary Knudsen number. *Appl. Phys. Lett.* **2008**, *93*, 253103. [CrossRef]
15. Zhu, J.; Xu, Y.X.; Hang, X. A Non-Newtonian Magnetohydrodynamics (MHD) Nanofluid Flow and Heat Transfer with Nonlinear Slip and Temperature Jump. *Mathematics* **2019**, *7*, 1199. [CrossRef]
16. Almutairi, F.; Khaled, S.M.; Ebaid, A. MHD Flow of Nanofluid with Homogeneous-Heterogeneous Reactions in a Porous Medium under the influence of Second-Order Velocity. *Mathematics* **2019**, *7*, 220. [CrossRef]
17. Noeiaghdam, S.; Dreglea, A.; He, J.H.; Avazzadeh, Z.; Suleman, M.; Araghi, M.A.F.; Sidorov, D.N.; Sidorov, N. Error Estimation of the Homotopy Perturbation Method to Solve Second Kind Volterra Integral Equations with Piecewise Smooth Kernels: Application of the CADNA Library. *Symmetry* **2020**, *12*, 1730. [CrossRef]
18. Nobari, M.R.H.; Gharali, K. A numerical study of flow and heat transfer in internally finned rotating straight pipes and stationary curved pipes. *Int. J. Heat Mass Transf.* **2005**, *49*, 1185–1194. [CrossRef]
19. Ganga, B.; Ansari, S.M.Y.; Ganesh, N.V.; Abdul Hakeem, A.K. MHD flow of Boungiorno model nanofluid over a vertical plate with internal heat generation/absorption. *Propuls. Power Res.* **2016**, *5* 211–222. [CrossRef]
20. Zhu, J.; Wang, S.N.; Zheng, L.C.; Zhang, X.X. Heat transfer of nanofluids considering nanoparticle migration and second-order slip velocity. *Appl. Math. Mech.* **2016**, *38*, 125–136. [CrossRef]
21. Moein, S.; Mohsen, K. Study of water based nanofluid flows in annular tubes using numerical simulation and sensitivity analysis. *Heat Mass Transf.* **2018**, *54*, 2995–3014.
22. Liao, S.J. On the homotopy analysis method for nonlinear problems. *Appl. Math. Comput.* **2004**, *147*, 499–513. [CrossRef]
23. Fan, T. Applications of Homotopy Analysis Method in Boundary Layer Flow and Nanofluid Flow Problems. Ph.D. Thesis, Shanghai Jiao Tong University, Shanghai, China, 2012. (In Chinese)
24. Yang, C.; Li, W.; Sano, Y.; Mochizuki, M.; Nakayama, A. On the anomalous convective heat transfer enhancement in nanofluids: a theoretical answer to the nanofluids controversy. *J. Heat Transf.* **2013**, *135*, 054504. [CrossRef]

Article

Matrix Method by Genocchi Polynomials for Solving Nonlinear Volterra Integral Equations with Weakly Singular Kernels

Elham Hashemizadeh [1,*], Mohammad Ali Ebadi [2] and Samad Noeiaghdam [3]

1. Department of Mathematics, Karaj Branch, Islamic Azad University, Karaj 3149968111, Iran
2. Young Researchers and Elite Club, Karaj Branch, Islamic Azad University, Karaj 3149968111, Iran; ma_ebadi268@stumail.liau.ac.ir
3. Department of Applied Mathematics and Programming, South Ural State University, Lenin Prospect 76, 454080 Chelyabinsk, Russia; noiagdams@susu.ru
* Correspondence: hashemizadeh@kiau.ac.ir

Received: 28 November 2020; Accepted: 14 December 2020; Published: 17 December 2020

Abstract: In this study, we present a spectral method for solving nonlinear Volterra integral equations with weakly singular kernels based on the Genocchi polynomials. Many other interesting results concerning nonlinear equations with discontinuous symmetric kernels with application of group symmetry have remained beyond this paper. In the proposed approach, relying on the useful properties of Genocchi polynomials, we produce an operational matrix and a related coefficient matrix to convert nonlinear Volterra integral equations with weakly singular kernels into a system of algebraic equations. This method is very fast and gives high-precision answers with good accuracy in a low number of repetitions compared to other methods that are available. The error boundaries for this method are also presented. Some illustrative examples are provided to demonstrate the capability of the proposed method. Also, the results derived from the new method are compared to Euler's method to show the superiority of the proposed method.

Keywords: nonlinear Volterra integral equation; weakly singular kernels; Abel's integral equations; the Genocchi polynomials; operational matrix

1. Introduction

Spectral schemes are invaluable tools for the numerical solution of fractional partial differential equations (FPDEs), ordinary differential equations (ODEs), integral equations (IEs), and integrodifferential equations (IDEs).

Spectral approaches are a class of schemes used in applied mathematics and scientific computing to numerically solve certain differential equations and nonlinear integral equations. In recent years, these approaches have been used in modeling of many problems of physical phenomena, engineering and chemical processes in chemical kinetics [1], super fluidity biology and economics [2], axially symmetric problems in the case of an elastic body containing an inclusion [3], and fluid dynamics [4], and the Hammerstein integral equation is employed for modeling nonlinear physical phenomena such as electromagnetic fluid dynamics reformulation of boundary value problems with a nonlinear boundary condition [5].

Various numerical approaches have been presented for solving a class of nonlinear singular integral equations including Abel's integral equation, Hammerstein integral equation, Volterra integral equation, etc. For example, Noeiaghdam et al. in [6] applied the Laplace homotopy analysis method to solve Abel's integral equation, and validation of this method was discussed in [7]. Also, the numerical studies on the Volterra integral equation with discontinuous kernels can be found in [8,9]. Allaei et al. in [10] presented an analytical and computational method for a class of nonlinear singular integral equations. Maleknejad et al. in [11] proposed a new numerical approach for solving the nonlinear integral equations of Hammerstein and Volterra–Hammerstein. In [12], the authors applied the operational Tau method (OTM) to find a numerical solution for weakly singular Volterra integral equations (WSVIEs) and Abel's equation.

Other researchers have attempted to solve nonlinear integral equations in recent years. Among them, in recent years, Mehdi Dehghan et al. in [13] solved nonlinear fractional integrodifferential equations (NFIDEs) by using the collocation numerical method. Li Zhu and Qibin Fan in [14] presented a spectral method based on the second Chebyshev wavelet (SCW) operational matrix for solving the fractional nonlinear Fredholm integrodifferential equation, and the Ferdholm and Volterra integral equations.

Nemati in [15] applied a numerical approach for solving nonlinear fractional integrodifferential equations with weakly singular kernels by using a modification of hat functions. Somveer et al. [16] presented an efficient spectral method based on shifted Legendre polynomials for solving nonlinear Volterra singular partial integrodifferential equations (PIDEs) which involve both integrals and derivatives of a function.

Recently, with the effort of other scientists, many of the nonlinear differential and integral equations which appear in different fields of physical phenomena and engineering were solved by using numerical methods, and nonlinear differential and integral equations have also been explored in delayed scaled consensus problems [17–24].

In the study of many nonlinear problems in heat conduction, boundary-layer heat transfer, chemical kinetics, and superfluidity, we are often led to singular Volterra integral equations for which real answers are hard to find [10]. In this article, we use efficient functions such as Genocchi polynomials and their operational matrices to solve nonlinear Volterra integral equations with weakly singular kernels of the following form:

$$y(t) = f(t) - \int_0^t \frac{s^\beta}{(t-s)^\alpha} g(y(s))ds, \quad t > 0, \tag{1}$$

where $f(t)$ is in $L^2(\mathfrak{R})$ on the interval $0 \leq t, s \leq T$; g is locally Lipchitz continuous, smooth, and a Hammerstein nonlinear function; and α, β are real positive numbers.

For future works, we can use other polynomials like Chebyshev, Lagger, etc. for implementation, and by comparing the archived results, we can expand the present method and implement it on the system of nonlinear Volterra integral equations or nonlinear Volterra integral equations of mixed type. Because of important applications of the first kind of Volterra integral equations with discontinuous kernels in load leveling problems and power engineering systems, the proposed method can also be used for future works.

The rest of the article is organized as follows: In Section 2, we state some necessary basic definitions and properties of Genocchi polynomials. Numerical implementation of the suggested technique based on Genocchi polynomials is shown in Section 3. Section 4 estimates the error analysis of our proposed technique. In Section 5, two examples with tables and graphs are presented to show the efficiency and accuracy of the proposed scheme. Section 6 provides some discussion and concluding remarks.

2. Genocchi Polynomials and Their Properties

2.1. Definition of the Genocchi Polynomials

Genocchi polynomials and Genocchi numbers have been widely applied in many branches of mathematics and physics such as complex analytic number theory, homotopy theory, differential topology, and quantum physics (quantum groups) [25,26]. The Genocchi polynomials

$G_n(x)$ and numbers G_n are usually expressed by using the exponential generating functions $Q(t,x)$ and $Q(t)$ respectively as follows:

$$Q(t) = \frac{2t}{e^t + 1} = \sum_{n=0}^{\infty} G_n \frac{t^n}{n!}, \quad (|t| < \pi), \tag{2}$$

$$Q(t,x) = \frac{2te^{xt}}{e^t + 1} = \sum_{n}^{\infty} G_n(x) \frac{t^n}{n!}, \quad (|t| < \pi), \tag{3}$$

where $G_n(x)$ is the well-known Genocchi polynomials of order n. Also, we note that the Genocchi polynomials can be determined as follows:

$$G_n(x) = \sum_{k=0}^{n} \binom{n}{k} G_{n-k} x^k = 2B_n(x) - 2^{n+1} B_n(x), \tag{4}$$

where the Genocchi number G_{n-k} is obtained by the following relation:

$$G_n = 2(1 - 2^n) B_n, \tag{5}$$

B_n is the famous Bernoulli number.

The first few Genocchi numbers are given in the table below:

n	0	1	2	4	6
G_n	0	1	-1	1	-3

We also have to pay attention that $G_{2n+1} = 0$, $n = 1, 2, 3, \ldots$. We list the first few Genocchi polynomials that are given as follows:

$$\begin{aligned} G_0(x) &= 0, \\ G_1(x) &= 1, \\ G_2(x) &= 2x - 1, \\ G_3(x) &= 3x^2 - 3x, \\ G_4(x) &= 4x^3 - 6x^2 + 1, \\ G_5(x) &= 5x^4 - 10x^3 + 5x. \end{aligned} \tag{6}$$

The Genocchi polynomials are depicted in Figure 1 for different n:

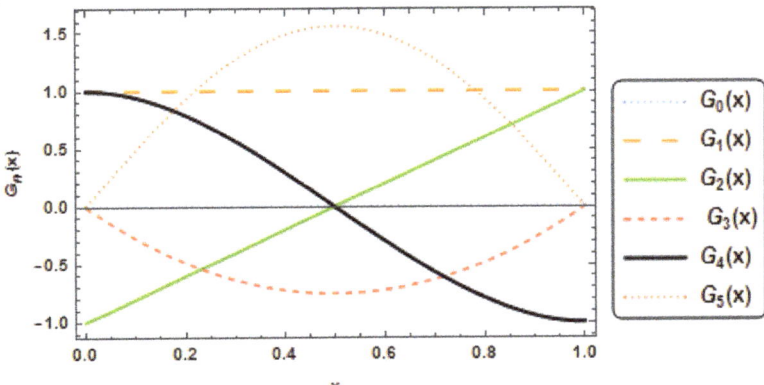

Figure 1. The plots of the Genocchi polynomials.

Therefore, some of the important basic properties of the Genocchi polynomials are as follows:

$$\int_0^1 G_n(x)G_m(x)dx = \frac{2(-1)^n n!m!}{(n+m)!} G_{m+n}, \quad n,m \geq 1, \tag{7}$$

$$\frac{dG_n(x)}{dx} = nG_{n-1}(x), \quad n \geq 1, \tag{8}$$

$$G_n(1) + G_n(0) = 0, \quad n > 1, \tag{9}$$

Also, by using them in Relations (5) and (9), we can write the following:

$$G_n(x) = \int_0^x nG_{n-1}(x)dx + G_n, \quad n \geq 1. \tag{10}$$

For more information, you can refer to References [27] and [28], which discuss the Genocchi polynomials extensively.

2.2. Approximation of Arbitrary Function by Applying Genocchi Polynomials

The approximation theory plays an important role in solving a variety of differential equations. The main goal of this section is to approximate the arbitrary function $f(x) \in L^2[0,1]$ by Genocchi polynomials. Let $\{G_1(x), G_2(x), \ldots, G_N(x)\} \subseteq L^2[0,1]$ be the set of Genocchi polynomials and $P = \text{span}\{G_1(x), G_2(x), \ldots, G_N(x)\}$. Since P is a finite dimensional subspace of the $L^2[0,1]$ space, therefore $f(x)$ as an arbitrary element of the $L^2[0,1]$ space has a unique best approximation in P, say $f^*(x)$, such that

$$\|f(x) - f^*(x)\|_2 \leq \|f(x) - y(x)\|_2 : \forall y(t) \in P. \tag{11}$$

Therefore, inequality (11) requires that the following equation to be true.

$$\langle f(x) - f^*(x), y(t) \rangle = 0 : \forall y(t) \in P. \tag{12}$$

where $\langle .,. \rangle$ denotes the inner product.

Any arbitrary function $f(x) \in L^2[0,1]$ can be expanded in the finite series to the number of the Genocchi polynomials as follows:

$$f(x) \approx f^*(x) = \sum_{n=1}^N c_n G_n(x) = C^T G(x), \tag{13}$$

where T means transpose and the Genocchi coefficient vector C and Genocchi vector $G(x)$ are given by the following:

$$C = [c_1, c_2, \ldots, c_N]^T, \quad G(x) = [G_1(x), G_2(x), \ldots, G_N(x)]^T. \tag{14}$$

Hence, the coefficient c_n can be obtained using the Genocchi polynomials as follows:

$$c_n = \frac{1}{2n!}\left(f^{(n-1)}(0) + f^{(n-1)}(1)\right), \quad n = 1, \ldots, N. \tag{15}$$

Of course, we have to note the important fact that calculating the approximation coefficient by the Genocchi polynomials in Equation (15) for a function that is not $(n-1)$ differentiable at the points $x = 0, x = 1$ leads to failure. The following example illustrates the problem.

Let $N = 3$, $f(x) = x^{3/2}$, $f(x) = \sum_{n=1}^{3} c_n G_n(x) = c_1 G_1(x) + c_2 G_2(x) + c_3 G_3(x)$;

$$c_3 = \frac{1}{2 \times 3!} \left[\frac{d^2}{dx^2} x^{3/2} \bigg|_{x=0} + \frac{d^2}{dx^2} x^{3/2} \bigg|_{x=1} \right]$$
$$= \frac{1}{2 \times 3!} \left[\frac{1}{4\sqrt{x}} \bigg|_{x=0} + \frac{1}{4\sqrt{x}} \bigg|_{x=1} \right]. \tag{16}$$

To avoid this problem for functions that are not $(n-1)$ differentiable at points $x = 0$, $x = 1$, we use the matrix approach taken in the next section to compute the unknown approximation coefficients.

2.3. Using the Matrix Approach to Compute the Genocchi Approximation Coefficients

In this section, we compute the Genocchi coefficient vector C using the matrix method. Before we apply this approach, we need to demonstrate and verify the following theorems. We first introduce Theorem 1, which gives the expression and proof of integration of the two Genocchi polynomials on arbitrary interval $[a, b]$, $0 \leq a \leq b$ which will be used to prove Theorem 2. Therefore, the proof of Theorem 1 is of particular important.

Theorem 1. *Let us assume that $G_n(x)$ and $G_m(x)$ are two Genocchi polynomials for $x \geq 0$:*

$$\begin{aligned} \gamma_{n,m}(x) &= \int_0^x G_n(x) G_m(x) dx \\ &= \sum_{r=0}^{n-1} (-1)^r \frac{n_{(r)}}{(m+1)^{(r+1)}} (G_{n-r}(x) G_{m+1+r}(x) - G_{n-r}(0) G_{m+1+r}(0)), \end{aligned} \tag{17}$$

where $n_{(r)}$, $(m+1)^{(r+1)}$ are respectively the falling and rising factorials. In particular, we have the following relations for $[a, b]$, $0 \leq a \leq b$:

$$\begin{aligned} \gamma_{n,m}^{(a,b)} &= \int_a^b G_n(x) G_m(x) dx = \gamma_{n,m}(b) - \gamma_{n,m}(a) \\ &= \sum_{r=0}^{n-1} (-1)^r \frac{n_{(r)}}{(m+1)^{(r+1)}} (G_{n-r}(b) G_{m+1+r}(b) - G_{n-r}(a) G_{m+1+r}(a)), \\ \gamma_{n,m}^{(0,1)} &= \sum_{r=0}^{n-1} (-1)^r \frac{n_{(r)}}{(m+1)^{(r+1)}} (G_{n-r}(1) G_{m+1+r}(1) - G_{n-r}(0) G_{m+1+r}(0)), \end{aligned} \tag{18}$$

Proof. See [26]. □

On the other hand, by applying Theorem 1, we can calculate the arbitrary function approximation coefficients with the matrix approach by using the following theorem.

Theorem 2. *Suppose that $f(x) \in L^2[0,1]$ is an arbitrary function and $\{G_i(x) : i = 1, \ldots, N\}$ is the set of the Genocchi polynomials up to order N. Let $Y = \text{span}\{G_1, \ldots, G_N\}$. Since Y is a finite dimensional closed subspace of $L^2[0,1]$, then $\exists f^*(x) \in Y$ is the unique best approximation in the Genocchi polynomials such that any arbitrary function $f(x)$ can be expressed in terms of the Genocchi polynomials by unique coefficient c_n, $n = 0, 1, \ldots, N$:*

$$f(x) \approx f^*(x) = \sum_{n=1}^{N} c_n G_n(x) = C^T G(x), \tag{19}$$

where C consisting of the unique coefficient is called the Genocchi coefficient matrix C given by the following:

$$C^T = F^T T^{(0,1)^{-1}}, \tag{20}$$

where $F = \int_0^1 f(x)G_m(x)dx$, $m = 0, 1, \ldots, N$ and $T^{(0,1)} = \left[\int_0^1 G_n(x)G_m(x)dx\right]_{N \times N}$ is the matrix derived in Theorem 1.

Proof. Assume that $f(x) \in L^2[0,1]$. Therefore, this arbitrary function can be approximated using Equation (13) as follows:

$$f(x) \approx \sum_{n=1}^{N} c_n G_n(x) = C^T G(x), \tag{21}$$

Therefore,

$$\begin{aligned}
\int_0^1 f(x)G_m(x)dx &= \int_0^1 \left(\sum_{n=1}^{N} c_n G_n(x)\right) G_m(x)dx \\
&= \sum_{n=1}^{N} c_n \int_0^1 G_n(x)G_m(x)dx.
\end{aligned} \tag{22}$$

Let the first side of Equation (22) have $f_m = \int_0^1 f(x)G_m(x)dx$ alternatives; thus, we have the following:

$$\begin{aligned}
f_m &= \int_0^1 \left(\sum_{n=1}^{N} c_n G_n(x)\right) G_m(x)dx \\
&= \sum_{n=1}^{N} c_n \int_0^1 G_n(x)G_m(x)dx, \\
m &= 1, \ldots, N.
\end{aligned} \tag{23}$$

In fact, we can construct Equation (23) as a system of N equations for which the matrix representation of the device is as follows:

$$\begin{bmatrix} f_1 \\ \cdot \\ \cdot \\ \cdot \\ f_N \end{bmatrix} = [c_1, \ldots, c_N] \begin{bmatrix} \gamma_{1,1}^{(0,1)} & \cdots & \gamma_{1,N}^{(0,1)} \\ \gamma_{2,1}^{(0,1)} & \cdots & \gamma_{2,N}^{(0,1)} \\ \cdot & \cdots & \cdot \\ \cdot & \cdots & \cdot \\ \gamma_{N,1}^{(0,1)} & \cdots & \gamma_{N,N}^{(0,1)} \end{bmatrix}, \tag{24}$$

$$F^T = C^T T^{(0,1)}.$$

Therefore, we have the Genocchi coefficient matrix C as follows:

$$C^T = F^T {T^{(0,1)}}^{-1}, \tag{25}$$

where $\gamma_{i,j}$ can be calculated by using Theorem 1. □

3. Implementation of the Genocchi Polynomial Method for Solving Nonlinear Volterra Integral Equations with Weakly Singular Kernels

In this section, we implement a new spectral approach based on the Genocchi polynomials to solve the following equation:

$$y(t) = f(t) - \int_0^t \frac{s^\beta}{(t-s)^\alpha} g(y(s))ds, \quad t > 0,$$

where $f(t)$ is in $L^2(\mathcal{R})$ on the interval $0 \le t, s \le T$; g is locally Lipchitz continuous, smooth, and a Hammerstein nonlinear function; and α, β are real positive numbers.

Let us assume that function $f(x) \in L^2[0,1]$ is arbitrary; then, we can approximate it, as follows:

$$f(x) \approx \sum_{n=1}^{N} c_n G_n(x) = C^T G(x) = C^T G X_x, \qquad (26)$$

where $C = [c_1, c_1, \ldots, c_N]^T$ is a vector of unknown coefficient; $X_x = [1, x, x^2, \ldots, x^n]^T$; and $G(x) = [G_1(x), G_2(x), \ldots, G_N(x)]^T = G X_x$, where G is a $n \times n$ matrix of coefficients that can be approximated by X_x.

Thus, we need to compute the following integral before applying the new approach to solve Equation (1).

$$\int_0^x \frac{t^m}{(x-t)^\alpha} dt = \frac{\Gamma(1-\alpha)\Gamma(m+1)}{\Gamma(m-\alpha+2)} x^{(m-\alpha+1)}, \; m = 0, 1, \ldots. \qquad (27)$$

Therefore, by considering Relation (27), we let

$$z(s) = g(y(s)), \; 0 \leq s \leq 1. \qquad (28)$$

since we have

$$y(t) = f(t) - \int_0^t \frac{s^\beta}{(t-s)^\alpha} g(y(s)) ds, \; t > 0. \qquad (29)$$

By substituting Equation (29) into Equation (28), we have

$$z(t) = g(f(t) - \int_0^t \frac{s^\beta}{(t-s)^\alpha} g(y(s)) ds), \; 0 \leq t \leq 1. \qquad (30)$$

We approximate Equation (30) as follows:

$$C^T G(t) = g(f(t) - \int_0^t \frac{s^\beta}{(t-s)^\alpha} C^T G X_s ds), \; 0 \leq t \leq 1, \qquad (31)$$

and

$$C^T G(t) = g(f(t) - C^T G \int_0^t \frac{s^\beta}{(t-s)^\alpha} X_s ds), \; 0 \leq t \leq 1. \qquad (32)$$

Thus, we need to convert the integral part of Equation (32) to the matrix form. Therefore, by assuming $X_s = [1, s, s^2, \ldots, s^n]^T$, we can write the following:

$$\begin{aligned}\int_0^t \frac{s^\beta}{(t-s)^\alpha} \cdot X_s ds &= \left[\int_0^t \frac{s^\beta}{(t-s)^\alpha} ds, \int_0^t \frac{s^\beta}{(t-s)^\alpha} \cdot s ds, \ldots, \int_0^t \frac{s^\beta}{(t-s)^\alpha} \cdot s^n ds, \ldots\right]^T \\ &= \left[\int_0^t \frac{s^\beta}{(t-s)^\alpha} ds, \int_0^t \frac{s^{\beta+1}}{(t-s)^\alpha} ds, \ldots, \int_0^t \frac{s^{\beta+n}}{(t-s)^\alpha} ds, \ldots\right]^T,\end{aligned} \qquad (33)$$

and using Equation (27), we have

$$\int_0^t \frac{s^{\beta+m}}{(t-s)^\alpha} ds = \frac{\Gamma(1-\alpha)\Gamma(\beta+m+1)}{\Gamma(\beta+m-\alpha+2)} t^{(\beta+m-\alpha+1)}, \; m = 0, 1, 2, \ldots. \qquad (34)$$

Therefore, by using Relation (34), we can rewrite Equation (33) as follows:

$$\int_0^t \frac{s^\beta}{(t-s)^\alpha} \cdot X_s ds = \left[\frac{\Gamma(1-\alpha)\Gamma(\beta+1)}{\Gamma(\beta-\alpha+2)} t^{(\beta-\alpha+1)}, \frac{\Gamma(1-\alpha)\Gamma(\beta+2)}{\Gamma(\beta-\alpha+3)} t^{(\beta-\alpha+2)}, \ldots \right. \\ \left. , \frac{\Gamma(1-\alpha)\Gamma(\beta+m+1)}{\Gamma(\beta+m-\alpha+2)} t^{(\beta+m-\alpha+1)}, \ldots\right]^T. \qquad (35)$$

If we consider $\gamma_{m,m} = \frac{\Gamma(1-\alpha)\Gamma(\beta+m+1)}{\Gamma(\beta+m-\alpha+2)}$, $m = 0, 1, 2, \ldots$, then, we can reconstruct Equation (35) in the matrix form as follows:

$$\int_0^t \frac{s^\beta}{(t-s)^\alpha} \cdot X_s ds = \begin{bmatrix} \gamma_{0,0} & 0 & 0 & \cdots & 0 \\ 0 & \gamma_{1,1} & 0 & 0 & 0 \\ 0 & 0 & \gamma_{2,2} & 0 & 0 \\ \vdots & \vdots & \vdots & \ddots & \vdots \\ 0 & 0 & 0 & \cdots & \gamma_{m,m} \end{bmatrix} \begin{bmatrix} t^{\beta-\alpha+1} \\ t^{\beta-\alpha+2} \\ \vdots \\ t^{\beta+m-\alpha+1} \\ \vdots \end{bmatrix} = \Omega\Pi, \quad (36)$$

where Ω is an infinite diagonal matrix and

$$\Pi = \left[t^{\beta-\alpha+1}, t^{\beta-\alpha+2}, \cdots, t^{\beta+m-\alpha+1}, \cdots \right]^T. \quad (37)$$

Now, each element of infinite vector Π can be approximated by using the Genocchi polynomials as follows:

$$t^{\beta+m-\alpha+1} = \sum_{i=1}^{\infty} a_{m,i} G_i(t) = \partial_m G X_t, \quad \partial_m = [a_{m,1}, a_{m,2}, \ldots], \quad m = 0, 1, \ldots, \quad (38)$$

and we obtain

$$\Pi = [\partial_1 G X_t, \partial_2 G X_t, \ldots, \partial_m G X_t, \ldots]^T = A G X_t, \quad A = [\partial_1, \partial_2, \ldots, \partial_m, \ldots]^T. \quad (39)$$

Substituting (39) in (32), we have

$$\int_0^t \frac{s^\beta}{(t-s)^\alpha} \cdot X_s ds == \Omega A G X_t. \quad (40)$$

By using Equations (40) and (39), we get

$$C^T G(t) = g(f(t) - C^T G \Omega A G X_t), \quad 0 \le t \le 1. \quad (41)$$

We select N nodal points of the Newton–Cotes rule for finding vector C as follows:

$$x_p = \frac{2p-1}{2N}, \quad p = 1, 2, \ldots, N, \quad (42)$$

By collocating Equation (41) at the points x_p, we have

$$\begin{aligned} C^T G(x_p) &= g(f(x_p) - C^T G \Omega A G X_{x_p}), \quad 0 \le t \le 1, \\ p &= 1, 2, \ldots, N. \end{aligned} \quad (43)$$

We can solve the nonlinear system (43) by using the Newton iteration scheme to calculate unknown vector C. After calculating unknown vector C by solving the nonlinear Equation (43), we use Equations (29), (31), and (32) to obtain the approximate solution of Equation (1), as follows:

$$y_n(t) = f(t) - C^T G \Omega A G(t), \quad 0 \le t \le 1. \quad (44)$$

4. Error Analysis

In this section, we perform error estimation of the approximation solution to find the error boundaries of the new numerical approach by applying the Genocchi polynomials. Consider the nonlinear Volterra integral equations with weakly singular kernels of the form Equation (1),

We suppose that $\Omega = L^2[0,1]$, $\{G_1(t), G_2(t), \ldots, G_n(t)\} \subset \Omega$, and $T = Span\{G_1(t), G_2(t), \ldots, G_n(t)\}$. Here, we let $y(t)$ be an arbitrary function of Ω, so, it has the best approximation of T. Let $y_n \in T$, that is,

$$\exists y_n \in T : \forall h \in T \ \|y - y_n\|_2 \le \|y - h\|_2, \tag{45}$$

where $\|y(t)\|_2^2 = \int_0^1 |y(t)|^2 dt$. $y(t)$ is approximated by using the truncated Genocchi polynomials:

$$y(t) \simeq y_n = \sum_{n=1}^{N} c_n G_n(t) = C^T G(t), \tag{46}$$

where $C^T = [c_1, c_2, \ldots, c_N]$ and $G(x) = [G_1(x), G_2(x), \ldots, G_N(x)]^T$.

In the following study, we present an upper bound for the error of Equation (45). Let $e_n(t) = y(t) - y_n(t)$ be the error function of Equation (1), where $y(t), y_n(t)$ are the exact and approximate solutions

Therefore, the mean error bound is presented as follows:

$$\begin{aligned} \|e_n(t)\|_2^2 &= \|y(t) - y_n(t)\|_2 = \int_0^1 |y(t) - y_n(t)|^2 dt \\ &= \int_0^1 \left| \left(f(t) - \int_0^t \frac{s^\beta}{(t-s)^\alpha} g(y(s))ds\right) - \left(f(t) - \int_0^t \frac{s^\beta}{(t-s)^\alpha} g(y_n(s))ds\right) \right|^2 dt \\ &= \int_0^1 \left| \int_0^t \frac{s^\beta}{(t-s)^\alpha} (g(y(s)) - g(y_n(s)))ds \right|^2 dt. \end{aligned} \tag{47}$$

On the other hand, $g(s)$ is continuous on the interval $[0,1]$ and locally Lipchitz continuous in $s \in R$; therefore, there is a constant $C_1 > 0$ such that

$$|g(y(s)) - g(y_n(s))| \le C_1 |y(s) - y_n(s)|. \tag{48}$$

Then, by using Equations (47) and (48), we have

$$\begin{aligned} \|e_n(t)\|_2^2 &\le \int_0^1 \left(\int_0^t \frac{s^\beta}{(t-s)^\alpha} \cdot C_1 |y(s) - y_n(s)| ds \right)^2 dt \\ &= \int_0^1 \left(\int_0^t \frac{s^\beta}{(t-s)^\alpha} \cdot C_1 \left| y(s) - \sum_{n=1}^{N} c_n G_n(s) \right| ds \right)^2 dt \\ &= \int_0^1 \left(\int_0^t \frac{s^\beta}{(t-s)^\alpha} \cdot C_1 \left| \sum_{n=N+1}^{\infty} c_n G_n(s) \right| ds \right)^2 dt \\ &\le \int_0^1 \left(\int_0^t \frac{s^\beta}{(t-s)^\alpha} \cdot C_1 \sum_{n=N+1}^{\infty} |c_n| |G_n(s)| ds \right)^2 dt. \end{aligned} \tag{49}$$

By substituting (4) into (49), we get

$$\begin{aligned} \|e_n(t)\|_2^2 &\le \int_0^1 \left(\int_0^t \frac{s^\beta}{(t-s)^\alpha} \cdot C_1 \sum_{n=N+1}^{\infty} |c_n| \sum_{k=0}^{n} \binom{n}{k} G_{n-k} s^k \middle| ds \right)^2 dt \\ &\le \int_0^1 \left(\int_0^t \frac{s^\beta}{(t-s)^\alpha} \cdot C_1 \sum_{n=N+1}^{\infty} |c_n| \sum_{k=0}^{n} \binom{n}{k} |G_{n-k}| s^k ds \right)^2 dt \\ &= \int_0^1 \left(\sum_{k=0}^{n} \sum_{n=N+1}^{\infty} \cdot C_1 |c_n| \binom{n}{k} |G_{n-k}| \int_0^t \frac{s^{\beta+k}}{(t-s)^\alpha} ds \right)^2 dt, \end{aligned} \tag{50}$$

where $\gamma(t, \beta, \alpha)$ is defined by

$$\gamma(t, \beta, \alpha) = \int_0^t \frac{s^\beta}{(t-s)^\alpha} ds = B(1-\alpha, 1+\beta) t^{1-\alpha+\beta}. \tag{51}$$

On the other hand, $B(\alpha, \beta)$ is the beta function that is usually defined by

$$B(\alpha, \beta) = \int_0^1 \tau^{\alpha-1}(1-\tau)^{\beta-1} d\tau, \quad (\text{Re}(\alpha) > 0, \text{Re}(\beta) > 0). \tag{52}$$

Therefore, by using Inequality (51) and Equation (52), we get

$$\|e_n(t)\|_2^2 \leq \int_0^1 \left(\sum_{k=0}^n \sum_{n=N+1}^{\infty} \cdot C_1 |c_n| \binom{n}{k} |G_{n-k}| B(1-\alpha, 1+\beta+k) t^{1-\alpha+\beta+k} \right)^2 dt. \tag{53}$$

and

$$\|e_n(t)\|_2 \leq \sqrt{\sum_{k=0}^n \sum_{n=N+1}^{\infty} \frac{1}{2(-\alpha+\beta+k)+3} \left(C_1 |c_n| \binom{n}{k} |G_{n-k}| B(1-\alpha, 1+\beta+k) \right)^2}. \tag{54}$$

5. Illustrative Examples

In this section, two numerical examples are performed to check the perfection of the proposed method as well as the accuracy and efficiency of the Genocchi polynomials scheme.

In order to demonstrate the error of a new numerical approach based on Genocchi polynomials, we define the notations as follows:

$$e_2(N) = \|y - y_n\|_\infty = \max_{0 \leq t \leq 1} |y - y_n|,$$
$$e_n(t) = |y - y_n|,$$
$$\xi_n = \left(\int_0^T w(t) e_n^2(t) dt \right)^{\frac{1}{2}}, \tag{55}$$

where $y(t)$ is the exact solution and $y_n(t)$ is the approximate function to the proposed method and we have $w(t) = 1$. In our implementation, the calculations are done on a personal computer with core-i5 processor, 2.67 GHZ frequency, and 4 GB memory, and the codes were written in Mathematica 11 software.

Example 1. *We consider the following nonlinear Volterra integral equation which was proposed in [10]:*

$$y(t) = t^{\frac{1}{3}} + \frac{4\Gamma\left(\frac{4}{3}\right)\Gamma\left(\frac{13}{6}\right)}{\sqrt{\pi}} t^{\frac{3}{2}} - \int_0^t \frac{s^{1/2} y^2(s)}{(t-s)^{2/3}} ds, \quad t \in [0, 1]. \tag{56}$$

The exact solution of this equation is $y(t) = t^{1/3}$.

We solved this equation with the proposed numerical method by using different values of N. The diagonal matrix Ω with elements $\frac{\Gamma[1-\alpha]\Gamma[m+1+\beta]}{\Gamma[m-\alpha+2+\beta]}$, $m = 0, 1, \ldots, N$, and vector Π for $N = 5$ are obtain in the following forms:

$$\Omega = \begin{pmatrix} 2.52393 & 0 & 0 & 0 & 0 & 0 \\ 0 & 2.06503 & 0 & 0 & 0 & 0 \\ 0 & 0 & 1.82209 & 0 & 0 & 0 \\ 0 & 0 & 0 & 1.66364 & 0 & 0 \\ 0 & 0 & 0 & 0 & 1.54891 & 0 \\ 0 & 0 & 0 & 0 & 0 & 1.4604 \end{pmatrix},$$

$$\Pi = [x^{5/6}, x^{11/6}, x^{17/6}, x^{23/6}, x^{29/6}, x^{35/6}]^T.$$

Also, the unknown vector elements C are as follows:

$$c_0 = 0.529883, c_1 = 0.598039, c_2 = -0.351362, c_3 = 0.343517,$$
$$c_4 = -0.104667, c_5 = 0.0717268$$

After numerical computations, a system of algebraic nonlinear equations is obtained under the proposed method. Therefore, by solving this system, we obtain the approximate solution for $N = 5$ as follows:

$$\begin{aligned} y_5(t) &= f(t) - C^T G \Omega A G(t) = t^{1/3} - 0.151891 t^{5/6} + 2.18117 t^{3/2} - 3.56596 t^{11/6} \\ &+ 3.71576 t^{17/6} - 4.02725 t^{23/6} + 2.47707 t^{29/6} - 0.628499 t^{35/6}. \end{aligned}$$

According to the error boundaries in Relation (55), we have

$$\|e_5(t)\|_2 \leq 0.000598532.$$

Figure 2 is devoted to comparing the exact solution with the approximate solution obtained from the proposed method for $N = 5$. Observing Figure 2, overlap of the exact and approximate solutions shows the exactness and correctness of the proposed method. The absolute error functions with $N = 5, 10, 18, 20$ are shown in Figures 3–6. Therefore, these plots quickly explain that the proposed approach has small absolute errors.

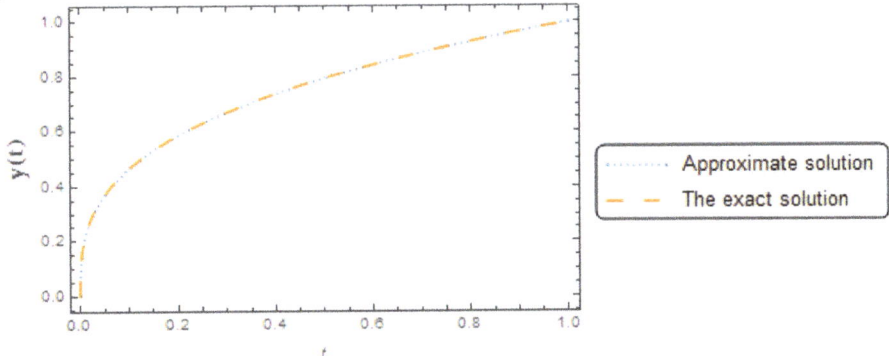

Figure 2. Plot of comparison between the exact and approximate solutions of Example 1 for $N = 5$.

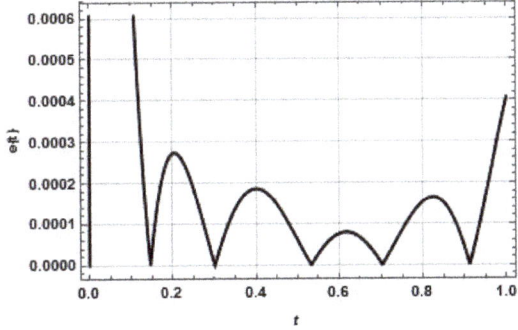

Figure 3. Plot of the absolute error with $N = 5$ for Example 1.

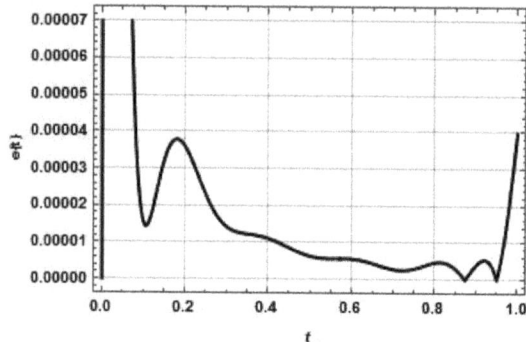

Figure 4. Plot of the absolute error with $N = 10$ for Example 1.

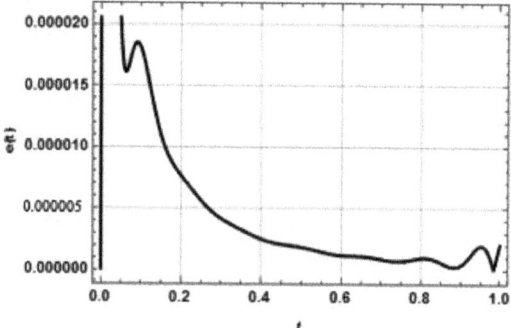

Figure 5. Plot of the absolute error with $N = 15$ for Example 1.

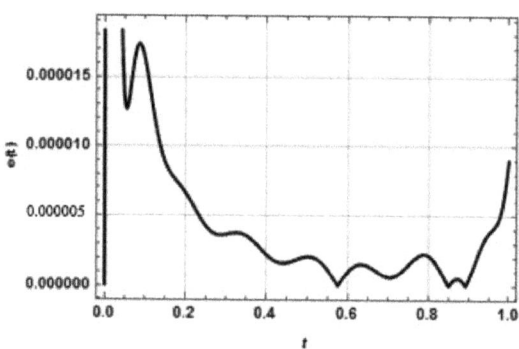

Figure 6. Plot of the absolute error with $N = 20$ for Example 1.

We reported the numerical results of the exact and approximate solutions for various values N on the interval $[0, 1]$ in Table 1. On the other hand, numerical results are showed for different values N in Table 2. The absolute error functions are displayed for various values of N on the interval $[0, 1]$ for this problem in Table 3. Also, Table 4 compares the numerical results of a new proposed numerical approach with Euler's method [10] for different values of N. Also, Table 4 indicates that the new numerical method has better accuracy and efficiency compared to the old method.

Table 1. Approximate and exact values of nonlinear Volterra integral equations with $N = 5, 10, 15, 20$ for Example 1.

	$N = 5$	$N = 10$	$N = 15$	$N = 20$	y_{Exact}
0.0	0.000000	0.000000	0.000000	0.000000	0.000000
0.2	0.585076	0.584768	0.584793	0.584797	0.584804
0.4	0.736620	0.736795	0.736802	0.736804	0.736806
0.6	0.843508	0.843427	0.843431	0.843434	0.843433
0.8	0.928164	0.928313	0.928317	0.928319	0.928318
1.0	1.00041	0.99996	1.000001	1.000001	1.000000

Table 2. Numerical results of ζ_N for different values N on the interval $[0, 1]$ for Example 1.

N	ζ_N	Computing Time (s)
5	5.98532×10^{-4}	0.321
10	1.14944×10^{-4}	0.357
15	4.48214×10^{-5}	0.420
20	2.85973×10^{-5}	0.451

Table 3. The absolute error function of various values N on the interval $[0, 1]$ for Example 1.

t	$e_5(t)$	$e_{10}(t)$	$e_{15}(t)$	$e_{20}(t)$
0.0	0.000000000	0.0000000000	0.00000000	0.00000000
0.2	0.000272294	0.0000359627	0.00001089	6.67632×10^{-6}
0.4	0.000185829	0.0000111075	3.8548×10^{-6}	2.25757×10^{-6}
0.6	0.000075540	5.5081×10^{-6}	1.83763×10^{-6}	9.72768×10^{-6}
0.8	0.000153622	4.62581×10^{-6}	1.11576×10^{-6}	2.18163×10^{-6}
1.0	0.000406512	0.0000397521	8.73408×10^{-6}	9.0017×10^{-6}

Table 4. Comparison of maximum absolute errors between a new approach approximate solution and Euler's method on $[0, \varepsilon]$ for Example 1.

	Euler's Method [10]					Our Method (Genocchi Polynomials)			
N	$\varepsilon = 0$	$\varepsilon = 0.01$	$\varepsilon = 0.02$	$\varepsilon = 0.03$	N	$\varepsilon = 0$	$\varepsilon = 0.01$	$\varepsilon = 0.02$	$\varepsilon = 0.03$
	$e_\infty(N)$	$e_\infty(N)$	$e_\infty(N)$	$e_\infty(N)$		$e_\infty(N)$	$e_\infty(N)$	$e_\infty(N)$	$e_\infty(N)$
80	0.67×10^{-2}	6.60×10^{-3}	6.50×10^{-3}	6.30×10^{-3}	5	0.000	1.851×10^{-3}	2.343×10^{-3}	2.427×10^{-3}
160	3.21×10^{-3}	3.10×10^{-3}	$3,10 \times 10^{-3}$	3.03×10^{-3}	10	0.000	6.595×10^{-4}	6.704×10^{-4}	6.704×10^{-4}
320	1.55×10^{-3}	1.50×10^{-3}	1.50×10^{-3}	1.50×10^{-3}	15	0.000	3.193×10^{-4}	3.193×10^{-4}	3.178×10^{-4}
640	753×10^{-4}	7.40×10^{-4}	7.20×10^{-4}	7.20×10^{-4}	20	0.000	2.305×10^{-4}	2.305×10^{-4}	2.305×10^{-4}

Example 2. *Next, we discuss the following Lighthill's equation which was proposed in [10] and extensively studied in [10,29,30]. The authors employed the iterative method and schemes to solve this integral equation.*

$$y(t) = 1 - \frac{\sqrt{3}}{\pi} \int_0^t \frac{s^{\frac{1}{3}} y^4(s)}{(t-s)^{\frac{2}{3}}} ds, \ t \in [0, 1]., \tag{57}$$

The numerical results for this example are obtained by the presented approach for different values of N and are given in Tables 5 and 6. Also, in Table 7, the maximum absolute errors can be compared with those that were achieved by Euler's method in [10] by different values of N on the interval $[0, \varepsilon]$. We can see that our proposed method is very fast compared to Euler's method. Figure 7 displays the convergence approximate solutions using our method (Genocchi polynomials) and the Picard iteration y_2 with different values of N on the interval $[0, \varepsilon]$ with $\varepsilon = 0.002$ for this problem.

Table 5. Numerical results on the interval $[0, \varepsilon]$, with $\varepsilon = 0.002$ for Example 2.

N	ζ_N	Computing Time (s)
5	1.912914×10^{-4}	0.351
10	1.087754×10^{-4}	0.402
15	9.106063×10^{-5}	0.457
20	7.200394×10^{-5}	0.530

Table 6. The approximate solutions by different values of N and M for Example 2.

	$\|y_N - y_M\|_\infty$			
t	$N = 5; \mathbf{M} = 7$	$N = 7; \mathbf{M} = 10$	$N = 10; \mathbf{M} = 12$	$N = 12; \mathbf{M} = 13$
0.0000	1.11022×10^{-16}	1.12022×10^{-16}	0.000000000	0.000000000
0.0004	0.000507344	0.000477524	0.000214749	0.0000875408
0.0008	0.000800212	0.000750437	0.000336102	0.0001366621
0.0012	0.001041861	0.000973492	0.000434214	0.0001761062
0.0016	0.001254042	0.001167461	0.000518585	0.0002097873
0.002	0.001445845	0.001341082	0.000593241	0.000239374

Table 7. Comparison of maximum absolute errors $e_\infty(N) = \|y_2 - y_N\|_\infty$ between our method (Genocchi polynomials) and Euler's method: the Picard iterate y_2 was used on $[0, \varepsilon]$ for Example 2.

	Euler's Method [10]				Our Method (Genocchi Polynomials)		
N	$\varepsilon = 0.002$	$\varepsilon = 0.003$	$\varepsilon = 0.008$	N	$\varepsilon = 0.002$	$\varepsilon = 0.003$	$\varepsilon = 0.008$
	$e_\infty(N)$	$e_\infty(N)$	$e_\infty(N)$		$e_\infty(N)$	$e_\infty(N)$	$e_\infty(N)$
40	3.60×10^{-2}	3.00×10^{-2}	1.70×10^{-2}	5	6.165×10^{-3}	7.441×10^{-3}	9.878×10^{-3}
80	2.10×10^{-2}	1.7×10^{-2}	9.10×10^{-3}	10	3.378×10^{-3}	3.386×10^{-3}	4.162×10^{-3}
160	1.01×10^{-2}	8.4×10^{-3}	4.00×10^{-3}	12	2.785×10^{-3}	3.113×10^{-3}	3.222×10^{-3}
320	4.00×10^{-3}	3.00×10^{-3}	1.30×10^{-3}	15	2.151×10^{-3}	2.311×10^{-3}	2.232×10^{-3}

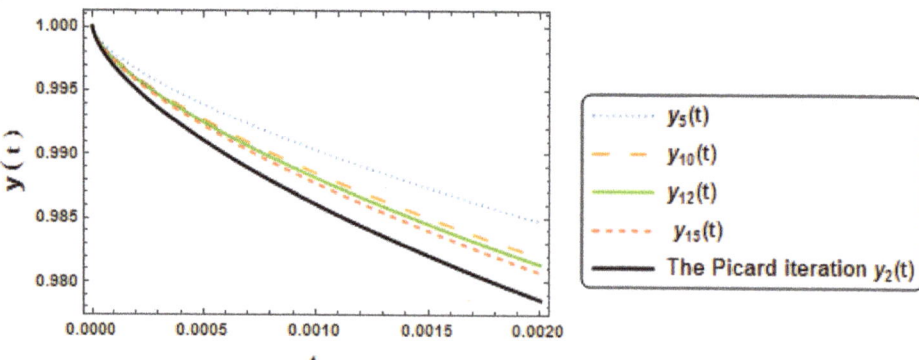

Figure 7. Plot of approximate solutions by our method (Genocchi polynomials) with different values of N on the interval $[0, \varepsilon]$ with $\varepsilon = 0.002$ for Example 1.

6. Conclusions and Future Work

In the study of many nonlinear problems in heat conduction, boundary-layer heat transfer, chemical kinetics, and superfluidity, we are often led to singular Volterra integral equations that are difficult to solve analytically. In this article, a spectral method based on Genocchi polynomials is presented for solving nonlinear Volterra integral equations with weakly singular kernels. An error analysis of the spectral approach has been done. Two numerical examples are provided to confirm the applicability and accuracy of the scheme. Also, the proposed method results have been compared with Euler's method to show the superiority of the present method with better results in smaller N. For future works, we can use other polynomials like Chebyshev, Lagger, etc. for implementation, and by comparing the archived results, we can expand the present method and implement it on the system of nonlinear Volterra integral equations and nonlinear Volterra integral equations of mix type or the first kind of Volterra integral equations with discontinuous kernels.

Author Contributions: Conceptualization, E.H. and M.A.E. and S.N.; methodology, E.H. and M.A.E.; software, E.H. and M.A.E.; validation, E.H. and M.A.E. and S.N.; formal analysis, E.H.; investigation, E.H. and M.A.E. and S.N.; resources, E.H and M.A.E.; data curation, E.H. and M.A.E. and S.N.; writing—original draft preparation, E.H. and M.A.E.; writing—review and editing, E.H. and S.N.; visualization, M.A.E.; supervision, E.H.; project administration, E.H.; funding acquisition, S.N. All authors have read and agreed to the published version of the manuscript.

Funding: This research received no external funding.

Conflicts of Interest: The authors declare no conflict of interest.

References

1. Abdou, M.A. On a symptotic methods for Fredholm–Volterra integral equation of the second kind in contact problems. *J. Comput. Appl. Math.* **2003**, *154*, 431–446. [CrossRef]
2. Datta, K.B.; Mohan, B.M. *Orthogonal Functions in Systems and Control*; World Scientific: Singapore, 1995.
3. Smetanin, B.I. On an integral equation for axially-symmetric problems in the case of an elastic body containing an inclusion. *J. Appl. Math. Mech.* **1991**, *55*, 371–375. [CrossRef]
4. Ramos, J.I.; Canuto, C.; Hussaini, M.Y.; Quarteroni, A.; Zang, T.A. *Spectral Methods in Fluid Dynamics*; Springer: New York, NY, USA, 1988.
5. Atkinson, K.E. *The Numerical Solution of Integral Equations of the Second Kind*; Cambridge University Press: Cambridge, UK, 1996.
6. Noeiaghdam, S.; Zarei, E.; Kelishami, H.B. Homotopy analysis transform method for solving Abel's integral equations of the first kind. *Ain Shams Eng. J.* **2016**, *7*, 483–495. [CrossRef]
7. Noeiaghdam, S.; Araghi, M.A.; Abbasbandy, S. Finding optimal convergence control parameter in the homotopy analysis method to solve integral equations based on the stochastic arithmetic. *Numer. Algorithms* **2019**, *81*, 237–267. [CrossRef]
8. Noeiaghdam, S.; Sidorov, D.; Sizikov, V.; Sidorov, N. Control of accuracy on Taylor-collocation method to solve the weakly regular Volterra integral equations of the first kind by using the CESTAC method. *Appl. Comput. Math. Int. J.* **2020**, *19*, 81–105.
9. Noeiaghdam, S.; Dreglea, A.; He, J.; Avazzadeh, Z.; Suleman, M.; Fariborzi Araghi, M.A.; Sidorov, D.N.; Sidorov, N. Error Estimation of the Homotopy Perturbation Method to Solve Second Kind Volterra Integral Equations with Piecewise Smooth Kernels: Application of the CADNA Library. *Symmetry* **2020**, *12*, 1730. [CrossRef]
10. Allaei, S.S.; Diogo, T.; Rebelo, M. Analytical and computational methods for a class of nonlinear singular integral equations. *Appl. Numer. Math.* **2017**, *114*, 2–17. [CrossRef]
11. Maleknejad, K.; Hashemizadeh, E.; Basirat, B. Numerical solvability of Hammerstein integral equations based on hybrid Legendre and Block-Pulse functions. In Proceedings of the 2010 International Conference on Parallel and Distributed Processing Techniques and Applications, Las Vegas, NV, USA, 12–15 July 2010; pp. 172–175.
12. Pourgholi, R.; Tahmasebi, A.; Azimi, R. Tau approximate solution of weakly singular Volterra integral equations with Legendre wavelet basis. *Int. J. Comput. Math.* **2017**, *94*, 1337–1348. [CrossRef]
13. Eslahchi, M.R.; Dehghan, M.; Parvizi, M. Application of the collocation method for solving nonlinear fractional integro-differential equations. *Comput. Appl. Math.* **2014**, *257*, 105–128. [CrossRef]
14. Zhu, L.; Fan, Q. Numerical solution of nonlinear fractional-order Volterra integro-differential equations by SCW. *Commun. Nonlinear Sci. Numer. Simul.* **2013**, *18*, 1203–1213. [CrossRef]
15. Nemati, S.; Lima, P.M. Numerical solution of nonlinear fractional integro-differential equations with weakly singular kernels via a modification of hat functions. *Appl. Math. Comput.* **2018**, *327*, 79–92. [CrossRef]

16. Singh, S.; Patel, V.K.; Singh, V.K. Operational matrix approach for the solution of partial integro-differential equation. *Appl. Math. Comput.* **2016**, *283*, 195–207. [CrossRef]
17. Garg, M.; Sharma, A. Solution of space-time fractional telegraph equation by Adomian decomposition method. *J. Inequalities Spec. Funct.* **2011**, *2*, 1–7.
18. Ray, S.S.; Bera, R.K. An approximate solution of a nonlinear fractional differential equation by Adomian decomposition method. *Appl. Math. Comput.* **2005**, *167*, 561–571. [CrossRef]
19. Wu, G.C. A fractional variational iteration method for solving fractional nonlinear differential equations. *Comput. Math. Appl.* **2011**, *61*, 2186–2190. [CrossRef]
20. Khan, Y.; Faraz, N.; Yildirim, A.; Wu, Q. Fractional variational iteration method for fractional initial-boundary value problems arising in the application of nonlinear science. *Comput. Math. Appl.* **2011**, *62*, 2273–2278. [CrossRef]
21. Arikoglu, A.; Ozkol, I. Solution of fractional differential equations by using differential transform method. *Chaos Solitons Fractals* **2007**, *34*, 1473–1481. [CrossRef]
22. Nazari, D.; Shahmorad, S. Application of the fractional differential transform method to fractional-order integro-differential equations with nonlocal boundary conditions. *J. Comput. Appl. Math.* **2010**, *234*, 883–891. [CrossRef]
23. Ebadi, M.A.; Hashemizadeh, E. A new approach based on the Zernike radial polynomials for numerical solution of the fractional diffusion-wave and fractional Klein–Gordon equations. *Phys. Scr.* **2018**, *93*, 125202. [CrossRef]
24. Shang, Y. On the delayed scaled consensus problems. *Appl. Sci.* **2017**, *7*, 713. [CrossRef]
25. Isah, A.; Phang, C. On Genocchi operational matrix of fractional integration for solving fractional differential equations. *AIP Conf. Proc.* **2017**, *1795*, 020015.
26. Loh, J.R.; Phang, C.; Isah, A. New operational matrix via Genocchi polynomials for solving Fredholm-Volterra fractional integro-differential equations. *Adv. Math. Phys.* **2017**, *2017*, 1–12. [CrossRef]
27. Isah, A.; Phang, C.; Phang, P. Collocation method based on Genocchi operational matrix for solving generalized fractional pantograph equations. *Int. J. Differ. Equ.* **2017**, *2017*, 1–10. [CrossRef]
28. Sadeghi Roshan, S.; Jafari, H.; Baleanu, D. Solving FDEs with Caputo-Fabrizio derivative by operational matrix based on Genocchi polynomials. *Math. Methods Appl. Sci.* **2018**, *41*, 9134–9141. [CrossRef]
29. Diogo, T.; Lima, P.; Rebelo, M. Numerical solution of a nonlinear Abel type Volterra integral equation. *Commun. Pure Appl. Anal.* **2006**, *5*, 277–288. [CrossRef]
30. Diogo, M.T.; Lima, P.M.; Rebelo, M.S. Comparative study of numerical methods for a nonlinear weakly singular Volterra integral equation. *Hermis J.* **2006**, *7*, 1–20.

Publisher's Note: MDPI stays neutral with regard to jurisdictional claims in published maps and institutional affiliations.

© 2020 by the authors. Licensee MDPI, Basel, Switzerland. This article is an open access article distributed under the terms and conditions of the Creative Commons Attribution (CC BY) license (http://creativecommons.org/licenses/by/4.0/).

Article

A Numerical Method for Weakly Singular Nonlinear Volterra Integral Equations of the Second Kind

Sanda Micula

Department of Mathematics, Babeş-Bolyai University, 400084 Cluj-Napoca, Romania; smicula@math.ubbcluj.ro

Received: 21 October 2020; Accepted: 10 November 2020; Published: 12 November 2020

Abstract: This paper presents a numerical iterative method for the approximate solutions of nonlinear Volterra integral equations of the second kind, with weakly singular kernels. We derive conditions so that a unique solution of such equations exists, as the unique fixed point of an integral operator. Iterative application of that operator to an initial function yields a sequence of functions converging to the true solution. Finally, an appropriate numerical integration scheme (a certain type of product integration) is used to produce the approximations of the solution at given nodes. The resulting procedure is a numerical method that is more practical and accessible than the classical approximation techniques. We prove the convergence of the method and give error estimates. The proposed method is applied to some numerical examples, which are discussed in detail. The numerical approximations thus obtained confirm the theoretical results and the predicted error estimates. In the end, we discuss the method, drawing conclusions about its applicability and outlining future possible research ideas in the same area.

Keywords: weakly singular Volterra integral equations; Picard iteration; product integration; numerical approximation

MSC: 65R20; 45D05; 45E10; 37C25; 65D30

1. Introduction

Many fields in the area of Applied Mathematics rely on knowledge of integral equations, as they arise naturally in various applications in Mathematics, Engineering, Physics, and Technology. They can be used to model a wide range of physical problems such as heat conduction, diffusion, continuum mechanics, geophysics, electricity, magnetism, neutron transport, traffic theory, and many more. Integral equations provide solutions in designing efficient parametrization algorithms for algebraic curves, surfaces, and hypersurfaces. Many initial and boundary value problems associated with ordinary and partial differential equations can be reformulated as integral equations.

Singular and weakly singular integral equations are of particular interest, since they are used to solve inverse boundary value problems whose domains are fractal curves, where classical calculus cannot be used. Abel equations and other fractional order integral equations were studied extensively and are used in modeling various phenomena in biophysics, viscoelasticity, electrical circuits, etc.

Solvability and properties of singular Volterra integral equations were studied using various analytical and approximating methods. We mention existence (and uniqueness) results [1–3], resolvent methods [4], Laplace transforms [2,5,6], fixed point theorems [3,7], etc. Numerical solutions have been found, using product integration [8], collocation and iterated collocation [9–12], homotopy perturbation transform method [2,13], Tau method based on Jacobi functions [14], Nyström methods [8], quadrature schemes [15], variational iteration methods [6], block-pulse wavelets [16], modified quadratic spline approximation [17], reproducing kernel method [18], etc.

Researchers around the world have studied properties of the solutions, such as regularity [9,19], properties of the resolvent [4], monotonicity [20] and others [21–23].

In many applications modeled by integral equations, the kernels are not smooth, making it difficult both to find a solution and to approximate it numerically, as the convergence of approximate methods depends in general on the smoothness of the solution. Thus, classical analytical methods, such as projection methods perform poorly in such cases, as the linear system they lead to is generally badly conditioned and difficult to solve. Proof of convergence and error estimation can also be laborious, when classical calculus cannot be used. Oftentimes, they also have a high implementation cost. Hence, there is a high need for speedy, easy to use numerical methods for these types of equations. The method we propose is based on a classical fixed point result, adapted appropriately. Then, for the approximation of the integrals involved, the product integration numerical scheme we use is also quite efficient, since most of the computations can be done only once, not at each iteration.

In this paper, we consider a Volterra integral equation of the type

$$u(t) = \int_0^t K(t,s,u(s))\, ds + f(t), \quad t \in [0,T], \tag{1}$$

with the kernel of the form

$$K(t,s,u(s)) = a(t,s,u(s))(t-s)^{\alpha-1} \tag{2}$$

where $0 < \alpha < 1$ and $a : [0,T] \times [0,T] \times \mathbb{R} \to \mathbb{R}$, $f : (0,T] \to \mathbb{R}$ are continuous functions. Later on, other smoothness assumptions will be made on a and f.

We derive conditions under which results from fixed point theory will provide the existence of a unique solution of this equation, as well as a sequence of successive iterations to approximate it. We briefly summarize the main results for the existence of fixed points of an operator on a Banach space.

Definition 1. *Let $(X, ||\cdot||)$ be a Banach space. A mapping $T : X \to X$ is called a q-**contraction** if there exists a constant $0 \leq q < 1$ such that*

$$||Tx - Ty|| \leq q||x - y||,$$

for all $x, y \in X$.

On Banach spaces, the well known contraction principle holds:

Theorem 1. *Consider a Banach space $(X, ||\cdot||)$ and let $T : X \to X$ be a q-contraction. Then*

(a) *T has exactly one fixed point, which means equation $x = Tx$ has exactly one solution $x^* \in X$;*
(b) *the sequence of successive approximations $x_{n+1} = Tx_n$, $n \in \mathbb{N}$, converges to the solution x^*, where x_0 can be any arbitrary point in X;*
(c) *for every $n \in \mathbb{N}$, the following error estimate*

$$||x_n - x^*|| \leq \frac{q^n}{1-q} ||Tx_0 - x_0||$$

holds.

Remark 1. *Theorem 1 remains valid when X is replaced by a closed subset $Y \subseteq X$, satisfying $T(Y) \subseteq Y$.*

We use Banach's theorem to establish, under certain conditions, the existence and uniqueness of a solution of Equation (1) and to approximate it by applying the operator successively. Then we use a suitable numerical integration scheme to approximate the values of the solution at given nodes.

The numerical method thus resulted is quite easy to use and implement, while giving accurate approximations.

The paper is organized as follows. In Section 2 we derive necessary conditions for the existence and uniqueness of the solution and discuss its regularity. In Section 3 the numerical method is described, by use of a special type of product integration. The convergence and error analysis of the method are also discussed in details. Numerical examples are given in Section 4, illustrating the applicability of the proposed method. In Section 5, the advantages of this new method are summarized and future possible work ideas in the same area are discussed.

2. Existence and Uniqueness of the Solution

To solve Equation (1), we apply the contraction principle to the associated integral operator

$$Fu(t) = \int_0^t a(t,s,u(s))(t-s)^{\alpha-1}\,ds + f(t). \tag{3}$$

Remark 2. *Since $a \in C([0,T] \times [0,T] \times \mathbb{R})$ and $f \in C(0,T]$, it is well known that the operator $F : C[0,T] \to C[0,T]$ is well defined, i.e., $F(C[0,T]) \subseteq C[0,T]$ (for the proof, see e.g., [7]).*

Then we solve the integral Equation (1) by finding a fixed point for the operator F:

$$u = Fu. \tag{4}$$

We consider the space $X = C[0,T]$ equipped with the Bielecki norm

$$||u||_\tau := \max_{t \in [0,T]} |u(t)|\, e^{-\tau t}, \quad u \in X,$$

for some suitable constant $\tau > 0$. Then, as is well known, $(X, ||\cdot||_\tau)$ is a Banach space (see e.g., [24]) and we have the following result.

Theorem 2. *Let $F : (X, ||\cdot||_\tau) \to (X, ||\cdot||_\tau)$ be defined by Equation (3). Assume that there exists a constant $L > 0$ such that*

$$|a(t,s,u) - a(t,s,v)| \leq L|u-v|, \tag{5}$$

for all $t, s \in [0,T]$ and all $u, v \in \mathbb{R}$. Then

(a) *Equation (4) has a unique solution $u^* \in X$;*
(b) *the sequence of successive approximations*

$$u_{n+1} = Fu_n, \quad n = 0, 1, \ldots \tag{6}$$

converges to the solution u^* for any $u_0 \in X$;
(c) *for every $n \in \mathbb{N}$, the following error estimate*

$$||u_n - u^*||_\tau \leq \frac{q^n}{1-q} ||Fu_0 - u_0||_\tau \tag{7}$$

holds, where $q := \dfrac{L\Gamma(\alpha)}{\tau^\alpha}$ is the contraction constant.

Proof of Theorem 2. Let $t \in [0, T]$ be fixed. By Equation (5), we have

$$
\begin{aligned}
|Fu(t) - Fv(t)| &\leq \int_0^t |a(t,s,u(s)) - a(t,s,v(s))|(t-s)^{\alpha-1}\, ds \\
&\leq L \int_0^t |u(s) - v(s)|(t-s)^{\alpha-1}\, ds \\
&= L \int_0^t |u(s) - v(s)| e^{-\tau s} e^{\tau s}(t-s)^{\alpha-1}\, ds \\
&\leq L ||u - v||_\tau \int_0^t e^{\tau s}(t-s)^{\alpha-1}\, ds \\
&= L ||u - v||_\tau e^{\tau t} \int_0^{\tau t} e^{-y} \left(\frac{1}{\tau}y\right)^{\alpha-1} \frac{1}{\tau}\, dy \\
&\leq \frac{L\Gamma(\alpha)}{\tau^\alpha} ||u - v||_\tau e^{\tau t},
\end{aligned}
$$

where the change of variables $y = \tau(t-s)$, $0 \leq y \leq \tau t$ was used and $\Gamma(\alpha) = \int_0^\infty e^{-x} x^{\alpha-1}\, dx$ denotes Euler's Gamma function. Then

$$|Fu(t) - Fv(t)| e^{-\tau t} \leq \frac{L\Gamma(\alpha)}{\tau^\alpha} ||u - v||_\tau,$$

for every $t \in [0, T]$ and, so,

$$||Fu - Fv||_\tau \leq \frac{L\Gamma(\alpha)}{\tau^\alpha} ||u - v||_\tau.$$

We can choose $\tau > 0$ such that $q := \frac{L\Gamma(\alpha)}{\tau^\alpha} < 1$, so F is a q-contraction. The conclusions now follow from Theorem 1. □

Next, we address the question of smoothness of the solution of Equation (1). The following result holds:

Theorem 3. *Let the conditions of Theorem 2 hold. If, in addition, $f \in C^{2,1-\alpha}(0,T]$ and $a \in C^2([0,T] \times [0,T] \times \mathbb{R})$, then $u^* \in C^{2,1-\alpha}(0,T]$, also.*

Remark 3. *For the proof, see e.g., [19] (with $i = j = k = 0$ and $\nu = 1 - \alpha$).*

The Lipschitz condition in Theorem 2 can be very prohibitive if required on the entire space. To be able to use it on a wider range of applications, we restrict it to a closed subset. Let $||\cdot||$ denote the Chebyshev norm on $C[0,T]$ (which is equivalent to the Bielecki norm) and consider the closed ball $B_R := \{u \in C[0,T] \mid ||u - f|| \leq R\}$, for some $R > 0$. Then $B_R \subseteq X$ and we have the following result.

Theorem 4. *Let us suppose that there exists a constant $L > 0$ such that*

$$|a(t,s,u) - a(t,s,v)| \leq L|u - v|, \tag{8}$$

for all $t, s \in [0, T]$ and all $u, v \in [R_1 - R, R_2 + R]$, where $R_1 := \min_{t \in [0,T]} f(t)$, $R_2 := \max_{t \in [0,T]} f(t)$. Further assume that

$$\frac{MT^\alpha}{\alpha} \leq R, \qquad (9)$$

where $M := \max |a(t, s, u)|$ over all $t, s \in [0, T]$ and all $u, v \in [R_1 - R, R_2 + R]$. Then the conclusions of Theorem 2 hold on B_R.

Proof of Theorem 4. By Remark 1, all we need to show is that $F(B_R) \subseteq B_R$. Let $u \in B_R$. Then, for all $t \in [0, T]$,

$$R_1 - R \leq u(t) \leq R_2 + R,$$

so, for $u \in B_R$, conditions in Equations (8) and (9) hold.

Fix $t \in [0, T]$. We have

$$
\begin{aligned}
|Fu(t) - f(t)| &\leq \int_0^t |a(t, s, u(s))| (t-s)^{\alpha-1} \, ds \\
&\leq M \int_0^t (t-s)^{\alpha-1} \, ds \\
&\leq \frac{MT^\alpha}{\alpha} \leq R.
\end{aligned}
$$

Thus, $\|Fu - f\| \leq R$ and $F(B_R) \subseteq B_R$. □

3. Numerical Method

We have now established that under the conditions of Theorem 4 a unique solution of Equation (1) exists and that it can be obtained as the limit of the sequence of successive approximations given in Equation (6). Still, the integrals involved in the iteration process cannot be computed exactly, so they have to be approximated numerically. We now proceed to approximate the values of the solution $u^*(t)$ at a given set of nodes $0 = t_0 < \ldots < t_m = T$. That means the singular integrals in Equation (6) have to be approximated numerically at the nodes.

3.1. Product Integration

For the numerical solution, we use product integration (see [24]). The idea is to approximate the integral

$$I(\varphi) = \int_a^b \varphi(s) w(s) \, ds,$$

for φ a smooth function and a singular weight function w, using a sequence of functions φ_m such that $\|\varphi - \varphi_m\| \to 0$ as $m \to \infty$ and the integrals

$$I_m(\varphi) = \int_a^b \varphi_m(s) w(s) \, ds$$

can be easily computed. Then

$$|I(\varphi) - I_m(\varphi)| \leq \|\varphi - \varphi_m\| \int_a^b |w(s)| \, ds, \qquad (10)$$

so $I_m(\varphi) \to I(\varphi)$ as $m \to \infty$, at least as fast as $\varphi_m \to \varphi$. Hence, for a set of nodes $a = s_0 < \ldots < s_m = b$, we use the approximation formula

$$I(\varphi) = \sum_{k=1}^{m} \int_{s_{k-1}}^{s_k} \varphi(s)w(s)\,ds$$

$$\approx \sum_{k=1}^{m} \int_{s_{k-1}}^{s_k} \varphi_k(s)w(s)\,ds = \sum_{k=0}^{m} w_k\,\varphi(s_k), \qquad (11)$$

with the error given in Equation (10).

One of the easiest (in terms of keeping the algebra simple) product integration methods is the so-called *product trapezoidal rule*. The name comes from the fact that the idea is the same as the one used to produce the trapezoidal rule, i.e., start with piecewise linear interpolation of the function φ, in order to obtain the sequence φ_m.

Next, we derive the formulas for approximating

$$I(\varphi) = \int_0^b \varphi(s)(b-s)^{\alpha-1}\,ds,$$

for $\varphi \in C^2[0,b]$ and $w(s) = (b-s)^{\alpha-1}, 0 < \alpha < 1$. Let $s_k = kh = k\dfrac{b}{m}$, for $k = 0, \ldots, m$. Let

$$\varphi_m(s) = \frac{1}{h}\left[(s_j - s)\varphi(s_{j-1}) + (s - s_{j-1})\varphi(s_j)\right], \text{ for } s \in [s_{j-1}, s_j],\ j = 1, \ldots, m.$$

Then

$$\|\varphi - \varphi_m\| \leq \frac{h^2}{8}\|\varphi''\|_\infty \text{ and}$$

$$|I(\varphi) - I_m(\varphi)| \leq \frac{h^2}{8}\frac{b^\alpha}{\alpha}\|\varphi''\|_\infty = \frac{b^2}{8m^2}\frac{b^\alpha}{\alpha}\|\varphi''\|_\infty. \qquad (12)$$

Now,

$$I(\varphi) \approx \sum_{k=0}^{m} w_k\,\varphi(s_k), \qquad (13)$$

where

$$w_0 = \frac{1}{h}\int_{s_0}^{s_1}(s_1 - s)w(s)\,ds, \qquad w_m = \frac{1}{h}\int_{s_{m-1}}^{s_m}(s - s_{m-1})w(s)\,ds,$$

$$w_j = \frac{1}{h}\left[\int_{s_{j-1}}^{s_j}(s - s_{j-1})w(s)\,ds + \int_{s_j}^{s_{j+1}}(s_{j+1} - s)w(s)\,ds\right],\ j = 1, \ldots, m-1.$$

To simplify the computations, we make the substitution $s - s_{j-1} = hy, 0 \leq y \leq 1$. We get

$$w_0 = h \int_0^1 (1-y) w(s_0 + hy) \, dy = h \int_0^1 (1-y)(b-hy)^{\alpha-1} dy$$

$$w_m = h \int_0^1 y w(s_{m-1} + hy) \, dy = h \int_0^1 y (b - h(m-1+y))^{\alpha-1} dy$$

$$w_j = h \int_0^1 y w(s_{j-1} + hy) \, dy + h \int_0^1 (1-y) w(s_j + hy) \, dy$$

$$= h \int_0^1 (1-y)(b - h(j-1+y))^{\alpha-1} dy + h \int_0^1 y (b - h(j+y))^{\alpha-1} dy.$$

Let

$$\psi_1(i) = \int_0^1 y \bigl(b - h(i+y) \bigr)^{\alpha-1} dy,$$

$$\psi_2(i) = \int_0^1 (1-y) \bigl(b - h(i+y) \bigr)^{\alpha-1} dy, \quad i = 0, 1, \ldots, m-1. \tag{14}$$

Then the coefficients in Equation (13) can be written as

$$\begin{aligned} w_0 &= h\psi_2(0), \quad w_m = h\psi_1(m-1), \\ w_j &= h\psi_1(j-1) + h\psi_2(j), \quad j = 1, \ldots, m-1. \end{aligned} \tag{15}$$

Next, we apply these formulas to the integrals in Equation (6), i.e., to

$$Fu_n(t_k) = \int_0^{t_k} a(t_k, s, u_n(s))(t_k - s)^{\alpha-1} ds,$$

for $h = T/m$ and $t_k = kh$, $k = 0, 1, \ldots, m$. For a fixed $k \in \{0, \ldots, m\}$, let $w^{(k)}(s) = (t_k - s)^{\alpha-1}$ denote the weight function. On each interval $[0, t_k]$, we use the nodes $\{t_0, \ldots, t_k\}$. Please note that on each subinterval $[0, t_k]$, we still have the same step size $\frac{t_k}{k} = \frac{kh}{k} = h$. We now have

$$\int_0^{t_k} a(t_k, s, u_n(s))(t_k - s)^{\alpha-1} ds = \int_0^{t_k} a(t_k, s, u_n(s)) w^{(k)}(s) \, ds$$

$$= \sum_{j=0}^k w_{j,k} a(t_k, t_j, u_n(t_j)) + R_{n,k}. \tag{16}$$

In analogy to Equation (14), for $i = 0, 1, \ldots$, let

$$\psi_{1,k}(i) = \int_0^1 y \bigl(t_k - h(i+y) \bigr)^{\alpha-1} dy = h^{\alpha-1} \int_0^1 y (k-i-y)^{\alpha-1} dy, \tag{17}$$

$$\psi_{2,k}(i) = \int_0^1 (1-y) \bigl(t_k - h(i+y) \bigr)^{\alpha-1} dy = h^{\alpha-1} \int_0^1 (1-y)(k-i-y)^{\alpha-1} dy.$$

Now, the coefficients in Equation (16) can be expressed as

$$\begin{aligned} w_{0,k} &= h\psi_{2,k}(0), \quad w_{k,k} = h\psi_{1,k}(k-1), \\ w_{j,k} &= h\psi_{1,k}(j-1) + h\psi_{2,k}(j), \quad j=1,\ldots,k-1. \end{aligned} \qquad (18)$$

Remark 4. *It is worth mentioning that by Equation (17), the functions $\psi_{1,k}$ and $\psi_{2,k}$ can be computed once, for $k = 0, \ldots, m$ and then be used in Equation (18) to find the coefficients $w_{j,k}$ at every step, they do not have to be computed at each iteration n. This makes the implementation of the method very efficient.*

By Equation (12), the error bound satisfies

$$|R_{n,k}| \leq \frac{h^2}{8} \frac{T^\alpha}{\alpha} \|a(t,s,u_n(s))''_s\| \qquad (19)$$

Let us notice that this bound does not depend on k, thus, we will simply write R_n, not $R_{n,k}$. Also, let us note the following thing that will be useful in the next subsection: for a fixed $k \in \{0, \ldots, m\}$, we have

$$\begin{aligned} \sum_{j=0}^{k} w_{j,k} &= h \sum_{j=0}^{k-1} \left(\psi_{1,k}(j) + \psi_{2,k}(j)\right) \\ &= \frac{h^\alpha}{\alpha} \sum_{j=0}^{k} \left[(k-j)^\alpha - (k-j-1)^\alpha\right] \\ &= \frac{(hk)^\alpha}{\alpha} \leq \frac{T^\alpha}{\alpha}. \end{aligned} \qquad (20)$$

3.2. Convergence and Error Analysis

Assuming the conditions of Theorems 3 and 4 hold, one can choose $u_0 \in B_R \cap C^{2,1-\alpha}(0,T]$, such that $u_n \in B_R \cap C^{2,1-\alpha}(0,T]$. To analyze the convergence and give an error estimate, we make the following notations. Let

$$\begin{aligned} M_a &= \max_{r \leq 2} \left| \frac{\partial^r a(t,s,u)}{\partial t^{r_1} \partial s^{r_2} \partial u^{r_3}} \right|, \quad r = r_1 + r_2 + r_3, \\ M_f &= \max\{\|f\|, \|f'\|, \|f''\|\}, \end{aligned}$$

over all $t \in [0,T], s \in [0,t)$ and $u \in [R_1 - R, R_2 + R]$. If $f \in C^{2,1-\alpha}(0,T]$ and $a \in C^2([0,T] \times [0,T] \times [R_1 - R, R_2 + R])$, one can find a constant $M_0 > 0$ such that the remainder in Equation (19) satisfies

$$|R_n| \leq \frac{h^2}{8} \frac{T^\alpha}{\alpha} M_0, \quad n = 0, 1, \ldots \qquad (21)$$

The constant M_0 may depend on M_a, M_f or τ, but not on m, k or n.
To simplify the writing, we make the following notation. Let

$$\gamma = \frac{LT^\alpha}{\alpha}.$$

Next we define our numerical method using Equation (5) iteratively, with initial point $u_0 \equiv f$. For every $k = \overline{0,m}$, we have

$$u_0(t_k) = f(t_k), \tag{22}$$
$$u_{n+1}(t_k) = \int_0^{t_k} a(t_k,s,u_n(s))(t_k-s)^{\alpha-1}ds + f(t_k), \quad n=0,1,\ldots$$

We will approximate $u_n(t_k)$ by $\tilde{u}_n(t_k)$, obtained by applying Equation (16) to the integrals above:

$$\begin{aligned}
u_1(t_k) &= \int_0^{t_k} a(t_k,s,f(s))(t_k-s)^{\alpha-1}ds + f(t_k) \\
&= \sum_{j=0}^{k} w_{j,k} a(t_k,t_j,f(t_j)) + R_1 + f(t_k) \\
&= \tilde{u}_1(t_k) + \tilde{R}_1,
\end{aligned} \tag{23}$$

with

$$\tilde{u}_1(t_k) = \sum_{j=0}^{k} w_{j,k} a(t_k,t_j,f(t_j)) + f(t_k).$$

Denote the error at the nodes by

$$e(u_n,\tilde{u}_n) := \max_{t_k \in [0,T]} |u_n(t_k) - \tilde{u}_n(t_k)|.$$

By Equation (21), we have

$$e(u_1,\tilde{u}_1) = |\tilde{R}_1| = |R_1| \leq \frac{h^2}{8} \frac{T^\alpha}{\alpha} M_0. \tag{24}$$

Similarly, we get

$$\begin{aligned}
u_2(t_k) &= \int_0^{t_k} a(t_k,s,u_1(s))(t_k-s)^{\alpha-1}ds + f(t_k) \\
&= \sum_{j=0}^{k} w_{j,k} a(t_k,t_j,u_1(t_j)) + R_2 + f(t_k) \\
&= \sum_{j=0}^{k} w_{j,k} a(t_k,t_j,\tilde{u}_1(t_j) + \tilde{R}_1) + R_2 + f(t_k) \\
&= \tilde{u}_2(t_k) + \tilde{R}_2,
\end{aligned}$$

where

$$\tilde{u}_2(t_k) = \sum_{j=0}^{k} w_{j,k} a(t_k,t_j,\tilde{u}_1(t_j)) + f(t_k).$$

We have, by Equations (21) and (24),

$$
\begin{aligned}
e(u_2, \tilde{u}_2) = |\tilde{R}_2| &\leq L|\tilde{R}_1| \sum_{j=0}^{k} w_{j,k} + |R_2| \\
&\leq L|\tilde{R}_1| \frac{T^\alpha}{\alpha} + |R_2| \\
&\leq L \frac{h^2}{8} \frac{T^\alpha}{\alpha} M_0 \frac{T^\alpha}{\alpha} + \frac{h^2}{8} \frac{T^\alpha}{\alpha} M_0 \\
&= \frac{h^2}{8} \frac{T^\alpha}{\alpha} M_0 (1 + \gamma).
\end{aligned}
\tag{25}
$$

In a similar fashion, we get

$$
\begin{aligned}
u_n(t_k) &= \int_0^{t_k} a(t_k, s, u_{n-1}(s)) (t_k - s)^{\alpha-1} ds + f(t_k) \\
&= \sum_{j=0}^{k} w_{j,k} a(t_k, t_j, u_{n-1}(t_j)) + R_n + f(t_k) \\
&= \sum_{j=0}^{k} w_{j,k} a(t_k, t_j, \tilde{u}_{n-1}(t_j) + \tilde{R}_{n-1}) + R_n + f(t_k) \\
&= \tilde{u}_n(t_k) + \tilde{R}_n,
\end{aligned}
$$

with

$$
\tilde{u}_n(t_k) = \sum_{j=0}^{k} w_{j,k} a(t_k, t_j, \tilde{u}_{n-1}(t_j)) + f(t_k).
\tag{26}
$$

The values $\tilde{u}_n(t_k)$ can always be computed from the values at the previous step and, for the error, by induction, we have

$$
\begin{aligned}
e(u_n, \tilde{u}_n) = |\tilde{R}_n| &\leq L|\tilde{R}_{n-1}| \sum_{j=0}^{k} w_{j,k} + |R_n| \\
&\leq L|\tilde{R}_{n-1}| \frac{T^\alpha}{\alpha} + |R_n| \\
&\leq \frac{h^2}{8} \frac{T^\alpha}{\alpha} M_0 \gamma \left(1 + \cdots + \gamma^{n-1}\right) + \frac{h^2}{8} \frac{T^\alpha}{\alpha} M_0 \\
&= \frac{h^2}{8} \frac{T^\alpha}{\alpha} M_0 (1 + \cdots + \gamma^n).
\end{aligned}
\tag{27}
$$

Now we can give the following error estimate for our numerical method.

Theorem 5. *Assume the conditions of Theorem 4 hold with $f \in C^{2,1-\alpha}(0, T]$ and $a \in C^2([0, T] \times [0, T] \times [R_1 - R, R_2 + R])$. Furthermore, assume that*

$$
\gamma := \frac{LT^\alpha}{\alpha} < 1.
\tag{28}
$$

Then the following error estimate holds

$$
e(u^*, \tilde{u}_n) \leq \frac{q^n}{1-q} \|Fu_0 - u_0\|_\tau + \frac{T^2}{8m^2} \frac{T^\alpha}{\alpha} M_0 \frac{1}{1-\gamma},
\tag{29}
$$

for all $n = 1, 2, \ldots$ and any $m \in \mathbb{N}^*$, where u^* is the true solution of Equation (4) and $\tilde{u}_n$ are the approximations given by Equation (26).

Proof of Theorem 5. By Equations (27) and (28),

$$e(u_n, \tilde{u}_n) \leq \frac{T^2}{8m^2} \frac{T^\alpha}{\alpha} M_0 \frac{1}{1-\gamma}.$$

Then, by Theorem 4,

$$\begin{aligned} e(u^*, \tilde{u}_n) &\leq e(u^*, u_n) + e(u_n, \tilde{u}_n) \\ &\leq \frac{q^n}{1-q} \|Fu_0 - u_0\|_\tau + \frac{T^2}{8m^2} \frac{T^\alpha}{\alpha} M_0 \frac{1}{1-\gamma}, \end{aligned}$$

where q is the contraction constant given in Theorem 2. □

4. Numerical Experiments

In this section, we give numerical examples of nonlinear weakly singular integral equations, to show the applicability of the method proposed.

Example 1. *Consider the integral equation*

$$u(t) = \frac{1}{12} \int_0^t u^2(s)(t-s)^{-1/2}\, ds + t^{1/2}\left(1 - \frac{1}{9}t\right), \ t \in [0,1],$$

with exact solution $u^*(t) = \sqrt{t}$.

We have $\alpha = 1/2$, $a(t, s, u) = \frac{1}{12}u^2$ and $f(t) = t^{1/2}\left(1 - \frac{1}{9}t\right)$. Let us check that all our theoretical assumptions are met.

For the function f, $R_1 = 0$, $R_2 = 8/9$ and we choose $R = 1$. Then $M = \frac{1}{12}\left(\frac{17}{9}\right)^2 \approx 0.2973$ and for all $u \in [R_1 - R, R_2 + R]$,

$$\frac{MT^\alpha}{\alpha} \approx 0.5947 \leq R.$$

We have $\frac{\partial a}{\partial u} = \frac{1}{6}u$, so taking $L = \max\left|\frac{\partial a}{\partial u}\right|$, over $u \in [R_1 - R, R_2 + R]$, $L \approx 0.3148$ and

$$\gamma \approx 0.6296 < 1.$$

Also, choosing $\tau = 1$, we have

$$\frac{L\Gamma(\alpha)}{\tau^\alpha} \approx 0.5580 < 1.$$

Thus, all conditions of Theorem 5 are satisfied.

We apply the product trapezoidal rule for the values $m = 12$ and $m = 24$, with the corresponding nodes $t_k = \frac{1}{m}k$, $k = \overline{0, m}$. In Table 1 we give the errors $e(u^*, \tilde{u}_n)$, with initial approximation $u_0(t) = f(t)$. Figure 1 displays the graphs of the true solution $u^*(t)$ and of the approximate solution $\tilde{u}_n$, for $n = 10$ iterations and $m = 24$ nodes, for the values $t \in [0, 1]$. As both the errors in Table 1 and the graphs in Figure 1 show, there is very good agreement between the true values and the approximate ones of the solution at the nodes $t_0, \ldots, t_m$.

Table 1. Errors $e(u^*, \tilde{u}_n)$ for Example 1.

n	m	
	12	24
1	1.084348×10^{-1}	1.882162×10^{-2}
5	2.799553×10^{-4}	5.567188×10^{-6}
10	6.813960×10^{-7}	4.690204×10^{-9}

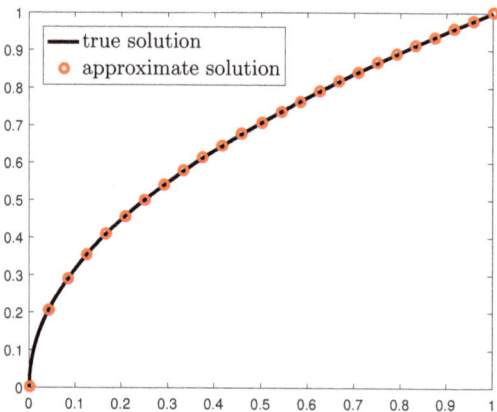

Figure 1. Example 1, $n = 10$, $m = 24$.

Example 2. *Next, consider the equation*

$$u(t) = \frac{1}{18} \int_0^t \left(\sin^2 s + u^2(s) \right) (t-s)^{-2/3} \, ds + \cos t - \frac{1}{6} t^{1/3}, \ t \in \left[0, \frac{\pi}{4}\right],$$

whose exact solution is $u^*(t) = \cos t$.

Now $\alpha = 1/3$, $a(t, s, u) = 1/18 \left(\sin^2 s + u^2 \right)$ and $f(t) = \cos t - t^{1/3}/6$. We check the applicability of the method, by verifying all the theoretical assumptions. Here, $R_1 = f(\pi/4) \approx 0.5533$, $R_2 = 1$ and taking $R = 1/2$, we have, for $u \in [R_1 - R, R_2 + R]$, $M = 11/72 \approx 0.1528$ and

$$\frac{MT^\alpha}{\alpha} \approx 0.4229 \leq R.$$

Again, taking $L = \max \left| \frac{\partial a}{\partial u} \right|$, over $u \in [R_1 - R, R_2 + R]$, we get $L = 1/6$ and

$$\gamma \approx 0.4613 < 1.$$

For $\tau = 5$, we have

$$\frac{L\Gamma(\alpha)}{\tau^\alpha} \approx 0.7227 < 1.$$

So all conditions in Theorem 5 are verified.

Table 2 contains the errors $e(u^*, \tilde{u}_n)$, with initial approximation $u_0(t) = f(t)$, for the values $m = 12$ and $m = 24$, with nodes $t_k = \frac{\pi}{4m} k, k = \overline{0, m}$. Figure 2 shows the graphs of the true solution $u^*(t)$ and of the approximate solution $\tilde{u}_n$, for $n = 10$ iterations and $m = 24$ nodes, for $t \in [0, \pi/4]$.

Table 2. Errors $e(u^*, \tilde{u}_n)$ for Example 2.

n	m	
	12	24
1	1.002977×10^{-1}	3.014020×10^{-2}
5	2.315358×10^{-4}	4.412851×10^{-5}
10	9.363611×10^{-7}	5.525447×10^{-9}

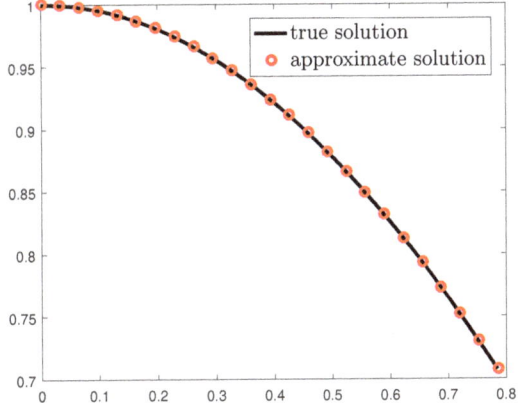

Figure 2. Example 2, $n = 10$, $m = 24$.

As seen in the examples above, the proposed method produces approximations that are in very good agreement with the exact values of the solution, thus confirming the theoretical results and error estimates given in the previous section.

5. Conclusions

In this paper, we presented a numerical iterative method for approximating solutions of nonlinear Volterra integral equations of the second kind, with weakly singular kernels. We used Banach's fixed point theorem to establish the existence and uniqueness of the solution and to find a sequence converging to it (Picard iteration). Then we employed the product trapezoidal rule to approximate each iteration at a given set of nodes in the domain. The present method is fairly simple to use, its convergence is based on a classical fixed point result. It is also quite efficient and inexpensive in (the cost of) implementation, most of the computations can be done only once, not at each iteration (see Remark 4). Thus, when only values of the solution at some points are needed (as is the case in many applications), this method is more practical and accessible than other classical methods.

Yet, the method converges quite fast, with order $O(q^n)$ with respect to the number of successive approximations and order $O\left(\frac{1}{m^2}\right)$ with respect to the number of nodes. As the examples show, it gives good approximations even with a relatively small number of iterations and of quadrature nodes.

In future works, other types of singularity of the kernel can be explored for Volterra or Fredholm integral equations. Also, more complicated kernels can be considered, such as kernels containing modified (or delayed) arguments, or other special types of kernels. Various other iteration techniques for fixed point successive approximations can be employed, such as Mann iteration, Krasnoselskii iteration, and others.

Funding: This research received no external funding.

Conflicts of Interest: The author declares no conflict of interest.

References

1. Agarwal, R.P.; O'Regan, D. Singular Volterra integral equations. *Appl. Math. Lett.* **2000**, *13*, 115–120. [CrossRef]
2. Gorenflo, R.; Vessella, S. *Abel Integral Equations: Analysis and Applications, Lecture Notes in Mathematics (1461)*; Springer: Berlin, Germany, 1991.
3. Wang, J.; Zhu, C.; Fečkan, M. Analysis of Abel-type nonlinear integral equations with weakly singular kernels. *Bound. Value Probl.* **2014**, *2014*, 20. [CrossRef]
4. Becker, L.C. Properties of the resolvent of a linear Abel integral equation: Implications for a complementary fractional equation. *Electron. J. Qual. Theory* **2016**, *64*, 1–38. [CrossRef]
5. Aghili, A.; Zeinali, H. Solution to Volterra singular integral equations and non homogenous time. *Gen. Math. Notes* **2013**, *14*, 6–20.
6. Wu, G.C.; Baleanu, D. Variational iteration method for fractional calculus—A universal approach by Laplace transform. *Adv. Differ. Equ.* **2013**, *2013*, 18. [CrossRef]
7. Andras, S. Weakly singular Volterra and Fredholm-Volterra integral equations. *Stud. Univ. Babeş-Bolyai Math.* **2003**, *48*, 147–155.
8. Bertram, B.; Ruehr, O. Product integration for finite-part singular integral equations: Numerical asymptotics and convergence acceleration. *J. Comput. Anal. Appl.* **1992**, *41*, 163–173. [CrossRef]
9. Brunner, H. The numerical solution of weakly singular Volterra integral equations by collocation on graded meshes. *Math. Comput.* **1985**, *45*, 417–437. [CrossRef]
10. Diogo, T. Collocation and iterated collocation methods for a class of weakly singular Volterra integral equations. *J. Comput. Appl. Math.* **2009**, *229*, 363–372. [CrossRef]
11. Assari, P. Solving weakly singular integral equations utilizing the meshless local discrete collocation technique. *Alexandria Eng. J.* **2018**, *57*, 2497–2507. [CrossRef]
12. Rehman, S.; Pedas, A.; Vainikko, G. Fast solvers of weakly singular integral equations of the second kind. *Math. Mod. Anal.* **2018**, *23*, 639–664. [CrossRef]
13. Kumar, S.; Kumar, A.; Kumar, D.; Singh, J.; Singh, A. Analytical solution of Abel integral equation arising in astrophysics via Laplace transform. *J. Egypt. Math. Soc.* **2015**, *23*, 102–107. [CrossRef]
14. Mokharty, P.; Ghoreishi, F. Convergence analysis of the operational Tau method for Abel-type Volterra integral equations. *Electron. Trans. Numer. Anal.* **2014**, *41*, 289–305.
15. Diogo, T.; Ford, N.J.; Lima, P.; Valtchev, S. Numerical methods for a Volterra integral equation with non-smooth solutions. *J. Comput. Appl. Math.* **2006**, *189*, 412–423. [CrossRef]
16. Ali, M.R.; Mousa, M.M.; Ma, W.-X. Solution of nonlinear Volterra integral equations with weakly singular kernel by using the HOBW method. *Adv. Math. Phys.* **2019**. [CrossRef]
17. Nadir, M.; Gagui, B. Quadratic numerical treatment for singular integral equations with logarithmic kernel. *Int. J. Comput. Sci. Math.* **2019**, *10*, 288–296.
18. Alvandi, A.; Paripour, M. Reproducing kernel method for a class of weakly singular Fredholm integral equations. *J. Taibah Univ. Sci.* **2018**, *12*, 409–414. [CrossRef]
19. Brunner, H.; Pedas, A.; Vainikko, G. The piecewise polynomial collocation method for nonlinear weakly singular Volterra equations. *Math. Comput.* **1999**, *68*, 1079–1095. [CrossRef]
20. Darwish, M.A. On monotonic solutions of an integral equation of Abel type. *Math. Bohem.* **2008**, *133*, 407–420. [CrossRef]
21. Sidorov, N.A.; Sidorov, D.N.; Krasnik, A.V. Solution of Volterra operator-integral equations in the nonregular case by the successive approximation method. *Diff. Equ.* **2010**, *46*, 882–891. [CrossRef]
22. Sidorov, D.N.; Sidorov, N.A. Convex majorants method in the theory of nonlinear Volterra equations. *Banach J. Math. Anal.* **2012**, *6*, 1–10. [CrossRef]

23. Noeiaghdam, S.; Dreglea, A.; He, J.; Avazzadeh, Z.; Suleman, M.; Fariborzi Araghi, M.A.; Sidorov, D.N.; Sidorov, N. Error Estimation of the Homotopy Perturbation Method to Solve Second Kind Volterra Integral Equations with Piecewise Smooth Kernels: Application of the CADNA Library. *Symmetry* **2020**, *12*, 1730. [CrossRef]
24. Atkinson, K.E. *An Introduction to Numerical Analysis*, 2nd ed.; John Wiley & Sons: New York, NY, USA, 1989.

Publisher's Note: MDPI stays neutral with regard to jurisdictional claims in published maps and institutional affiliations.

© 2020 by the author. Licensee MDPI, Basel, Switzerland. This article is an open access article distributed under the terms and conditions of the Creative Commons Attribution (CC BY) license (http://creativecommons.org/licenses/by/4.0/).

Article

Matrix Expression of Convolution and Its Generalized Continuous Form

Young Hee Geum [1,†], Arjun Kumar Rathie [2,†] and Hwajoon Kim [3,*,†]

[1] Department of Applied Mathematics, Dankook University, Cheonan 31116, Korea; conpana@empal.com
[2] Department of Mathematics, Vedant College of Engineering & Technology (Rajasthan Technical University), Kota 324010, India; arjunkumarrathie@gmail.com
[3] Department of IT Engineering, Kyungdong University, Yangju 11458, Korea
* Correspondence: math@kduniv.ac.kr or cellmath@gmail.com
† These authors contributed equally to this work.

Received: 30 September 2020; Accepted: 23 October 2020; Published: 29 October 2020

Abstract: In this paper, we consider the matrix expression of convolution, and its generalized continuous form. The matrix expression of convolution is effectively applied in convolutional neural networks, and in this study, we correlate the concept of convolution in mathematics to that in convolutional neural network. Of course, convolution is a main process of deep learning, the learning method of deep neural networks, as a core technology. In addition to this, the generalized continuous form of convolution has been expressed as a new variant of Laplace-type transform that, encompasses almost all existing integral transforms. Finally, we would, in this paper, like to describe the theoretical contents as detailed as possible so that the paper may be self-contained.

Keywords: matrix expression of convolution; Laplace-type transforms; convolution neural network; kernel

1. Introduction

Deep learning means the learning of deep neural networks, called deep and if multiple hidden layers exist. Deep learning allows computational models that are composed of multiple processing layers to learn representations of data with multiple levels of abstraction [1]. The convolution in convolutional deep neural network (CNN) is the tool for obtaining a feature map from the original image data, it sweeps the original image with a kernel matrix, and transforms the original data into a different shape. This distorted image is called a feature map. Therefore, in CNN, the convolution can be regarded as a tool that creates a feature map from the original image. Herein, the concept of convolution in artificial intelligence is demonstrated mathematically.

The core concept of CNN is the convolution which applies the weight to the receptive fields only, and it transforms the original data into a feature map. This process is called convolution. This is a similar principle to integral transform. The method of Integral transform maps from the original domain to another domain to solve a given problem more easily. Since the matrix expression of convolution is an essential concept in artificial intelligence, we believe that this study would certainly be meaningful. In addition to this, the generalized continuous form of convolution has also been studied, and thus this form is expressed as a new variant of Laplace-type transform.

On one hand, the transform theory is extensively utilized in fields involving medical diagnostic equipment, such as magnetic resonance imaging or computed tomography. Typically, a projection data are obtained by an integral transform, and an image using an inverse transform is produced. Although plausible integral transforms exist, almost all existing integral transforms are not sufficiently satisfied with fullness, and can be interpreted as a Laplace-type transform. One of us proposed a comprehensive form of the Laplace-type integral transform in [2]. The present study is being conducted to investigate the matrix expression of convolution and its generalized continuous form.

In [2], a Laplace-type integral transform was proposed, expressed as

$$G_\alpha(f) = G(f) = u^\alpha \int_0^\infty e^{-\frac{t}{u}} f(t) dt. \tag{1}$$

For values of α as $0, -1, 1$, and -2, we have, respectively, the Laplace [3], Sumudu [4], Elzaki [5], and Mohand transforms [6]. This form can be expressed in various manners. Replacing t by ut, we have

$$G(f) = u^\beta \int_0^\infty e^{-t} f(ut) dt,$$

where $\beta = \alpha + 1$. In the form, β values of $1, 0, 2$, and -1 correspond to the Laplace, Sumudu, Elzaki, and Mohand transforms, respectively. If we substitute $u = 1/s$ in (1), we then obtain the simplest form of the generalized integral transform as follows:

$$G(f) = s^\gamma \int_0^\infty e^{-st} f(t) dt,$$

where $\gamma = -\alpha$. In this form, the Laplace, Sumudu, Elzaki, and Mohand transforms have γ values of $0, 1, -1$, and 2, respectively. It is somewhat paved, but essentially a simple way to derive the Sumudu transform is to multiply the Laplace transform by s. Similarly, it can be obtained multiply by s^{-1} to obtain the Elzaki transform, and multiply by s^2 to obtain the Mohand transform. The natural transform [7] can be obtained by substituting $f(t)$ with $f(ut)$. Additionally, by substituting $t = \ln x$, the Laplace-type transform $G(f)$ can be expressed as

$$s^{-\alpha} \int_1^\infty f(\ln x) x^{-s-1} dx.$$

As a similar form, there is a Mellin transform [8] of the form

$$\int_0^\infty f(x) x^{s-1} dx.$$

As shown above, many integral transforms have their own fancy masks, but most of them can essentially be interpreted as Laplace-type transforms. From a different point of view, a slight change in the kernel results in a significant difference in the integral transform theory. Meanwhile, plausible transforms exist, such as the Fourier, Radon, and Mellin transforms. Typically, if the interval of integration and the power of kernel are different, it can be interpreted as a completely different transform. Studies using Laplace transform were conducted in [9,10]. The generalized solutions of the third-order Cauchy–Euler equation in the space of right-sided distributions has found [9], studied the solution of the heat equation without boundary conditions [10], and investigated further properties of Laplace-type transform [11]. As an application, a new class of Laplace-type integrals involving generalized hypergeometric functions has been studied [12,13]. As for research related to the integral equation, Noeiaghdam et al. [14] presented a new scheme based on the stochastic arithmetic. The scheme is presented to guarantee the validity and accuracy of the homotopy analysis method. Different kinds of integral equations such as singular and first kind are considered to find the optimal results by applying the proposed algorithms.

The main objective of this study is to investigate the matrix expression of convolution and its generalized continuous form. The generalized continuous form of the matrix expression was carried out in the form of a new variant of Laplace-type transform. The obtained result are as follows:

(1) If the matrix representing the function (image) f is A and the matrix representing the function g is B, then the convolution $f * g$ is represented by *the sum of all elements of $A \circ B$* and this is the same as $tr(AB^T)$ where $\circ$ is array multiplication, T is the transpose, and tr is the trace. Thus, the convolution in artificial intelligence (AI) is the same as $tr(AB^T)$.

(2) The generalized continuous form of the convolution in AI can be represented as

$$V(f) = \Phi(u) \int_0^\infty e^{-t\Delta} f(t) \, dt,$$

where $\Phi(u)$ is an arbitrary bounded function and

$$\Delta = \Delta(\delta, u) = \frac{\ln[1 + \frac{\delta - 1}{u}]}{\delta - 1}.$$

2. Matrix Expression of Convolution in Convolutional Neural Network (CNN)

Note that functions can be interpreted as images in artificial intelligence (AI). The convolution is changed from $\int_0^t f(\tau)g(t-\tau)d\tau$ to

$$\sum_{\tau=0}^{t} f(\tau)g(t-\tau)$$

by the discretization. The convolution in CNN is the tool for obtaining a feature map from the original image data, plays a role to sweeping the original image with kernel matrices (or filter), and it transforms original data into a different shape. In order to calculate the convolution, each $n \times n$ part of the original matrix is element-wise multiplied by the kernel matrix and all its components are added. Typically, the kernel matrix is using by 3×3 matrix. On the one hand, the pooling (or sub-sampling) is a simple job, reducing the size of the image made by convolution. It is the principle that the resolution is increased when the screen is reduced.

Let the matrix representing the function f is A and the matrix representing the function g is B. For two matrices A and B of the same dimension, the array multiplication (or sweeping) $A \circ B$ is given by

$$(A \circ B)_{ij} = (A)_{ij}(B)_{ij}.$$

For example, the array multiplication for 2×2 matrices is

$$\begin{pmatrix} a & b \\ c & d \end{pmatrix} \circ \begin{pmatrix} e & f \\ g & h \end{pmatrix} = \begin{pmatrix} ae & bf \\ cg & dh \end{pmatrix}.$$

The array multiplication appears in lossy compression such as joint photographic experts group and the decoding step. Let us look at an example.

Example 1. *In the classification field of AI, the pixel is treated as a matrix. When the original image is*

$$\begin{pmatrix} 1 & 2 & 3 & 0 & 0 \\ 0 & 1 & 2 & 1 & 0 \\ 3 & 0 & 1 & 0 & 1 \\ 1 & 0 & 2 & 1 & 0 \\ 0 & 1 & 1 & 2 & 0 \end{pmatrix},$$

array multiplying the kernel matrix

$$\begin{pmatrix} 2 & 0 & 1 \\ 0 & 1 & 2 \\ 2 & 0 & 1 \end{pmatrix}$$

on the first 3 × 3 matrix, we obtain the matrix

$$\begin{pmatrix} 2 & 0 & 3 \\ 0 & 1 & 4 \\ 6 & 0 & 1 \end{pmatrix}.$$

Now, adding all of its components, we obtain 17. Next, if we array-multiply the kernel matrix to the 3 × 3 matrix

$$\begin{pmatrix} 2 & 3 & 0 \\ 1 & 2 & 1 \\ 0 & 1 & 0 \end{pmatrix}$$

on the right and add all the components, we get 8 by stride 1. If we continue this process to the final matrix

$$\begin{pmatrix} 1 & 0 & 1 \\ 2 & 1 & 0 \\ 1 & 2 & 0 \end{pmatrix},$$

we get 6. Consequently, the original matrix changes to

$$\begin{pmatrix} 17 & 8 & 10 \\ 8 & 5 & 10 \\ 12 & 8 & 6 \end{pmatrix}$$

by using the convolution kernel. This is called the convolved feature map.

This is just an example for understanding, and in perceptron the output uses a value between −1 and 1 using the activation function. Note that the perceptron is an artificial network designed to mimic the brain's cognitive abilities. Therefore, the output of neuron (or node) Y can be represented as

$$Y = sign[\sum_{i=1}^{n} x_i w_i - \Theta] = \begin{cases} 1 & X \geq \Theta \\ -1 & X < \Theta \end{cases},$$

where w is a weight, Θ is the threshold value, and X is the activation function with $X = \sum_{i=1}^{n} x_i w_i$. In the backpropagation algorithm of deep neural network, the sigmoid function

$$Y^{sigmoid} = \frac{1}{1 + e^{-x}}$$

is used as the activation function [15]. This function is easy to differentiate and ensures neuron output is in $[0, 1]$. If max-pulling is applied to the above convolved feature map, the resulting matrix becomes 1×1 matrix $(17) = 17$.

As discussed above, convolution in AI can be obtained by array multiplication. We would like to associate this definition with matrix multiplication in mathematics.

Definition 1. *(Convolution in AI) If the matrix representing the function (image) f is A and the matrix representing the function g is B, then the convolution $f * g$ is represented by the sum of all elements of $A \circ B$ and this is the same as $tr(AB^T)$ where $\circ$ is array multiplication, T is the transpose, and tr is the trace. Thus, the convolution in AI is the same as $tr(AB^T)$, the sum of all elements on the diagonal with the right side facing down in AB^T.*

Typically, the convolution kernel is used as a 3 × 3 matrix, but for easy understanding, let us consider a 2 × 2 matrix.

Example 2. If
$$A = \begin{pmatrix} a & b \\ c & d \end{pmatrix}$$

and
$$B = \begin{pmatrix} e & f \\ g & h \end{pmatrix},$$

then the convolution in AI is calculated as $ae + bf + cg + dh$ by the sweeping. On the other hand,

$$AB^T = \begin{pmatrix} a & b \\ c & d \end{pmatrix} \begin{pmatrix} e & g \\ f & h \end{pmatrix} = \begin{pmatrix} ae+bf & ag+bh \\ ce+df & cg+dh \end{pmatrix},$$

and
$$tr(AB^T) = ae + bf + cg + dh$$

for T is the transpose and tr is the trace. This is the same result as in AI.

3. Generalized Continuous Form of Matrix Expression of Convolution

If the matrix representing a function f is A and the matrix representing a function g is B, then the convolution of the functions f and g can be denoted by $tr(AB^T)$. Intuitively, the diagonal part of B^T corresponds to a graph of $g(t-\tau)$. The overlapping part of the graph can be interpreted as the concept of intersection, that is, the concept of multiplication. Thus, the generalized continuous form of the convolution in AI can be represented in a variant of Laplace-type transform given by

$$G_\alpha(f) = G(f) = u^\alpha \int_0^\infty e^{-\frac{t}{u}} f(t) dt$$

$$= u^\alpha \int_0^\infty \lim_{\delta \to 1} (1 + \frac{\delta-1}{u})^{-\frac{t}{\delta-1}} f(t) dt.$$

If $f(t)$ is a function defined for all $t \geq 0$, an integral of Laplace-type transform $vi(f)$ is given by

$$F(\Delta) = vi(f) = \int_0^\infty e^{-t\Delta(\delta,u)} f(t) \, dt$$

$$= \int_0^\infty [1 + \frac{\delta-1}{u}]^{-\frac{t}{\delta-1}} f(t) \, dt$$

for $\delta > 1$ with

$$\Delta(\delta, u) = \frac{\ln[1 + \frac{\delta-1}{u}]}{\delta - 1}.$$

Additionally, let $\Phi(u)$ be an arbitrary bounded function and let $V(f)$ be a variant of Laplace-type transform of $f(t)$. If $f(t)$ is a function defined for all $t \geq 0$, $V(f)$ is defined by

$$V(f) = \Phi(u) \int_0^\infty e^{-t\Delta} f(t) \, dt$$

$$= \Phi(u) \int_0^\infty [1 + \frac{\delta-1}{u}]^{-\frac{t}{\delta-1}} f(t) \, dt$$

for $\delta > 1$ with $\Delta = \Delta(\delta, u)$.

Based on the above two definitions, it is clear that the above variant of Laplace-type transform is represented as $V(f) = \Phi(u) \cdot vi(f)$ for an arbitrary function $\Phi(u)$. If so, let us see the relation with other integral transforms. Since

$$V(f) = \Phi(u) \int_0^\infty e^{-t\Delta} f(t)\, dt,$$

if $\delta \to 1+$ and $\Phi(u) = u^\alpha$, then it corresponds to the G_α transform. When we take $\delta \to 1$, $\Phi(u) = 1$, and $u = 1/s$, we get the Laplace transform. Similarly, when we take $\delta \to 1$, $\Phi(u) = u^{-1}$ ($\delta \to 1$, $\Phi(u) = u$), we get the Sumudu transform (Elzaki transform), respectively. In order to obtain a simple form of generalization, it is better to set $\phi(u)$ to u^α for an arbitrary integer α. However, it is judged that $\phi(u)$ is better than u^α as a suitable generalization, where $\phi(u)$ is a bounded arbitrary function. The reason is that $\phi(u)$ can express more integral transforms.

Lemma 1. *(Lebesgue dominated convergence theorem [16,17]). Let (X, M, μ) be a measure space and suppose $\{f_n\}$ is a sequence of extended real-valued measurable functions defined on X such that*
(a) $\lim_{n\to\infty} f_n(x) = f(x)$ exists μ-a.e.
(b) There is an integrable function g so that for each n, $|f_n| \leq g$ μ-a.e.
Then, f is integrable and

$$\lim_{n\to\infty} \int_X f_n d\mu = \int_X f d\mu.$$

Beppo Livi's theorem is a special form of Lemma 1. Its contents are as follows:

$$\int \sum_{n=1}^\infty g_n\, d\mu = \sum_{n=1}^\infty \int g_n\, d\mu$$

for (g_n) is a nondecreasing sequence. The details are can be found on page 71 in [16]. Note that the convolution of f and g is given by

$$(f * g)(t) = \int_0^\infty f(\tau)g(t-\tau)\, d\tau.$$

The following theorem is as follows. Since the proof is not difficult, we would like to cover just a few.

Theorem 1.

(1) *(Duality with Laplace transform) If $\mathcal{L}(f) = F_*(s)$ is the Laplace transform of a function $f(t)$, then it satisfies the relation of $V(f) = \Phi(u) \cdot F_*(\Delta)$.*

(2) *(Shifting theorem) If $f(t)$ has the transform $F(u)$, then $e^{at} f(t)$ has the transform $\Phi(u) \cdot F(\Delta - a)$. That is,*

$$V[e^{at} f(t)] = \Phi(u) \cdot F(\Delta - a).$$

Moreover, If $f(t)$ has the transform $F(u)$, then the shifted function $f(t-a)h(t-a)$ has the transform $e^{-a\Delta} \cdot \Phi(u)F(\Delta)$. In formula,

$$V[f(t-a)h(t-a)] = e^{-a\Delta} \cdot \Phi(u) F(\Delta)$$

for $h(t-a)$ is Heaviside function (We write h since we need u to denote u-space).

(3) *(Linearity) Let $V(f)$ be the variant of Laplace-type transform. Then $V(f)$ is a linear operation.*

(4) (Existence) If $f(t)$ is defined, piecewise continuous on every finite interval on the semi-axis $t \geq 0$ and satisfies
$$|f(t)| \leq Me^{kt}$$
for all $t \geq 0$ and some constants M and k, then the variant of Laplace-type transform $V(f)$ exists for all $\Delta > k$.

(5) (Uniqueness) If the variant of Laplace-type transform of a given function exists, then it is uniquely determined.

(6) (Heaviside function)
$$vi[\,h(t-a)] = \int_0^\infty e^{-t\Delta} h(t-a)\, dt = \int_a^\infty e^{-t\Delta} \cdot 1\, dt$$
$$= e^{-a\Delta}/\Delta,$$
where h is Heaviside function.

(7) (Dirac's delta function) We consider the function
$$f_k(t-a) = \begin{cases} 1/k & \text{if } a \leq t \leq a+k \\ 0 & \text{otherwise.} \end{cases}$$

In a similar way to Heaviside, taking the integral of Laplace-type transform, we get
$$vi[f_k(t-a)] = \int_0^\infty e^{-t\Delta} f_k(t-a)\, dt = -\frac{1}{k\Delta}[e^{-t\Delta}]_a^{a+k}$$
$$= -\frac{1}{k\Delta}(e^{-(a+k)\Delta} - e^{-a\Delta}) = -\frac{1}{k\Delta} e^{-a\Delta}(e^{-k\Delta} - 1).$$

If we denote the limit of f_k as $\delta(t-a)$, then
$$vi(\delta(t-a)) = \lim_{k \to 0} (vi)[f_k(t-a)] = e^{-a\Delta}.$$

(8) (Shifted data problems) For a given differential equation $y'' + ay' + by = r(t)$ subject to $y(t_0) = c_0$ and $y'(t_0) = c_1$, where $t_0 \neq 0$ and a and b are constant, we can set $t = t_1 + t_0$. Then $t = t_0$ gives $t_1 = 0$ and so, we have
$$y_1'' + ay_1' + by_1 = r(t_1 + t_0),\ y_1(0) = c_0,\ y_1'(0) = c_1$$
for input $r(t)$. Taking the variant, we can obtain the output $y(t)$.

(9) (Transforms of derivatives and integrals) Let a function f is n-th differentiable and integrable, and let us consider the fraction Δ as an operator. Then $V(f)$ of the n-th derivatives of $f(t)$ satisfies
$$vi(f^{(n)}) = \Delta^n vi(f) - \sum_{k=1}^n \Delta^{n-k} f^{(k-1)}(0) \qquad (2)$$
and
$$V[\int_0^t f(\tau)\, d\tau] = \Phi(u) \cdot \frac{1}{\Delta} V(f).$$

(10) (Convolution) If two functions f and g are integrable for $*$ is the convolution, then $V(f * g)$ satisfies
$$V(f * g) = \Phi(u) \cdot F(\Delta) G(\Delta)$$
for $V(f) = F(\Delta)$.

Proof. (5) Assume that $V(f)$ exists by $V(f_1)$ and $V(f_2)$ both. If $V(f_1) \neq V(f_2)$ for $f_1 = f_2$, then

$$\begin{aligned} V(f_1) - V(f_2) &= \Phi(u) \int_0^\infty e^{-t\Delta} f_1(t)\, dt - \Phi(u) \int_0^\infty e^{-t\Delta} f_2(t)\, dt \\ &= \Phi(u) \int_0^\infty e^{-t\Delta} (f_1(t) - f_2(t))\, dt \\ &= V(f_1 - f_2) = 0. \end{aligned}$$

This is a contradiction on $V(f_1) \neq V(f_2)$, and hence the transform is uniquely determined. Conversely, if two functions f_1 and f_2 have the same transform (i.e., if $V(f_1) = V(f_2)$), then

$$V(f_1) - V(f_2) = \Phi(u) \int_0^\infty e^{-t\Delta}(f_1(t) - f_2(t))\, dt = 0,$$

and so $f_1 = f_2$ a.e. Hence $f_1 = f_2$ excepting for the set of measure zero.

(9) Note that $vi(f) = \int_0^\infty e^{-t\Delta} f(t)\, dt$, and let us approach the proof by induction. In case of $n = 1$,

$$vi(f') = \int_0^\infty e^{-t\Delta} f'(t)\, dt.$$

Integrating by parts, we have

$$vi(f') = [e^{-t\Delta} f(t)]_0^\infty + \Delta \int_0^\infty e^{-t\Delta} f(t)\, dt$$

$$= -f(0) + \Delta vi(f).$$

which is true by (2).

Next, let us suppose that $n = m$ is valid for some m. Thus,

$$vi(f^{(m)}) = \Delta^m vi(f) - \sum_{k=1}^{m} \Delta^{m-k} f^{(k-1)}(0)$$

holds for $f^{(m)}$ is the m-th derivative of f. Let us show that

$$vi(f^{(m+1)}) = \Delta^{m+1} vi(f) - \sum_{k=1}^{m+1} \Delta^{m+1-k} f^{(k-1)}(0). \tag{3}$$

Now we start with the left-hand side of (2).

$$vi(f^{(m+1)}) = \int_0^\infty e^{-t\Delta} f^{(m)}(t)\, dt$$

$$= \Delta vi(f^{(m)}) - f^{(m)}(0)$$

$$= \Delta [\Delta^m vi(f) - \sum_{k=1}^{m} \Delta^{m-k} f^{(k-1)}(0)] - f^{(m)}(0)$$

$$= \Delta^{m+1} vi(f) - \sum_{k=1}^{m} \Delta^{m+1-k} f^{(k-1)}(0)] - f^{(m)}(0)$$

$$= \Delta^{m+1} vi(f) - \sum_{k=1}^{m+1} \Delta^{m+1-k} f^{(k-1)}(0).$$

Therefore, this theorem is valid for an arbitrary natural number n. Putting $g(t) = \int_0^t f(\tau)d\tau$,

$$vi(f(t)) = vi(g'(t)) = \Delta vi(g) - g(0) = \Delta vi(g)$$

follows. □

As the direct results of (9), $vi(f') = \Delta vi(f) - f(0)$ and $vi(f'') = \Delta^2 vi(f) - \Delta f(0) - f'(0)$ are follow.

For example, we consider $y'' - y = t$ subject to $y(0) = 1$ and $y'(0) = 1$. Taking the integral of Laplace-type transform on both sides, we have

$$\Delta^2 Y - \Delta y(0) - y'(0) - Y = 1/\Delta^2$$

for $Y = (vi)(y)$. Organizing this equation, we get $(\Delta^2 - 1)Y = \Delta + 1 + 1/\Delta^2$. Simplification gives

$$Y = \frac{1}{\Delta - 1} + \frac{1}{\Delta^2 - 1} - \frac{1}{\Delta^2}.$$

From the relation of $V(f) = \Phi(u) \cdot F_*(\Delta)$, we have the solution

$$y(t) = -t + 2\sin ht + \cos ht = e^t + \sin ht - t,$$

where h is hyperbolic function.

Example 3. *(Integral equations of Volterra type) Find the solution of*

$$(1)\ y(t) + \int_0^t (t - \tau) y(\tau) d\tau = 1.$$

$$(2)\ y(t) - \int_0^t y(\tau) \sin(t - \tau) d\tau = t.$$

$$(3)\ y(t) - \int_0^t (1 + \tau) y(t - \tau) d\tau = 1 - \sinh t.$$

Solution.

(1) Since this equation is $y + y * t = 1$, taking the integral of Laplace-type transform on both sides, we have

$$Y + (Y \cdot \frac{1}{\Delta^2}) = \frac{1}{\Delta}$$

for $Y = vi(y)$. Thus

$$Y = \frac{\Delta}{\Delta^2 + 1}$$

and so, we obtain the solution $y = \cos t$.

Let us do the check by expansion. Expanding, we get $y''(t) + y(t) = 0$. Since $\int_a^a f = 0$, we get $y(0) = 1$ and $y'(0) = 0$. Thus, we obtain $y = \cos t$.

(2) This is rewritten as a convolution

$$y(t) - y * \sin t = t.$$

Taking the integral of Laplace-type transform, we have

$$Y(u) - Y(u)\frac{1}{\Delta^2 + 1} = Y(u)(1 - \frac{1}{\Delta^2 + 1}) = \frac{1}{\Delta^2}$$

for $Y = vi(y)$. The solution is

$$Y(u) = \frac{1}{\Delta^2} + \frac{1}{\Delta^4}$$

and gives the answer

$$y(t) = t + \frac{1}{6}t^3.$$

(3) Note that the equation is the same as $y - (1+t) * y = 1 - \sinh t$. Taking the transform, we get

$$Y - (\frac{1}{\Delta} + \frac{1}{\Delta^2})Y = \frac{1}{\Delta} - \frac{1}{\Delta^2 - 1},$$

and hence

$$Y(1 - \frac{1}{\Delta} - \frac{1}{\Delta^2}) = \frac{\Delta^2 - \Delta - 1}{\Delta(\Delta^2 - 1)}.$$

Simplification gives

$$Y = \frac{\Delta}{\Delta^2 - 1},$$

and so, we obtain the answer

$$y(t) = \cosh t$$

by the relation of $V(f) = \Phi(u) \cdot F_*(\Delta)$.

Let us turn the topic to initial value problem of the convolution. The initial value problem

$$ay'' + by' + cy = f(t), \ y(0) = y_0, \ y'(0) = y_0'$$

gives

$$(a\Delta^2 + b\Delta + c)Y(\Delta) - (a\Delta + b)y(0) - ay'(0) = F(\Delta),$$

where $Y(\Delta) = vi(y)$ and $F(\Delta) = vi(f)$. Simplification gives

$$Y(\Delta) = \frac{1}{a\Delta^2 + b\Delta + c} \cdot F(\Delta) + y_0 \cdot \frac{a\Delta + b}{a\Delta^2 + b\Delta + c}$$

$$+ y_0' \cdot \frac{a}{a\Delta^2 + b\Delta + c}.$$

If we put the system function $H(\Delta) = (a\Delta^2 + b\Delta + c)^{-1}$, then

$$Y(\Delta) = H(\Delta)F(\Delta) + y_0(a\Delta + b)H(\Delta) + y_0'aH(\Delta).$$

Since $H(\Delta)F(\Delta) = vi(h)vi(f) = vi(h * f)$ for $H(\Delta) = vi(h(t))$, taking the inverse transform, we have

$$y = (h * f) + y_0 \, vi^{-1}\{(a\Delta + b)H(\Delta)\}$$

$$+ y_0' \, vi^{-1}\{aH(\Delta)\}.$$

Theorem 2. *(Differentiation and integration of transforms) Let us put $V = F(\Delta)$ and $Y = V(y)$. Then*

(1) $V(t^n f(t)) = \Phi(u) \cdot (-1)^n \dfrac{d^n}{ds^n} F(\Delta)$

(2) $V(f(t)/t) = \Phi(u) \displaystyle\int_\Delta^\infty F(\delta)d\delta.$

Proof. This is an immediate consequence of $V(f) = \Phi(u) \cdot F_*(\Delta)$ and $V(f) = \Phi(u) \cdot vi(f)$. For this reason, detailed proofs are omitted.

The statements below are the immediate results of Theorem 2.

(1) $V(tf(t)) = -\Phi(u)F'(\Delta)$

(2) $V(tf'(t)) = -\Phi(u)(Y + \Delta\dfrac{dY}{d\Delta})$

(3) $V(tf''(t)) = \Phi(u) \cdot (-2\Delta Y - \Delta^2 \dfrac{dY}{d\Delta} + y(0))$

Let us check examples for temperature in an infinite bar and displacement in a semi-infinite string by the variant of Laplace-type transform. □

Example 4. *(Semi-infinite string) Find the displacement $w(x,t)$ of an elastic string subject to the following conditions [3].*

(a) The string is initially at rest on the x-axis from $x = 0$ to ∞.
(b) For $t > 0$ the left end of the string is moved in a given fashion, namely, according to a single sine wave

$$w(0,t) = f(t) = \begin{cases} \sin 2t & \text{if } 0 \le t \le \pi \\ 0 & \text{otherwise.} \end{cases}$$

(c) Furthermore, $w(x,t) \to 0$ as $x \to \infty$ for $t \ge 0$.

Then the displacement w is

$$w(x,t) = f(t - \frac{x}{c})h(t - \frac{x}{c}) = \begin{cases} \sin 2(t - \frac{x}{c}) & \text{if } x/c < t < x/c + \pi \\ 0 & \text{otherwise,} \end{cases}$$

where h is Heaviside function.

The proof is simple, and the interchangeability of limit and integral in the proof process guarantees its validity by the Lebesgue dominated convergence theorem.

Example 5. *(Temperature in an infinite bar) Find the temperature w in an infinite bar if the initial temperature is*

$$f(x) = w(x,0) = \begin{cases} k_0 \text{ (constant)} & \text{if } |x| < 1 \\ 0 & \text{otherwise} \end{cases}$$

with $w(0,t) = 0$.

Solution. Taking the integral of Laplace-type transform on both sides of $w_t = c^2 w_{xx}$, we have

$$\Delta F - w(x,0) = c^2 \frac{\partial^2 F}{\partial x^2}$$

for $F(x,u) = vi[w(x,t)]$. Organizing the equality, we get

$$\frac{\partial^2 F}{\partial x^2} - \frac{\Delta}{c^2}F = -\frac{1}{c^2}w(x,0).$$

Organizing this equality, we get

$$F(x,u) = A(u)e^{-\sqrt{\Delta}x/c} + B(u)e^{\sqrt{\Delta}x/c}$$

$$+ e^{-\sqrt{\Delta}x/c} \int \frac{c}{2\sqrt{\Delta}} e^{\sqrt{\Delta}x/c} \frac{1}{c^2} w(x,0) \, dx$$

$$- e^{\sqrt{\Delta}x/c} \int \frac{c}{2\sqrt{\Delta}} e^{-\sqrt{\Delta}x/c} \frac{1}{c^2} w(x,0) \, dx, \tag{4}$$

where the Wronskian $W = 2\sqrt{\Delta}/c$. The value $\lim_{x \to \infty} f(x) = 0$ gives $\lim_{x \to \infty} F(x,u) = 0$, and hence $B(u) = 0$. Thus, from (4), we get

$$F(x,u) = A(u)e^{-\sqrt{\Delta}x/c}$$

$$+ \frac{w(x,0)}{2c\sqrt{\Delta}}(e^{-\sqrt{\Delta}x/c}\int e^{\sqrt{\Delta}x/c}\,dx - e^{\sqrt{\Delta}x/c}\int e^{-\sqrt{\Delta}x/c}\,dx).$$

By the direct calculation, we have

$$F(x,u) = A(u)e^{-\sqrt{\Delta}x/c} + \frac{w(x,0)}{\Delta}.$$

From the formula of $vi(f) = F(\Delta)$ for $F(s) = \pounds(f)$ and $s = \Delta$, we know

$$vi\left(\frac{k}{2\sqrt{\pi t^3}}e^{-\frac{k^2}{4t}}\right) = e^{-k\sqrt{\Delta}}\ (k>0)$$

and $vi(1) = 1/\Delta$. Taking the inverse transform, we obtain the temperature $w(x,t)$ as follows:

$$w(x,t) = A(t) * \frac{x}{2c\sqrt{\pi t^3}} e^{-\frac{x^2}{4c^2 t}} + k_0$$

on $|x| < 1$, and $*$ is the convolution. In case of $|x| > 1$, we have the solution

$$w(x,t) = A(t) * \frac{x}{2c\sqrt{\pi t^3}} e^{-\frac{x^2}{4c^2 t}}.$$

In the above equality, we note that

$$vi^{-1}\left[A(u)e^{-\sqrt{\Delta}x/c}\right] = vi^{-1}\left[vi(A(t))\cdot vi\left(\frac{x}{2c\sqrt{\pi t^3}}e^{-\frac{x^2}{4c^2 t}}\right)\right]$$

$$= vi^{-1}\left[vi\left\{A(t) * \frac{x}{2c\sqrt{\pi t^3}}e^{-\frac{x^2}{4c^2 t}}\right\}\right] = A(t) * \frac{x}{2c\sqrt{\pi t^3}}e^{-\frac{x^2}{4c^2 t}}$$

because $vi(f*g) = F(\Delta)g(\Delta)$ for $vi(f) = F(\Delta)$.

4. Conclusions

In this study, the concept of convolution in convolutional neural networks (CNN) was presented mathematically and tried to connect with the concept of convolution in mathematics. As a continuous form of convolution in CNN, a new form of Laplace-type transform has been proposed. In the future, we will study the change of convolution in CNN by changing the stride. In addition to this, we shall also explore the possibility of our applying our newly defined Laplace-type transform in obtaining certain new and interesting results involving generalized hypergeometric functions that would certainly unify and generalized the results available in the literature and may be potentially useful from an applications point of view.

Author Contributions: Conceptualization, A.K.R.; validation, Y.H.G.; and writing, H.K. All authors have read and agreed to the published version of the manuscript.

Funding: This research received no external funding.

Acknowledgments: The corresponding author (H.K.) acknowledges the support of the Kyungdong University Research Fund, 2021. The authors are also grateful to the anonymous referees whose valuable suggestions and comments significantly improved the quality of this paper.

Conflicts of Interest: The authors declare no conflict of interest.

References

1. LeCun, Y.; Bengio, Y.; Hinton, G. Deep learning. *Nature* **2015**, *521*. [CrossRef] [PubMed]
2. Kim, H. The intrinsic structure and properties of Laplace-typed integral transforms. *Math. Probl. Eng.* **2017**, *2017*, 1–8. [CrossRef]

3. Kreyszig, E. *Advanced Engineering Mathematics*; Wiley: Singapore, 2013.
4. Watugula, G.K. Sumudu Transform: A new integral transform to solve differential equations and control engineering problems. *Integr. Educ.* **1993**, *24*, 35–43. [CrossRef]
5. Elzaki, T.M.; Ezaki, S.M.; Hilal, E.M.A. ELzaki and Sumudu Transform for Solving some Differential Equations. *Glob. J. Pure Appl. Math.* **2012**, *8*, 167–173.
6. Mohand, M.; Mahgoub, A. The New Integral Transform 'Mohand Transform. *Adv. Theor. Appl. Math.* **2017**, *12*, 113–120.
7. Belgacem, F.B.M.; Silambarasan, R. Theory of natural transform. *Math. Eng. Sci. Aerosp.* **2012**, *3*, 105–135.
8. Bertrand, J.; Bertrand, P.; Ovarlez, J.P. *The Mellin Transform, The Transforms and Applications*; Poularkas, A.D., Ed.; CRC Press: Boca Raton, FL, USA, 1996.
9. Jhanthanam, S.; Nonlaopon, K.; Orankitjaroen, S. Generalized Solutions of the Third-Order Cauchy-Euler Equation in the Space of Right-Sided Distributions via Laplace Transform. *Mathematics* **2019**, *7*, 376. [CrossRef]
10. Kim, H. The solution of the heat equation without boundary conditions. *Dyn. Syst. Appl.* **2018**, *27*, 653–662.
11. Supaknaree, S.; Nonlaopon, K.; Kim, H. Further properties of Laplace-type integral transforms. *Dyn. Syst. Appl.* **2019**, *28*, 195–215.
12. Koepf, W.; Kim, I.; Rathie, A.K. On a New Class of Laplace-Type Integrals Involving Generalized Hypergeometric Functions. *Axioms* **2019**, *8*, 87. [CrossRef]
13. Sung, T.; Kim, I.; Rathie, A.K. On a new class of Eulerian's type integrals involving generalized hypergeometric functions. *Aust. J. Math. Anal. Appl.* **2019**, *16*, 1–15.
14. Noeiaghdam, S.; Fariborzi Araghi, M.A.; Abbasbandy, S. Finding optimal convergence control parameter in the homotopy analysis method to solve integral equations based on the stochastic arithmetic. *Numer. Algorithms* **2019**, *81*. [CrossRef]
15. Negnevitsky, M. *Artificial Intelligence*; Addison-Wesley: Essex, England, 2005.
16. Cohn, D.L. *Measure Theory*; Birkhäuser: Boston, MA, USA, 1980.
17. Jang, J.; Kim, H. An application of monotone convergence theorem in PDEs and Fourier analysis. *Far East J. Math. Sci.* **2015**, *98*, 665–669.

Publisher's Note: MDPI stays neutral with regard to jurisdictional claims in published maps and institutional affiliations.

© 2020 by the authors. Licensee MDPI, Basel, Switzerland. This article is an open access article distributed under the terms and conditions of the Creative Commons Attribution (CC BY) license (http://creativecommons.org/licenses/by/4.0/).

Article

Detecting Optimal Leak Locations Using Homotopy Analysis Method for Isothermal Hydrogen-Natural Gas Mixture in an Inclined Pipeline

Sarkhosh S. Chaharborj [1], Zuhaila Ismail [1] and Norsarahaida Amin [1,2,*]

[1] Department of Mathematical Sciences, Universiti Teknologi Malaysia, Johor Bahru 81310, Malaysia; sseddighi2014@yahoo.com.my (S.S.C.); zuhaila@utm.my (Z.I.)
[2] Department of Mathematics, Universitas Airlangga, Surabaya 60115, Indonesia
* Correspondence: norsarahaida@utm.my

Received: 18 August 2020; Accepted: 29 September 2020; Published: 26 October 2020

Abstract: The aim of this article is to use the Homotopy Analysis Method (HAM) to pinpoint the optimal location of leakage in an inclined pipeline containing hydrogen-natural gas mixture by obtaining quick and accurate analytical solutions for nonlinear transportation equations. The homotopy analysis method utilizes a simple and powerful technique to adjust and control the convergence region of the infinite series solution using auxiliary parameters. The auxiliary parameters provide a convenient way of controlling the convergent region of series solutions. Numerical solutions obtained by HAM indicate that the approach is highly accurate, computationally very attractive and easy to implement. The solutions obtained with HAM have been shown to be in good agreement with those obtained using the method of characteristics (MOC) and the reduced order modelling (ROM) technique.

Keywords: hydrogen; natural gas; gas mixture; homotopy analysis method; method of characteristics; reduced order modelling; leak locations

1. Introduction

One of the strategies to reduce gas transportation costs is the use of natural gas pipeline networks by petroleum companies [1]. These networks are capable of supplying gas in long distances under high pressure and through compression stations [2]. Changes in pipeline pressure are a function of gas velocity, valve closure time, and arrangement of the closing valve [3].

When the valve is closed at the end of the pipeline, there is the possibility of the occurrence of maximum pressure, which can be decreased in short times during its closure. It is of utmost importance to control factors affecting transient pressure, such as initial pressure and mass ratio. This is because the damage caused by this pressure is not evident shortly after the event [4–6].

Several studies have been conducted on transient flow in the mixtures of hydrogen and natural gas with the use of isothermal flow and horizontal pipelines, which is not the case in reality [2,7–9]. Furthermore, another study has made an attempt to study the flow of these mixtures under high pressure through inclined pipelines [10]. In most pipelines working under high pressure, there are slow and fast fluid transients. As gas properties are not constant, a one-dimensional and non-isothermal gas flow model should be presented to simulate these transients [2].

The reason for proposing hydrogen and natural gas mixtures is their transportation through the same pipelines for the purpose of cost reduction. This is while the existing lines are just designed for natural gas, whose properties are significantly different from that of hydrogen [11,12]. The solution to this problem has been the mixture of the both with a great deal of care and attention, as hydrogen is a reactive gas with high pressure that can cause leakage [13,14]. This problem is of great importance since leakage can cause many economic, environmental, and safety problems and threaten industries and citizens by wasting natural gas [14].

According to the reports, two thirds of the 375 pipeline events between 1994 and 1999 were caused by leakage [4]. In addition, high-pressure wave celerity causes pipe splitting, and even, exploding, sometimes making intense holes that lead to inward collapse of pipes, necessitating the careful study of pressure wave celerity.

Studies have been done on leakage and its location for natural gas [15,16], leading to the introduction of methods [17,18], such as the acoustic method (AM) [18,19] and the negative pressure wave method (NPW) [20,21]. Means of transients and using unsteady-state tests, which give rise to small overpressure, can be considered as an appropriate method for detecting leaks locations in pressurised pipes [22]. Autocorrelation analysis of vibro-acoustic signals measured in a test field and amplitude distortion of measured leak noise signals caused by instrumentation have been used for water leak detection in [23,24]. In water-filled small-diameter polyethylene pipes by means of acoustic Emission Measurements, [25] has been used for detecting leak locations. However, there is paucity of research on this issue for hydrogen or its mixture with natural gas [15,16]. In this regard, isothermal and non-isothermal flow models have been proposed for hydrogen and natural gas mixtures [7,8,14].

There have been several studies on the detection of leakage location through novel approaches. For example, new leakage detection using AM [26] and new algorithm based on the attenuation of NPW in isothermal cases have been introduced in recent years [27].

Accordingly, the present study made an attempt to determine leakage location in an inclined pipe for isothermal flow containing hydrogen-natural gas mixture with the use of homotpy analysis method. This method is used for solving the governing equations, leading to quick and accurate analytical solutions for nonlinear transportation equations. Factors affecting pressure and celerity waves in inclined pipes, such as inclination angles and mass ratio of mixtures, have also been discussed. The obtained results are in good agreement for isothermal flow in a horizontal pipeline. Results showed that pressure drop and leak discharge are increased with an increase in the inclination angle, while the celerity wave and the leak location do not seem to be affected.

2. Mathematical Formulation

Figure 1 shows an inclined pipeline, which has a reservoir at the top and a valve at its bottom. The governing equations consist of three partial differential equations that are all coupled, non-linear and hyperbolic. The non-isothermal flow in the pipeline, a homogenous mixture of hydrogen and natural gas, was considered to be one-dimensional that is compressible and covers transient condition [7].

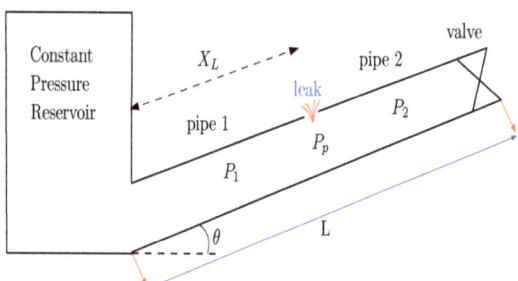

Figure 1. An inclined pipeline with a reservoir at the top and a valve at its bottom.

2.1. Governing Equation

The governing equations for the transport of hydrogen/natural gas mixture in an inclined pipeline from the principle of conserving mass and momentum are given by the following,

$$\frac{\partial \rho}{\partial t} + \frac{\partial \rho u}{\partial x} = 0, \tag{1}$$

$$\frac{\partial \rho u}{\partial t} + \frac{\partial (\rho u^2 + P)}{\partial x} + \frac{f\rho u|u|}{2D} + \rho g \sin(\theta) = 0, \tag{2}$$

with $u = Q/A$, $A = \pi D^2/4$, ρ is density, u is the gas velocity, P is the pressure, e is the gas internal energy per unit mass, D is the diameter of the pipeline, f is the coefficient of friction, g is the gravitational force and θ is an angle between the friction force and the direction.

Boundary conditions of this equations depend on the types of closure and the valve operational time. The boundary conditions at the initial point $x = 0$ and at the end point $x = L$, respectively are given by,

$$\rho(0,t) = \rho_0(t), \quad u(0,t) = u_0(t), \tag{3}$$
$$\rho(L,t) = \rho_L(t), \quad u(L,t) = u_L(t), \tag{4}$$

where ρ_0 and u_0 are defined as density and gas velocity at the inlet pipeline, respectively and ρ_L and u_L are defined as density and gas velocity at the outlet pipeline, respectively. The initial conditions that are assumed to be in a steady state condition at $t = 0$ are [7],

$$\frac{\partial \rho u}{\partial x}(x,0) = 0, \tag{5}$$

$$\frac{\partial (\rho u^2 + P)}{\partial x}(x,0) + \frac{f\rho u|u|}{2D} + \rho g \sin(\theta) = 0, \tag{6}$$

The commonly used equation of state for perfect gas is as follows:

$$P = \rho RT, \tag{7}$$

where,

R: is the specific gas constant.
T: is temperature.

The equation of state for the compressible flow, where there is a celerity pressure wave, is:

$$P = c^2 \rho, \tag{8}$$

The following equations are also achieved from ideal gas relation,

$$C_p - C_v = R, \quad \gamma = \frac{C_p}{C_v}, \quad C_v = \frac{R}{\gamma - 1}. \tag{9}$$

where,

C_v: is the specific heat at constant volume.
C_p: is the specific heat at constant pressure.
R: is the specific gas constant.
P: is pressure.
γ: is the flow process index.

2.2. Hydrogen-Natural Gas Mixture Equation

The mass ratio and the density of hydrogen-natural gas mixture are defined as,

$$\phi = \frac{m_h}{m_h + m_g}, \quad \frac{1}{\rho} = \frac{v_h + v_g}{m_h + m_g}, \quad (10)$$

with $\rho_h = \frac{m_h}{v_h}$, $\rho_g = \frac{m_g}{v_g}$, $\rho_h = \rho_{h_0}\left(\frac{P_0}{P}\right)^{\frac{1}{n_1}}$ and $\rho_g = \rho_{g_0}\left(\frac{P_0}{P}\right)^{\frac{1}{n_2}}$. Where m_g, m_h, V_g and V_h are defined as the mass of natural gas and hydrogen and volume of natural gas and hydrogen, respectively.

Therefore, the expression of the average density of the gas mixture is given by,

$$\rho = \left[\frac{\phi}{\rho_{h_0}}\left(\frac{P_0}{P}\right)^{\frac{1}{n_1}} + \frac{1-\phi}{\rho_{g_0}}\left(\frac{P_0}{P}\right)^{\frac{1}{n_2}}\right]^{-1}. \quad (11)$$

The celerity pressure wave for compressible flow is defined as,

$$c = \left(\frac{\partial \rho}{\partial P}\right)_s^{-\frac{1}{2}}, \quad (12)$$

where the subscript s is defined the constant entropy condition. The derivative of Equation (11) with respect to P, and substituting into Equation (12), then the celerity pressure wave yields [7],

$$c = \left[\frac{\phi}{\rho_{h_0}}\left(\frac{P_0}{P}\right)^{\frac{1}{n_1}} + \frac{1-\phi}{\rho_{g_0}}\left(\frac{P_0}{P}\right)^{\frac{1}{n_2}}\right] \times \left[\frac{1}{P}\left[\frac{\phi}{n_1\rho_{h_0}}\left(\frac{P_0}{P}\right)^{\frac{1}{n_1}} + \frac{1-\phi}{n_2\rho_{g_0}}\left(\frac{P_0}{P}\right)^{\frac{1}{n_2}}\right]\right]^{-\frac{1}{2}}. \quad (13)$$

The properties of hydrogen and natural gas used in the calculations are shown in the Table 1. For the simulation, the parameters are assumed as Table 2.

Table 1. Hydrogen properties in working conditions, $P = 35$ bar and $T = 15\ °C = 288$ K (See [7]).

Symbol	Fluid Properties	Values (J/kgK)	
		Hydrogen	Natural Gas
C_p	Specific heat at constant pressure	14,600	1497.5
C_v	Specific heat at constant volume	10,440	1056.8
R	Gas constant	4160	440.7

Table 2. Parameters used for the simulation (See [7]).

Symbols	Values	Symbols	Values
Pipe length	$L = 600$ m	Mass ratio	$\phi = 0, 0.5, 1$
Time	$t = 20$	Angle	$\theta = 0, \pi/6, \pi/4, \pi/3$
Pipe diameter	$D = 0.4$ m	Mass flow	$Q_0 = 55$ kg/s
Friction coefficient	$f = 0.03$	Absolute pressure	$P_0 = 35$ bar
Temperature	$T = 15\ °C = 288$ K		

3. Homotopy Analysis Method

A brief description of the standard homotopy analysis method (HAM) presented by [28–32]. This will be followed by a description of the algorithm of the homotopy analysis method (HAM). First, we consider the following differential equation,

$$\mathcal{N}[u(x,t)] = \mathcal{G}(x,t), \quad (14)$$

where $\mathcal{N}$ are nonlinear operators, x and t denotes the independent variable, $u(x,t)$ are unknown functions, and $\mathcal{G}(x,t)$ are known analytic functions. For $\mathcal{G}(x,t) = 0$, Equation (14) reduces to the homogeneous equation. By means of generalizing the traditional homotopy method, Liao [28] constructed the so-called zero-order deformation equation,

$$(1-q)\mathscr{L}\left[\Psi(x,t;q) - u_0(x,t)\right] = q\hbar \mathscr{H}(x,t)\left\{\mathscr{N}\left[\Psi(x,t;q)\right] - \mathscr{G}(x,t)\right\} \tag{15}$$

where $p \in [0,1]$ is an embedding parameter, $\hbar$ are nonzero auxiliary functions, $\mathscr{L}$ is an auxiliary linear operator, $u_0(x,t)$ are initial guesses of $u(x,t)$, $\mathscr{H}(x,t)$ denotes a nonzero auxiliary function and $\Psi(x,t;q)$ are unknown functions. It is important to note that one has great freedom to choose auxiliary objects such as $\hbar$ and $\mathscr{L}$ in HAM. Obviously, when $q=0$ and $q=1$, Equation (15) becomes,

$$\Psi(x,t;0) = u_0(x,t), \quad \Psi(x,t;1) = u(x,t), \tag{16}$$

Thus, as q increases from 0 to 1, the solution $\Psi(x,t;q)$ varies from the initial guesses $u_0(x,t)$ to the solutions $u(x,t)$. Expanding $\Psi(x,t;q)$ in Taylor series with respect to q, one has

$$\Psi(x,t;q) = u_0(x,t) + \sum_{m=1}^{\infty} u_m(x,t) q^m, \tag{17}$$

where,

$$u_m(x,t) = \frac{1}{m!} \frac{\partial^m \Psi(x,t;q)}{\partial q^m}\bigg|_{q=0}, \tag{18}$$

If the auxiliary linear operator, the initial guesses, the auxiliary parameters $\hbar$, and the auxiliary functions are so properly chosen, then series Equation (17) converges at $q=1$, and one has,

$$u(x,t) = u_0(x,t) + \sum_{m=1}^{\infty} u_m(x,t), \tag{19}$$

which must be one of the solutions of the original nonlinear equations, as proved by Liao [28]. As $\hbar = -1$ and $\mathscr{H}(x,t) = 1$, Equation (15) becomes,

$$(1-q)\mathscr{L}\left[\Psi(x,t;q) - u_0(x,t)\right] = q\left\{\mathscr{N}\left[\Psi(x,t;q)\right] - \mathscr{G}(x,t)\right\}, \tag{20}$$

which is used mostly in the homotopy perturbation method. Define the vectors,

$$\vec{u}_m = \left\{u_0(x,t), u_1(x,t), ..., u_m(x,t)\right\}, \tag{21}$$

Differentiate the zeroth-order deformation Equation (14) m-times with respect to q and then dividing them by $m!$ and finally setting $q=0$, we get the following mth-order deformation equation,

$$\mathscr{L}\left[u_m(x,t) - \chi_m u_{m-1}(x,t)\right] = \hbar \mathscr{R}_m\left(\vec{u}_{m-1}(x,t)\right), \tag{22}$$

where,

$$\mathscr{R}_m\left(\vec{u}_{m-1}(x,t)\right) = \frac{1}{(1-m)!} \frac{\partial^{m-1}\left\{\mathscr{N}\left[\Psi(x,t;q)\right] - \mathscr{G}(x,t)\right\}}{\partial q^{m-1}}, \tag{23}$$

with,

$$\chi_m = \begin{cases} 0, & m \leq 1 \\ 1, & m > 1 \end{cases} \tag{24}$$

It should be noted that the linear Equation (22), which has linear boundary conditions, governs $u_m(x,t)$ for $m \leq 1$ [33]. Boundary conditions stem from the main problem, the solution for which can be provided by Matlab, Maple, or Mathematica. The requirement for the limit of Equation (17) is that it should meet the conditions of the main equation $\mathscr{N}\left[u(x,t)\right] = 0$ when it is convergent at $q = 1$. It is noteworthy that drawing "$\hbar$-curves" or "curves for convergence-control parameter" aim to find a proper convergence-control parameter $\hbar$, a convergent series solution, or and accelerate its convergence rate. It is such that these curves with unknown quantities are drawn against $\hbar$ to approximately find the convergence region, though they are just graphical. This is because it is not possible to find which $\hbar_0 \in R_h$ provides the fastest convergent series (see Liao [28,34] for further reading). Another note to be made is that a unique solution is achieved when Equation (14) accepts a unique solution; otherwise, many possible solutions will be obtained from HAM.

3.1. Solving the Steady State Equations by High-Order Deformation HAM

We define the vectors,

$$\begin{cases} \vec{P}(x) = \{P_0(x), P_1(x), ..., P_m(x)\} \\ \vec{u}(x) = \{u_0(x), u_1(x), ..., u_m(x)\} \end{cases} \tag{25}$$

Differentiating Equations (5) and (6) m times with respect to the embedding parameter q and then setting $q = 0$ and finally dividing them by $m!$, we have the so-called mth-order deformation equations,

$$\begin{cases} \mathscr{L}_1\left[P_m(x) - \chi_m P_{m-1}(x)\right] = \hbar \mathscr{R}_m^1\left(\vec{P}_{m-1}(x), \vec{u}_{m-1}(x)\right), \\ \mathscr{L}_2\left[u_m(x) - \chi_m u_{m-1}(x)\right] = \hbar \mathscr{R}_m^2\left(\vec{P}_{m-1}(x), \vec{u}_{m-1}(x)\right), \end{cases} \tag{26}$$

with the initial conditions,

$$P(0) = P_0, \quad u(0) = u_0, \tag{27}$$

where,

$$\begin{cases} \mathscr{R}_m^1\left(\vec{P}_{m-1}(x), \vec{u}_{m-1}(x)\right) = \frac{dP_{m-1}(x)}{dx} + \sum_{i=0}^{m-1} P_{m-1-i}(x) \sum_{j=0}^{i} \frac{1}{u_j(x)} \frac{du_{i-j}(x)}{dx}, \\ \mathscr{R}_m^2\left(\vec{P}_{m-1}(x), \vec{u}_{m-1}(x)\right) = \frac{du_{m-1}(x)}{dx} + \sum_{i=0}^{m-1} u_{m-1-i}(x) \sum_{j=0}^{i} \frac{1}{P_j(x)} \frac{dP_{i-j}(x)}{dx} \\ \quad + \sum_{i=0}^{m-1} c_{m-1-i} \sum_{j=0}^{i} c_{i-j} \sum_{k=0}^{j} P_{j-k}(x) \sum_{l=0}^{k} \frac{1}{u_l(x)} \frac{dP_{k-l}(x)}{dx} \\ \quad + \frac{f}{2D} |u_{m-1}(x)| + u_{m-1}(x) g \sin(\theta), \end{cases} \tag{28}$$

with the celerity pressure wave c_i defined as follows,

$$c_i = \left[\frac{\phi}{\rho_{h_0}} \left(\frac{P_0}{P_i(x)}\right)^{\frac{1}{n_1}} + \frac{1-\phi}{\rho_{g_0}} \left(\frac{P_0}{P_i(x)}\right)^{\frac{1}{n_2}}\right]$$

$$\times \left[\frac{1}{P_i(x)} \left[\frac{\phi}{n_1 \rho_{h_0}} \left(\frac{P_0}{P_i(x)}\right)^{\frac{1}{n_1}} + \frac{1-\phi}{n_2 \rho_{g_0}} \left(\frac{P_0}{P_i(x)}\right)^{\frac{1}{n_2}}\right]\right]^{-\frac{1}{2}}.$$

with the following linear operators,

$$\mathscr{L}_1\left[\Psi_1(x;q)\right] = \frac{d\Psi_1(x;q)}{dx}, \quad \mathscr{L}_2\left[\Psi_2(x;q)\right] = \frac{d\Psi_2(x;q)}{dx}, \tag{29}$$

with the property that,

$$\mathscr{L}_1\left[C_1\right] = 0, \quad \mathscr{L}_2\left[C_2\right] = 0, \tag{30}$$

which implies that,

$$\mathscr{L}_1^{-1}(.) = \int_0^x (.) dx, \quad \mathscr{L}_2(.) = \int_0^x (.) d, \tag{31}$$

Now, the solution of the mth-order deformation Equations (5) and (6) becomes,

$$\begin{cases} P_m(x) = \chi_m P_{m-1}(x) + \hbar \mathscr{L}_1^{-1}\left[\mathscr{H}(x,t)\mathscr{R}_m^1\left(\vec{P}_{m-1}(x), \vec{u}_{m-1}(x)\right)\right], \\ u_m(x) = \chi_m u_{m-1}(x) + \hbar \mathscr{L}_2^{-1}\left[\mathscr{H}(x,t)\mathscr{R}_m^2\left(\vec{P}_{m-1}(x), \vec{u}_{m-1}(x)\right)\right], \end{cases} \quad (32)$$

which can be easily solved by a symbolic computation software such as Matlab, Maple, and Mathematica. Therefore, we will have $P(x)$ and $u(x)$ as follows,

$$P(x) \simeq P_M(x) = P_0(x) + \sum_{m=1}^{M} P_m(x), \quad (33)$$

$$u(x) \simeq u_M(x) = u_0(x) + \sum_{m=1}^{M} u_m(x). \quad (34)$$

Furthermore, to construct the zeroth-order deformation equations we can define the nonlinear operators $\mathscr{N}_1\left[\Psi_1(x;q)\right]$ and $\mathscr{N}_2\left[\Psi_2(x;q)\right]$ as follows,

$$\begin{cases} \mathscr{N}_1\left[\Psi_1(x;q)\right] = \Psi_1(x;q)\frac{d\Psi_2(x;q)}{dx} + \Psi_2(x;q)\frac{d\Psi_1(x;q)}{dx} \\ \mathscr{N}_2\left[\Psi_2(x;q)\right] = \dfrac{d\left[\Psi_1(x;q)\Psi_2(x;q)^2 + c^2\Psi_1(x;q)\right]}{dx} + \dfrac{f}{2D}\Psi_1(x;q)\Psi_2(x;q)|\Psi_2(x;q)| + \Psi_1(x;q)g\sin(\theta) \end{cases} \quad (35)$$

with the celerity pressure wave c defined as follows,

$$c = \left[\frac{\phi}{\rho_{h_0}}\left(\frac{P_0}{\Psi_1(x;q)}\right)^{\frac{1}{n_1}} + \frac{1-\phi}{\rho_{g_0}}\left(\frac{P_0}{\Psi_1(x;q)}\right)^{\frac{1}{n_2}}\right]$$

$$\times \left[\frac{1}{\Psi_1(x;q)}\left[\frac{\phi}{n_1\rho_{h_0}}\left(\frac{P_0}{\Psi_1(x;q)}\right)^{\frac{1}{n_1}} + \frac{1-\phi}{n_2\rho_{g_0}}\left(\frac{P_0}{\Psi_1(x;q)}\right)^{\frac{1}{n_2}}\right]\right]^{-\frac{1}{2}}. \quad (36)$$

3.2. Solving Isothermal Flow of Hydrogen-Natural Gas Mixture by HAM

We define the vectors,

$$\begin{cases} \vec{P}(x,t) = \{P_0(x,t), P_1(x,t), ..., P_m(x,t)\} \\ \vec{u}(x,t) = \{u_0(x,t), u_1(x,t), ..., u_m(x,t)\} \end{cases} \quad (37)$$

Differentiating Equations (1) and (2) m times with respect to the embedding parameter q and then setting $q = 0$ and finally dividing them by $m!$, we have the so-called mth-order deformation equations,

$$\begin{cases} \mathscr{L}_1\left[P_m(x,t) - \chi_m P_{m-1}(x,t)\right] = \hbar \mathscr{R}_m^1\left(\vec{P}_{m-1}(x,t), \vec{u}_{m-1}(x,t)\right), \\ \mathscr{L}_2\left[u_m(x,t) - \chi_m u_{m-1}(x,t)\right] = \hbar \mathscr{R}_m^2\left(\vec{P}_{m-1}(x,t), \vec{u}_{m-1}(x,t)\right), \end{cases} \quad (38)$$

with the initial and boundary conditions as follows,

$$\begin{cases} P(x,0) = P_0(x), \quad u(x,0) = u_0(x); \\ P(0,t) = P_0(t), \quad u(0,t) = u_0(t) \quad \text{or} \quad P(L,t) = P_L(t), \quad u(L,t) = u_L(t), \end{cases}$$

where,

$$\begin{cases} \mathcal{R}_m^1\left(\vec{P}_{m-1}(x,t), \vec{u}_{m-1}(x,t)\right) = \frac{\partial P_{m-1}(x,t)}{\partial t} + \sum_{i=0}^{m-1} u_i(x,t) \frac{\partial P_{m-1-i}(x,t)}{\partial x} \\ \qquad + \sum_{i=0}^{m-1} P_i(x,t) \frac{\partial u_{m-1-i}(x,t)}{\partial x}, \\ \mathcal{R}_m^2\left(\vec{P}_{m-1}(x), \vec{u}_{m-1}(x)\right) = \frac{\partial u_{m-1}(x,t)}{\partial t} + \sum_{i=0}^{m-1} u_{m-1-i}(x) \sum_{j=0}^{i} \frac{1}{P_j(x,t)} \frac{\partial P_{i-j}(x,t)}{\partial t} \\ \qquad + \sum_{i=0}^{m-1} u_i(x,t) \frac{\partial u_{m-1-i}(x,t)}{\partial x} \\ \qquad + \sum_{i=0}^{m-1} u_{m-1-i}(x,t) \sum_{j=0}^{i} u_{i-j}(x,t) \sum_{k=0}^{j} \frac{1}{P_k(x,t)} \frac{\partial P_{j-k}(x,t)}{\partial x} \\ \qquad + \sum_{i=0}^{m-1} c_{m-1-i}(x,t) \sum_{j=0}^{i} c_{i-j}(x,t) \sum_{k=0}^{j} \frac{1}{P_k(x,t)} \frac{\partial P_{j-k}(x,t)}{\partial x} \\ \qquad + \frac{f}{2D} \sum_{i=0}^{m-1} u_i(x,t) |u_{m-1-i}(x,t)| + g\sin(\theta) \end{cases} \quad (39)$$

with the celerity pressure wave c_i defined as follows,

$$c_i = \left[\frac{\phi}{\rho_{h_0}} \left(\frac{P_0}{P_i(x,t)} \right)^{\frac{1}{n_1}} + \frac{1-\phi}{\rho_{g_0}} \left(\frac{P_0}{P_i(x,t)} \right)^{\frac{1}{n_2}} \right]$$

$$\times \left[\frac{1}{P_i(x,t)} \left[\frac{\phi}{n_1 \rho_{h_0}} \left(\frac{P_0}{P_i(x,t)} \right)^{\frac{1}{n_1}} + \frac{1-\phi}{n_2 \rho_{g_0}} \left(\frac{P_0}{P_i(x,t)} \right)^{\frac{1}{n_2}} \right] \right]^{-\frac{1}{2}}.$$

with the following linear operators,

$$\mathcal{L}_1\left[\Psi_1(x,t;q)\right] = \frac{\partial \Psi_1(x,t;q)}{\partial t}, \quad \mathcal{L}_2\left[\Psi_2(x,t;q)\right] = \frac{\partial \Psi_2(x,t;q)}{\partial t}, \quad (40)$$

with the property that,

$$\mathcal{L}_1\left[C_1\right] = 0, \quad \mathcal{L}_2\left[C_2\right] = 0, \quad (41)$$

which implies that,

$$\mathcal{L}_1^{-1}(.) = \int_0^t (.)dt, \quad \mathcal{L}_2(.) = \int_0^t (.)dt, \quad (42)$$

Now, the solution of the mth-order deformation Equations (1) and (2) becomes,

$$\begin{cases} P_m(x,t) = \chi_m P_{m-1}(x,t) + \hbar \mathcal{L}_1^{-1}\left[\mathcal{H}(x,t) \mathcal{R}_m^1\left(\vec{P}_{m-1}(x,t), \vec{u}_{m-1}(x,t)\right)\right], \\ u_m(x,t) = \chi_m u_{m-1}(x,t) + \hbar \mathcal{L}_2^{-1}\left[\mathcal{H}(x,t) \mathcal{R}_m^2\left(\vec{P}_{m-1}(x,t), \vec{u}_{m-1}(x,t)\right)\right], \end{cases} \quad (43)$$

which can be easily solved by a symbolic computation software such as Matlab, Maple, and Mathematica. Therefore, we will have $P(x,t)$ and $u(x,t)$ as follows,

$$P(x,t) \simeq P_M(x,t) = P_0(x,t) + \sum_{m=1}^{M} P_m(x,t), \quad (44)$$

$$u(x,t) \simeq u_M(x,t) = u_0(x,t) + \sum_{m=1}^{M} u_m(x,t). \quad (45)$$

Furthermore, to construct the zeroth-order deformation equations we can define the nonlinear operators $\mathcal{N}_1\left[\Psi_1(x,t;q)\right]$ and $\mathcal{N}_2\left[\Psi_2(x,t;q)\right]$ as follows,

$$\begin{cases} \mathcal{N}_1\left[\Psi_1(x,t;q)\right] = \frac{\partial \Psi_1(x,t;q)}{\partial t} + \frac{\partial\left[\Psi_1(x,t;q)\Psi_2(x,t;q)\right]}{\partial x} \\ \mathcal{N}_2\left[\Psi_2(x,t;q)\right] = \frac{\partial\left[\Psi_1(x,t;q)\Psi_2(x,t;q)\right]}{\partial t} + \frac{\partial\left[\Psi_1(x,t;q)\Psi_2(x,t;q)^2 + c^2\Psi_1(x,t;q)\right]}{\partial x} \\ \quad + \frac{f}{2D}\Psi_1(x,t;q)\Psi_2(x,t;q)|\Psi_2(x,t;q)| + \Psi_1(x,t;q)g\sin(\theta) \end{cases}$$

with the celerity pressure wave c defined as follows,

$$c = \left[\frac{\phi}{\rho_{h_0}}\left(\frac{P_0}{\Psi_1(x,t;q)}\right)^{\frac{1}{n_1}} + \frac{1-\phi}{\rho_{g_0}}\left(\frac{P_0}{\Psi_1(x,t;q)}\right)^{\frac{1}{n_2}}\right] \\ \times \left[\frac{1}{\Psi_1(x,t;q)}\left[\frac{\phi}{n_1\rho_{h_0}}\left(\frac{P_0}{\Psi_1(x,t;q)}\right)^{\frac{1}{n_1}} + \frac{1-\phi}{n_2\rho_{g_0}}\left(\frac{P_0}{\Psi_1(x,t;q)}\right)^{\frac{1}{n_2}}\right]\right]^{-\frac{1}{2}}. \quad (46)$$

3.3. Results and Discussion

For solving the Equations (5) and (6) by suing the homotopy analysis method according the Equations (25)–(36) we can have,

$$\mathcal{R}_1^1 = 0,$$
$$\mathcal{R}_m^2 = \frac{fu_0}{2d} + \frac{g\sin(\theta)}{u_0},$$
$$P_1(x) = P_0,$$
$$u_1(x) = u_0 + \hbar\left(\frac{fu_0 x}{2d} + \frac{g\sin(\theta)x}{u_0}\right),$$
$$\mathcal{R}_2^1 = \frac{P_0\hbar}{u_0}\left(\frac{fu_0}{2d} + \frac{g\sin(\theta)}{u_0}\right),$$
$$\mathcal{R}_2^2 = \hbar\left(\frac{fu_0}{2d} + \frac{g\sin(\theta)}{u_0}\right) + 1102500\frac{\hbar}{P_0 u_0}\left(\frac{fu_0}{2d} + \frac{g\sin(\theta)}{u_0}\right)$$
$$+\frac{f}{2d}\left(u_0 + \hbar\left(\frac{fu_0 x}{2d} + \frac{g\sin(\theta)x}{u_0}\right)\right) + \left(\frac{1}{u_0} - \frac{\hbar x}{u_0^2}\left(\frac{fu_0}{2d} + \frac{g\sin(\theta)}{u_0}\right)\right)g\sin(\theta)$$
$$+\left(\frac{\hbar^2(2g\sin(\theta)d + fu_0^2)x^2}{2u_0^4 d}\left(\frac{fu_0}{2d} + \frac{g\sin(\theta)}{u_0}\right)\right)g\sin(\theta),$$
$$P_2(x) = P_0 + \frac{\hbar^2 P_0 x}{u_0}\left(\frac{fu_0}{2d} + \frac{g\sin(\theta)}{u_0}\right),$$
$$u_2(x) = u_0 + \hbar\left(\frac{fu_0 x}{2d} + \frac{g\sin(\theta)x}{u_0}\right)$$
$$+\hbar\left(\frac{\hbar^2(2g\sin(\theta)d + fu_0^2)g\sin(\theta)x^3}{6u_0^4 d}\left(\frac{fu_0}{2d} + \frac{g\sin(\theta)}{u_0}\right)\right)$$
$$+\frac{\hbar}{2}\left(\frac{f\hbar}{2d}\left(\frac{fu_0}{2d} + \frac{g\sin(\theta)}{u_0}\right) - \frac{\hbar g\sin(\theta)}{u_0^2}\left(\frac{fu_0}{2d} + \frac{g\sin(\theta)}{u_0}\right)\right)x^2$$
$$+\hbar\left(\hbar\left(\frac{fu_0}{2d} + \frac{g\sin(\theta)}{u_0}\right)x + 1102500\frac{\hbar x}{P_0 u_0}\left(\frac{fu_0}{2d} + \frac{g\sin(\theta)}{u_0}\right)\right)$$
$$+\hbar\left(\frac{fu_0 x}{2d} + \frac{g\sin(\theta)x}{u_0}\right),$$
$$\vdots$$

therefore, pressure $P(x)$ is as follows,

$$P(x) \simeq P_0 + \frac{3\hbar^2 P_0 xf}{4d} + \frac{3\hbar^2 P_0 xg\sin(\theta)}{2u_0^2} + \frac{5 P_0 x^3 \hbar^4 g^2 (\sin(\theta))^2 f}{24 u_0^4 d} \\ + \frac{P_0 x^3 \hbar^4 g^3 (\sin(\theta))^3}{6u_0^6} + \frac{P_0 x^3 \hbar^4 g\sin(\theta) f^2}{12 u_0^2 d^2} + \frac{P_0 x^3 \hbar^4 f^3}{96 d^3} \\ - \frac{P_0 x^2 \hbar^3 g^2 (\sin(\theta))^2}{4u_0^4} - \frac{P_0 x^2 \hbar^3 fg\sin(\theta)}{8u_0^2 d} + \frac{\hbar^3 P_0 xf}{4d} \\ + \frac{\hbar^3 P_0 xg\sin(\theta)}{2u_0^2} + \frac{275625 x\hbar^3 f}{2 du_0} + 275625 \frac{x\hbar^3 g\sin(\theta)}{u_0^3} + \dots, \quad (47)$$

Equation (47) is a approximation solution for pressure P to the problem Equations (25)–(36) in terms of the convergence parameters $\hbar$ and order $m = 12$ with $\mathscr{H}(x) = 1$. To find the valid region of $\hbar$, the $\hbar$-curves given by the 12th-order HAM approximation at different values of x are drawn in Figure 2; this figure shows the interval of $\hbar$ in which the value of P_{12} is constant at certain x, and M; we chose the horizontal line parallel to x-axis ($\hbar$) as a valid region which provides us with a simple way to adjust and control the convergence region.

Figure 3 is showing the comparison between the homotopy analysis method with Subani et al., 2017 and Elaoud et al., 2010 methods. In this comparison the order of homotopy analysis method have been used as $M = 5$ and $M = 12$. The auxiliary parameter $\hbar$ is chosen as $\hbar = -0.15$ from the convergence interval as showed in the Figure 2. As seen from this figure, with order $M = 12$ the homotopy analysis method is comparable with Subani et al., 2017 and Elaoud et al., 2010 methods. In this problem the auxiliary parameter $\mathscr{H}(x,t)$ is chosen equal 1.

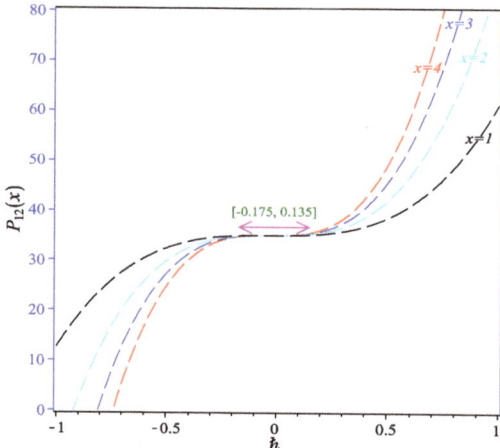

Figure 2. $\hbar$-curve for HAM approximation solution $P_{12}(x)$ of the problem Equations (5) and (6) at different values of x.

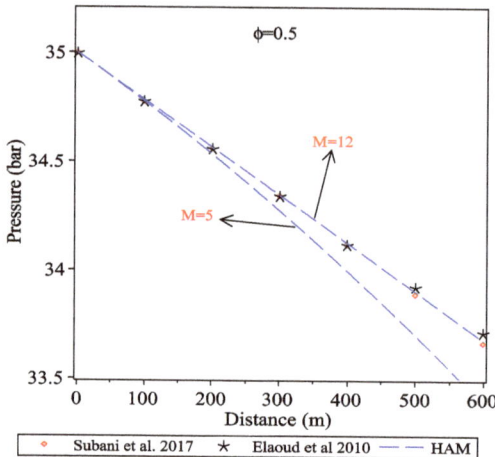

Figure 3. Comparison between homotopy analysis method of orders $M = 5, 12$ for $\hbar = -0.1$; with Subani et al., 2017 and Elaoud et al., 2010 methods.

Now we want to solve the Equations (1) and (2) with the homotopy analysis method (Equations (37)–(46)) using the following initial approximations,

$$P_0(x,t) = \frac{x(x-L)(1+t)}{(x+t)(x-L+t)} P_0(x) + \frac{t(x-L)(1+x)}{(x+t)(x-L+x)} P_0(t) + \frac{xt(1+x-L)}{xt+x-L} P_L(t), \quad (48)$$

$$u_0(x,t) = \frac{x(x-L)(1+t)}{(x+t)(x-L+t)} u_0(x) + \frac{t(x-L)(1+x)}{(x+t)(x-L+x)} u_0(t) + \frac{xt(1+x-L)}{xt+x-L} u_L(t), \quad (49)$$

we guessed the initial approximations Equations (48) and (49) using the initial and boundary conditions (for $x = 0, L$ results will be as Equations (3) and (4)). Therefore, using the homotopy analysis method for solving the Equations (1) and (2) with the initial approximation Equations (48) and (49) can obtain the following results,

$$
\begin{aligned}
P_0(x,t) &= -\frac{xa_0}{P_0^2 t} + \frac{x}{P_0 t} + \frac{1}{P_0} - \frac{xa_0}{P_0^2} - \frac{xa_0}{P_0^2 L} - \frac{x}{P_0 L} - \frac{x}{P_0} - \frac{xta_0}{P_0^2 L} - \frac{xta_0}{P_0^2 L^2} \\
&+ \frac{xtP_L}{P_0^2 L} - \frac{xtP_L}{P_0^2},
\end{aligned}
$$

$$
\begin{aligned}
u_0(x,t) &= -\frac{xb_0}{u_0^2 t} + \frac{x}{u_0 t} + \frac{1}{u_0} - \frac{xb_0}{u_0^2} - \frac{xb_0}{u_0^2 L} - \frac{x}{u_0 L} - \frac{x}{u_0} - \frac{xtb_0}{u_0^2 L} - \frac{xtb_0}{u_0^2 L^2} \\
&+ \frac{xtu_L}{u_0^2 L} - \frac{xtu_L}{u_0^2},
\end{aligned}
$$

$$
\begin{aligned}
\mathscr{R}_1^1 &= xP_L + u_0 a_0 + 2 u_0 P_0 + P_0 b_0 + 2 x u_L a_0 - 2 x u_L P_0 + 2 x b_0 P_L - 2 x u_0 P_L \\
&- \frac{xP_L}{L} + \frac{xa_0}{L} + \frac{xa_0}{L^2} + \frac{P_0 b_0}{L} + 2 x P_0 b_1 + 2 x b_0 a_0 + 2 x u_0 a_1 + 2 x u_0 a_0 \\
&+ 2 x u_0 P_0 + 2 x P_0 b_0 - 2 \frac{u_0 P_0}{t} + \frac{u_0 a_0}{t} + \frac{P_0 b_0}{t} + \frac{xP_0}{t^2} - \frac{xa_0}{t^2} + \frac{u_0 a_0}{L} \\
&+ 2 \frac{u_0 P_0}{L} + 4 \frac{x b_0 a_0}{tL} - 2 \frac{x P_0 b_0}{tL} - 2 \frac{x u_0 a_0}{tL} - 8 \frac{x u_0 P_0}{tL} - 4 \frac{x P_0 b_0}{t^2} \\
&+ 6 \frac{x u_0 P_0}{t^2} + 2 \frac{x b_0 a_0}{t^2} - 4 \frac{x u_0 a_0}{t^2} + 8 \frac{x b_0 a_0}{L} + 8 \frac{x u_0 P_0}{L} + 2 \frac{x P_0 b_1}{t} \\
&+ 4 \frac{x b_0 a_0}{t} + 2 \frac{x u_0 a_1}{t} - 2 \frac{x u_0 a_0}{t} - 8 \frac{x u_0 P_0}{t} - 2 \frac{x P_0 b_0}{t} + 2 \frac{x u_L P_0}{L} \\
&+ 2 \frac{x P_0 b_1}{L} + 2 \frac{x u_0 a_1}{L} + 10 \frac{x u_0 P_0}{L^2} + 6 \frac{x b_0 a_0}{L^2} - 2 \frac{x b_0 P_L}{L} + 2 \frac{x u_0 P_L}{L} \\
&- 2 \frac{x u_L a_0}{L},
\end{aligned}
$$

$$\begin{aligned}
\mathscr{R}_2^1 =\ & 1102501 - 1102500\,x + u_0 + \frac{u_0}{L} + 2\frac{xu_0 b_0 a_0}{t^2 P_0} + 8\frac{xu_0 b_0 a_0}{P_0 L} - 2\frac{xu_0 a_0}{P_0 tL} \\
& - 2\frac{xu_0 a_0^2}{P_0^2 tL} + 2\frac{xb_0 a_0}{P_0 tL} - 2\frac{xu_0^2 a_0}{P_0 tL} - 2\frac{xu_0^2 a_0^2}{P_0^2 tL} + \frac{xfu_0 b_0}{tD} + 4\frac{xu_0 b_0 a_0}{P_0 t} \\
& + 2\frac{xu_0 a_0 P_L}{P_0^2 L} + 2\frac{xu_0^2 a_0 P_L}{P_0^2 L} + 6\frac{xu_0 b_0 a_0}{P_0 L^2} - 2\frac{xu_0 b_0 P_L}{P_0 L} - 2\frac{xu_0 u_L a_0}{P_0 L} \\
& + \frac{xfu_0 b_0}{DL} + u_0^2 + 1102500\,L^{-1} - 1102500\,t^{-1} + g\sin(\theta) - 2\frac{xu_0^2 a_0 P_L}{P_0^2} \\
& + 2\frac{xu_0 b_0 P_L}{P_0} - 3\frac{xu_0 a_0}{P_0 L} - 4\frac{xu_0 a_0^2}{P_0^2 L} + 4\frac{xb_0 a_0}{P_0 L} - 2\frac{xu_0^2 a_0}{P_0 L} - 4\frac{xu_0^2 a_0^2}{P_0^2 L} \\
& + \frac{xfu_0 b_0}{D} - 2\frac{xu_0 a_0}{P_0 t} - 2\frac{xu_0 a_0^2}{P_0^2 t} + 2\frac{xu_0^2 a_1}{P_0 t} - 2205000\,\frac{xa_0}{P_0 tL} \\
& - 2205000\,\frac{xa_0^2}{P_0^2 tL} - 2\frac{xu_0^2 a_0^2}{P_0^2 t} + 2\frac{xb_0 a_0}{P_0 t} - 2\frac{xu_0^2 a_0}{P_0 t} + 2\frac{xu_0 a_1}{P_0 t} \\
& - \frac{fu_0^2 x}{tD} + 2\frac{xu_0 u_L a_0}{P_0} + 2205000\,\frac{xa_0 P_L}{P_0^2 L} + \frac{fu_0^2 x}{DL} + 2\frac{xu_0 a_1}{P_0 L} - 3\frac{xu_0 a_0^2}{P_0^2 L^2} \\
& - 2\frac{xu_0 a_0 P_L}{P_0^2} + 3\frac{xb_0 a_0}{P_0 L^2} - \frac{xb_0 P_L}{P_0 L} - \frac{xu_0 P_L}{P_0 L} - \frac{xu_L a_0}{P_0 L} + 2\frac{xu_0^2 a_1}{P_0 L} \\
& - 3\frac{xu_0^2 a_0^2}{P_0^2 L^2} + xb_0 + 3307500\,\frac{x}{L^2} + xu_0^2 - \frac{u_0^2}{t} + 1102500\,\frac{a_0}{P_0} + 4\frac{xu_0 b_0 a_0}{P_0 tL} \\
& + 4\frac{xu_0}{L^2} + 5\frac{xu_0^2}{L^2} - 2xu_0 u_L + 2xu_0 b_0 + 2\frac{xb_0}{L} + \frac{xb_0}{L^2} + 2205000\,\frac{xP_L}{P_0} \\
& + \frac{u_0 a_0}{P_0} + 3\frac{xu_0^2}{t^2} + 3\frac{xu_0}{t^2} + \frac{u_0^2}{L} + 1102500\,\frac{a_0}{P_0 L} + 2\frac{xu_0}{L} + 2205000\,\frac{xa_1}{P_0} \\
& - 2205000\,\frac{xa_0}{P_0} - 1102500\,\frac{xa_0^2}{P_0^2} + 4\frac{xu_0^2}{L} - 2\frac{xu_0}{t} - 4\frac{xu_0^2}{t} + 1102500\,\frac{a_0}{P_0 t} \\
& + \frac{u_0^2 a_0}{P_0} - 2\frac{xb_0}{t^2} + 1102500\,\frac{x}{t^2} + \frac{1}{2}\frac{fu_0^2}{D} - \frac{u_0}{t} - \frac{xu_0 a_0}{t^2 P_0} - \frac{xu_0 a_0^2}{t^2 P_0^2} + \frac{xb_0 a_0}{t^2 P_0} \\
& - 2\frac{xu_0^2 a_0}{t^2 P_0} - \frac{xu_0^2 a_0^2}{t^2 P_0^2} + 2\frac{xu_0 b_0 a_0}{P_0} + \frac{xb_0 P_L}{P_0} + \frac{xu_0 P_L}{P_0} + \frac{xu_L a_0}{P_0} \\
& + 2205000\,\frac{xa_1}{P_0 L} - 3307500\,\frac{xa_0^2}{P_0^2 L^2} - 2205000\,\frac{xa_0 P_L}{P_0^2} - 2205000\,\frac{xP_L}{P_0 L} \\
& + 2\frac{xu_0 u_L}{L} + 2\frac{xu_0 b_0}{L} - 2\frac{xu_0}{tL} + 2205000\,\frac{xa_1}{P_0 t} - 2205000\,\frac{xa_0}{P_0 t} \\
& - 2205000\,\frac{xa_0^2}{P_0^2 t} - 4\frac{xu_0^2}{tL} + \frac{u_0^2 a_0}{P_0 t} - 2\frac{xu_0 b_0}{t^2} - 1102500\,\frac{xa_0^2}{t^2 P_0^2} + \frac{u_0 a_0}{P_0 L} \\
& + \frac{u_0^2 a_0}{P_0 L} - \frac{xu_0 a_0}{P_0} - \frac{xu_0 a_0^2}{P_0^2} + 2\frac{xu_0^2 a_1}{P_0} - 4410000\,\frac{xa_0}{P_0 L} - 4410000\,\frac{xa_0^2}{P_0^2 L} \\
& - \frac{xu_0^2 a_0^2}{P_0^2} + \frac{xb_0 a_0}{P_0} + 2\frac{xu_0 a_1}{P_0} + \frac{u_0 a_0}{P_0 t} + \frac{fu_0^2 x}{D},
\end{aligned}$$

$$\begin{aligned}
P_1(x,t) = & \frac{xtP_L}{xt-L+x} - \frac{x^2Lta_1}{(x+t)(x-L+t)} - \frac{xLta_0}{(x+t)(x-L+t)} \\
& -\frac{P_0Lx}{(x+t)(2x-L)} + 2\frac{\hbar x^2 b_0 a_0}{tL} - \frac{\hbar x^2 P_0 b_0}{tL} - \frac{\hbar x^2 u_0 a_0}{tL} \\
& -4\frac{\hbar x^2 u_0 P_0}{tL} + \frac{1}{2}\hbar x^2 P_L + \frac{1}{2}\frac{\hbar x^2 P_0}{t^2} - \frac{1}{2}\frac{\hbar x^2 a_0}{t^2} + \hbar x^2 u_L a_0 \\
& -\hbar x^2 u_L P_0 + \hbar x^2 P_0 b_1 + \hbar x^2 b_0 P_L - \hbar x^2 u_0 P_L + \hbar x^2 b_0 a_0 \\
& +\hbar x^2 u_0 a_1 + \hbar x^2 u_0 a_0 + \hbar x^2 u_0 P_0 + \hbar x^2 P_0 b_0 - \frac{1}{2}\frac{\hbar x^2 P_L}{L} \\
& +\frac{1}{2}\frac{\hbar x^2 a_0}{L} + \frac{1}{2}\frac{\hbar x^2 a_0}{L^2} + \hbar x u_0 a_0 + 2\hbar x u_0 P_0 + \hbar x P_0 b_0 \\
& +\frac{x^3 a_1}{(x+t)(x-L+t)} + \frac{x^2 a_0}{(x+t)(x-L+t)} + \frac{x^2 t P_L}{xt-L+x} \\
& +\frac{\hbar x^2 u_0 a_1}{L} + 5\frac{\hbar x^2 u_0 P_0}{L^2} + 3\frac{\hbar x^2 b_0 a_0}{L^2} - \frac{\hbar x^2 b_0 P_L}{L} + \frac{\hbar x^2 u_0 P_L}{L} \\
& -\frac{\hbar x^2 u_L a_0}{L} + 4\frac{\hbar x^2 u_0 P_0}{L} + 2\frac{\hbar x^2 b_0 a_0}{t} + \frac{\hbar x^2 u_0 a_1}{t} + \frac{\hbar x^2 P_0 b_1}{t} \\
& -2\frac{\hbar x^2 P_0 b_0}{t^2} + \frac{\hbar x^2 b_0 a_0}{t^2} + 3\frac{\hbar x^2 u_0 P_0}{t^2} - 2\frac{\hbar x^2 u_0 a_0}{t^2} + \frac{\hbar x^2 u_L P_0}{L} \\
& +\frac{\hbar x^2 P_0 b_1}{L} + 4\frac{\hbar x^2 b_0 a_0}{L} - 4\frac{\hbar x^2 u_0 P_0}{t} + \frac{\hbar x P_0 b_0}{L} + \frac{\hbar x u_0 a_0}{L} \\
& +2\frac{\hbar x u_0 P_0}{L} + \frac{\hbar x u_0 a_0}{t} - 2\frac{\hbar x u_0 P_0}{t} + \frac{\hbar x P_0 b_0}{t} + \frac{x^3 t a_1}{(x+t)(x-L+t)} \\
& -\frac{x^2 L a_1}{(x+t)(x-L+t)} + \frac{x^2 t a_0}{(x+t)(x-L+t)} - \frac{x a_0 L}{(x+t)(x-L+t)} \\
& +\frac{P_0 t x^2}{(x+t)(2x-L)} - \frac{P_0 tL}{(x+t)(2x-L)} + \frac{P_0 tx}{(x+t)(2x-L)} \\
& -\frac{xtP_L L}{xt-L+x} - \frac{\hbar x^2 u_0 a_0}{t} - \frac{\hbar x^2 P_0 b_0}{t},
\end{aligned}$$

$$
\begin{aligned}
u_1(x,t) =\ & \frac{x^3 t b_1}{(x+t)(x-L+t)} - \frac{x^2 L b_1}{(x+t)(x-L+t)} + \frac{x^2 t b_0}{(x+t)(x-L+t)} \\
& - \frac{x L b_0}{(x+t)(x-L+t)} + \frac{u_0 t x^2}{(x+t)(2x-L)} - \frac{u_0 t L}{(x+t)(2x-L)} \\
& + \frac{u_0 t x}{(x+t)(2x-L)} + \frac{x t u_L}{xt-L+x} - \frac{x t u_L L}{xt-L+x} + 1102500\,\frac{\hbar\, x^2 a_1}{P_0 L} \\
& - 1653750\,\frac{\hbar\, x^2 a_0{}^2}{P_0{}^2 L^2} - 1102500\,\frac{\hbar\, x^2 a_0 P_L}{P_0{}^2} - 1102500\,\frac{\hbar\, x^2 P_L}{P_0 L} \\
& - 551250\,\frac{\hbar\, x^2 a_0{}^2}{t^2 P_0{}^2} - \frac{1}{2}\frac{\hbar\, x^2 u_0 a_0{}^2}{P_0{}^2} + \frac{\hbar\, x^2 u_0{}^2 a_1}{P_0} - 2205000\,\frac{\hbar\, x^2 a_0{}^2}{P_0{}^2 L} \\
& - \frac{\hbar\, x^2 u_0}{tL} - 2\,\frac{\hbar\, x^2 u_0{}^2}{tL} + 1102500\,\frac{\hbar\, x^2 a_1}{P_0 t} - 1102500\,\frac{\hbar\, x^2 a_0{}^2}{P_0{}^2 t} \\
& + \frac{1}{2}\frac{\hbar\, x^2 b_0 P_L}{P_0} + \frac{1}{2}\frac{\hbar\, x^2 u_0 P_L}{P_0} - \frac{1}{2}\frac{\hbar\, x^2 u_0{}^2 a_0{}^2}{P_0{}^2} - \frac{\hbar\, x^2 u_0 b_0}{t^2} \\
& + \frac{\hbar\, x^2 u_0 a_1}{P_0} + \frac{1}{2}\frac{\hbar\, x^2 b_0 a_0}{P_0} + \frac{\hbar\, x^2 u_0 u_L}{L} + \frac{\hbar\, x^2 u_0 b_0}{L} - \frac{1}{2}\frac{\hbar\, x^2 u_0 a_0}{P_0} \\
& - 2205000\,\frac{\hbar\, x^2 a_0}{P_0 L} - 1102500\,\frac{\hbar\, x^2 a_0}{P_0 t} + \frac{1}{2}\frac{\hbar\, x^2 f u_0{}^2}{D} + \frac{\hbar\, x u_0{}^2 a_0}{P_0} \\
& + \frac{1}{2}\frac{\hbar\, x^2 u_L a_0}{P_0} + 1102500\,\frac{\hbar\, x a_0}{P_0 t} + \frac{\hbar\, x u_0 a_0}{P_0} + 1102500\,\frac{\hbar\, x a_0}{P_0 L} \\
& + \frac{1}{2}\frac{\hbar\, f u_0{}^2 x}{D} - \frac{\hbar\, x^2 u_0{}^2 a_0 P_L}{P_0{}^2} + \frac{\hbar\, x^2 u_0 b_0 P_L}{P_0} + \frac{\hbar\, x^2 u_0 u_L a_0}{P_0} \\
& + 1102500\,\frac{\hbar\, x^2 a_0 P_L}{P_0{}^2 L} + \frac{1}{2}\frac{\hbar\, x^2 f u_0{}^2}{DL} - 2\,\frac{\hbar\, x^2 u_0{}^2 a_0{}^2}{P_0{}^2 L} + \frac{1}{2}\frac{\hbar\, x^2 f u_0 b_0}{D} \\
& - \frac{\hbar\, x^2 u_0 a_0{}^2}{P_0{}^2 t} + \frac{\hbar\, x^2 u_0{}^2 a_1}{P_0 t} - 1102500\,\frac{\hbar\, x^2 a_0{}^2}{P_0{}^2 tL} - \frac{\hbar\, x^2 u_0{}^2 a_0{}^2}{P_0{}^2 t} \\
& + \frac{\hbar\, x^2 u_0 a_1}{P_0 t} + \frac{\hbar\, x^2 b_0 a_0}{P_0 t} - 1102500\,\frac{\hbar\, x^2 a_0}{P_0 tL} - \frac{1}{2}\frac{\hbar\, x^2 f u_0{}^2}{tD} \\
& + \frac{\hbar\, x^2 u_0 a_1}{P_0 L} - \frac{3}{2}\frac{\hbar}{P_0{}^2 L^2}\, x^2 u_0 a_0{}^2 - \frac{\hbar\, x^2 u_0 a_0 P_L}{P_0{}^2} + \frac{3}{2}\frac{\hbar\, x^2 b_0 a_0}{P_0 L^2} \\
& - \frac{1}{2}\frac{\hbar\, x^2 b_0 P_L}{P_0 L} + \frac{\hbar\, x^2 u_0 b_0 a_0}{P_0} - \frac{1}{2}\frac{\hbar\, x^2 u_0 a_0}{t^2 P_0} - \frac{1}{2}\frac{\hbar\, x^2 u_0 a_0{}^2}{t^2 P_0{}^2} \\
& + \frac{1}{2}\frac{\hbar\, x^2 b_0 a_0}{t^2 P_0} - \frac{\hbar\, x^2 u_0{}^2 a_0}{t^2 P_0} - \frac{1}{2}\frac{\hbar\, x^2 u_0{}^2 a_0{}^2}{t^2 P_0{}^2} - 2\,\frac{\hbar\, x^2 u_0 a_0{}^2}{P_0{}^2 L} \\
& + 2\,\frac{\hbar\, x^2 b_0 a_0}{P_0 L} - \frac{\hbar\, x^2 u_0{}^2 a_0}{P_0 t} - \frac{3}{2}\frac{\hbar\, x^2 u_0 a_0}{P_0 L} - \frac{\hbar\, x^2 u_0{}^2 a_0}{P_0 L} - \frac{\hbar\, x^2 u_0 a_0}{P_0 t} \\
& + \frac{\hbar\, x u_0 a_0}{P_0 L} + \frac{\hbar\, x u_0{}^2 a_0}{P_0 L} + \frac{\hbar\, x u_0 a_0}{P_0 t} + \frac{\hbar\, x u_0{}^2 a_0}{P_0 t} - \frac{x^2 L t b_1}{(x+t)(x-L+t)} \\
& - \frac{x L t b_0}{(x+t)(x-L+t)} - \frac{u_0 t L x}{(x+t)(2x-L)} - \frac{1}{2}\frac{\hbar\, x^2 u_0 P_L}{P_0 L} - \frac{1}{2}\frac{\hbar\, x^2 u_L a_0}{P_0 L} \\
& + \frac{\hbar\, x^2 u_0{}^2 a_1}{P_0 L} - \frac{3}{2}\frac{\hbar\, x^2 u_0{}^2 a_0{}^2}{P_0{}^2 L^2} + 3\,\frac{\hbar\, x^2 u_0 b_0 a_0}{P_0 L^2} - \frac{\hbar\, x^2 u_0 b_0 P_L}{P_0 L} - \frac{\hbar\, x^2 u_0 u_L a_0}{P_0 L}
\end{aligned}
$$

$$+\frac{1}{2}\frac{\hbar x^2 f u_0 b_0}{DL} + 1102500\frac{\hbar x}{L} + \hbar u_0 x - 1102500\frac{\hbar x}{t} + \hbar x u_0^2 + \frac{1}{2}\hbar x^2 b_0$$

$$+\frac{1}{2}\hbar x^2 u_0^2 + 1653750\frac{\hbar x^2}{L^2} + 551250\frac{\hbar x^2}{t^2} - \frac{\hbar x u_0^2}{t} + \hbar g \sin(\theta) x$$

$$+\frac{\hbar x^2 b_0}{L} + \frac{\hbar x^2 u_0}{L} + \frac{5}{2}\frac{\hbar x^2 u_0^2}{L^2} - \hbar x^2 u_0 u_L + \frac{1}{2}\frac{\hbar x^2 b_0}{L^2} + 2\frac{\hbar x^2 u_0}{L^2}$$

$$+\hbar x^2 u_0 b_0 - 2\frac{\hbar x^2 u_0^2}{t} - 1102500\frac{\hbar x^2 a_0}{P_0} + 2\frac{\hbar x^2 u_0^2}{L} - \frac{\hbar x^2 u_0}{t}$$

$$+1102500\frac{\hbar x^2 P_L}{P_0} + 1102500\frac{\hbar x^2 a_1}{P_0} - 551250\frac{\hbar x^2 a_0^2}{P_0^2} - \frac{\hbar x^2 b_0}{t^2}$$

$$+\frac{3}{2}\frac{\hbar x^2 u_0}{t^2} + \frac{3}{2}\frac{\hbar x^2 u_0^2}{t^2} + \frac{\hbar u_0 x}{L} + 1102500\frac{\hbar x a_0}{P_0} + \frac{\hbar x u_0^2}{L} - \frac{\hbar u_0 x}{t}$$

$$+\frac{x^3 b_1}{(x+t)(x-L+t)} + \frac{x^2 b_0}{(x+t)(x-L+t)} + \frac{x^2 t u_L}{xt-L+x} + \frac{\hbar x^2 u_0 b_0 a_0}{t^2 P_0}$$

$$+4\frac{\hbar x^2 u_0 b_0 a_0}{P_0 L} - \frac{\hbar x^2 u_0^2 a_0^2}{P_0^2 t L} + \frac{1}{2}\frac{\hbar x^2 f u_0 b_0}{tD} + 2\frac{\hbar x^2 u_0 b_0 a_0}{P_0 t} - \frac{\hbar x^2 u_0 a_0^2}{P_0^2 t L}$$

$$+\frac{\hbar x^2 b_0 a_0}{P_0 t L} - \frac{\hbar x^2 u_0 a_0}{P_0 t L} - \frac{\hbar x^2 u_0^2 a_0}{P_0 t L} + \frac{\hbar x^2 u_0 a_0 P_L}{P_0^2 L} + \frac{\hbar x^2 u_0^2 a_0 P_L}{P_0^2 L}$$

$$+2\frac{\hbar x^2 u_0 b_0 a_0}{P_0 t L} - 551250\,\hbar x^2 + 1102501\,\hbar x,$$

therefore, pressure $P(x,t)$ is as follows,

$$\begin{aligned}
P(x,t) \simeq &\; \frac{2txP_L}{tx-L+x} - \frac{2x^2 Lt a_1}{(x+t)(x-L+t)} - \frac{2xLta_0}{(x+t)(x-L+t)} \\
&- \frac{2tP_0 Lx}{(x+t)(2x-L)} + \frac{2\hbar x^2 b_0 a_0}{tL} - \frac{\hbar x^2 P_0 b_0}{tL} - \frac{\hbar x^2 u_0 a_0}{tL} - \frac{4\hbar x^2 u_0 P_0}{tL} \\
&+ \frac{1}{2}\hbar x^2 P_L + \frac{1}{2}\frac{\hbar x^2 P_0}{t^2} - \frac{1}{2}\frac{\hbar x^2 a_0}{t^2} + \hbar x^2 u_L a_0 - \hbar x^2 u_L P_0 + \hbar x^2 P_0 b_1 \\
&+ \hbar x^2 b_0 P_L - \hbar x^2 u_0 P_L + \hbar x^2 b_0 a_0 + \hbar x^2 u_0 a_1 + \hbar x^2 u_0 a_0 + \hbar x^2 u_0 P_0 \\
&+ \hbar x^2 P_0 b_0 - \frac{1}{2}\frac{\hbar x^2 P_L}{L} + \frac{1}{2}\frac{\hbar x^2 a_0}{L} + \frac{1}{2}\frac{\hbar x^2 a_0}{L^2} + \hbar x u_0 a_0 + 2\hbar x u_0 P_0 \\
&+ \hbar x P_0 b_0 + \frac{2x^3 a_1}{(x+t)(x-L+t)} + \frac{2x^2 a_0}{(x+t)(x-L+t)} + \frac{2x^2 tP_L}{tx-L+x} \\
&+ \frac{\hbar x^2 u_0 a_1}{L} + \frac{5\hbar x^2 u_0 P_0}{L^2} + \frac{3\hbar x^2 b_0 a_0}{L^2} - \frac{\hbar x^2 b_0 P_L}{L} + \frac{\hbar x^2 u_0 P_L}{L} - \frac{\hbar x^2 u_L a_0}{L} \\
&+ \frac{4\hbar x^2 u_0 P_0}{L} + \frac{2\hbar x^2 b_0 a_0}{t} + \frac{\hbar x^2 u_0 a_1}{t} + \frac{\hbar x^2 P_0 b_1}{t} - \frac{2\hbar x^2 P_0 b_0}{t^2} + \frac{\hbar x^2 b_0 a_0}{t^2} \\
&+ \frac{3\hbar x^2 u_0 P_0}{t^2} - \frac{2\hbar x^2 u_0 a_0}{t^2} + \frac{\hbar x^2 u_L P_0}{L} + \frac{\hbar x^2 P_0 b_1}{L} + \frac{4\hbar x^2 b_0 a_0}{L} - \frac{4\hbar x^2 u_0 P_0}{t} \\
&+ \frac{\hbar x P_0 b_0}{L} + \frac{\hbar x u_0 a_0}{L} + \frac{2\hbar x u_0 P_0}{L} + \frac{\hbar x u_0 a_0}{t} - \frac{2\hbar x u_0 P_0}{t} + \frac{\hbar x P_0 b_0}{t} \\
&+ \frac{2x^3 t a_1}{(x+t)(x-L+t)} - \frac{2x^2 L a_1}{(x+t)(x-L+t)} + \frac{2x^2 t a_0}{(x+t)(x-L+t)} \\
&- \frac{2x a_0 L}{(x+t)(x-L+t)} + \frac{2tP_0 x^2}{(x+t)(2x-L)} - \frac{2tP_0 L}{(x+t)(2x-L)} \\
&+ \frac{2tP_0 x}{(x+t)(2x-L)} - \frac{2txP_L L}{tx-L+x} - \frac{\hbar x^2 u_0 a_0}{t} - \frac{\hbar x^2 P_0 b_0}{t} + \ldots,
\end{aligned} \qquad (50)$$

Equation (50) is a approximation solution for pressure $P(x,t)$ to the problem Equations (1) and (2) in terms of the convergence parameters $\hbar$ with $\mathcal{H}(x) = 1$. To find the valid region of $\hbar$, the $\hbar$-curves given by the 12th-order HAM approximation at different values of t and $x = 0$ are drawn in Figure 4; this figure shows the interval of $\hbar$ in which the value of $P_{12}(0,t)$ is constant at certain t, and M; we chose the horizontal line parallel to t-axis ($\hbar$) as a valid region which provides us with a simple way to adjust and control the convergence region.

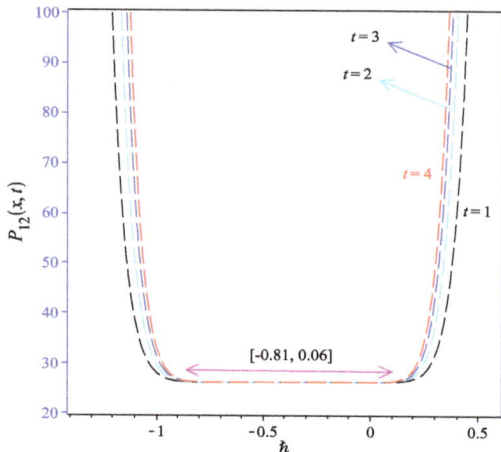

Figure 4. $\hbar$-curve for HAM approximation solution $P_{12}(x,t)$ of the problem Equations (1) and (2) at different values of t and $x = 0$.

3.4. Leak Detection Using Homotopy Analysis Method

Because of a small orifice between the high-pressure pipeline and the environment, the orifice of leak can be simulated leaning on the flow rate. The discharged flow from the orifice can be computed by the following Equation [7],

$$Q_l = \frac{\rho_l C_d A_l \sqrt{2P_l/\rho_l}}{X_L}, \tag{51}$$

where A_l is the leak orifice area with radius r_l, P_l is the pressure of gas mixture at the leak position and ρ_l is the density of gas mixture at the leak position respectively, C_d is a discharge coefficient and X_L is the distance of leak from the reservoir.

Analyzing transient pressure wave for hydrogen/natural gas mixtures is based on transmission and reflection properties of pressure wave effected by a downstream valves sudden closure. When the initial pressure wave reaches the leak, it will produce a reflection as it arrives back at the downstream end section. Then, the difference in time between the initial transient wave and the reflected wave is measured and the leakage position in the pipeline is computed by,

$$X_L = L - \frac{\Delta t_l c_{\Delta t_l}}{2}, \tag{52}$$

where X_L is defined as the distance between the leak and upstream end section, Δt_l is the difference of time between the initial transient wave and reflected wave and $c_{\Delta t_l}$ is defined as the transient celerity wave at time.

3.5. Results and Discussion

Figure 5 presents the transient pressure of hydrogen natural gas mixture for isothermal flow when leakage occurs at $X_L = L/3$ in horizontal pipeline. The homotopy analysis method of order $M = 12$ with $\hbar = -0.5$ has been used. This figure shows the comparison between homotopy analysis method from order 12 and Subani et al. method [7] when $\phi = 0.25$ and $\phi = 0.5$.

Figure 6 shows the transient pressure of hydrogen natural gas mixture ($\phi = 0.5$) for isothermal flow when leakage occurs at $X_L = L/3$ in an inclined pipeline with $\theta = 15$. Black line is homotopy analysis method from order 12 with $\hbar = -0.5$ and red line is Subani et al. method. As indicated in

Figures 5 and 6, the leak point are estimated at $t_s = 0.81$ and at $t_s = 0.808$ for Subani et al. method and HAM respectively.

The transient pressure of mixture of natural gas and hydrogen with a mass ratio of $\phi = 0.5$ is shown in Figure 7 in case of isothermal flow and leak location at $X_L = L/3$ with diverse angles. The homotopy analysis method from order 12 and $\hbar = -0.5$ has been used. Red line is for $\theta = 0$ and black line is for $\theta = 15$.

The celerity wave distribution is presented in Figure 8 as a function of time. In this case, the valve of the horizontal pipeline containing different mass ratios of a mixture of gas and hydrogen is abruptly closed when the leakage is at $X_L = L/3$. The values of celerity wave of the leak point for various mass rations are 819.20 ms^{-1}, 964.60 ms^{-1} and 1086.60 ms^{-1} for $\phi = 0.25, 0.5$ and 0.75, respectively.

As shown in Figures 6 and 7, the occurrence of the leakage is possible when Δt_l is equal to 0.808 s. Equation (53) can be used to calculate the leak location of the mixture of natural gas and hydrogen in case of an isothermal flow in a horizontal pipeline as follows:

$$X_L = 600 - \frac{0.808 \times 964.6}{2} \simeq 210.3. \tag{53}$$

As seen earlier, there are various mass ratios of the mixture and various angles of the pipeline each with a specific leak location at $X_L = L/3$, the values of which are presented in Table 3. It can be inferred that the leak location is not a function of pipe angle, it is rather a function of the mass ratio of the natural gas and hydrogen mixture. Therefore, mass ratio is of utmost importance here.

The real location of leak is 200 m, when the leak location is at $X_L = L/3$. The leak location calculations by Subani et al. and HAM turned out to be 211.10 m and 210.30 m, respectively. It is a mixture of natural gas and hydrogen with a mass ratio of 0.5. When the mass ration is increased to 0.75, the leakage location is less than 200 m. Therefore, when the mass ratio is decreased, the location is greater than 200 m. This is contrary to the calculations since the calculated value is less than 200 m when the mass ratio is 0.5. This is an indication of the dependence of leak location of mass ratio of the mixture considered. As Elaoud et al., (2010) state, the most important part in early determination of a leak close to the reservoir or compressor is the bottom of the pipeline.

Figure 9 shows the leak location with respect to the gas mixture (ϕ). As can be seen from this figure, there is a steep slope for the values $\phi \in [0, 0.25]$ and $\phi \in [0.75, 1]$, but for values $\phi \in [0.25, 0.75]$ there is a mild slope.

Table 3. Leak location for the hydrogen-natural gas mixture for isothermal flow at leakage $X_L = L/3$.

Gas Mixture	Pipeline's Angle	Leak Location (m)	
(ϕ)	(θ)	Subani et al., Method	HAM
0	0°	439.4	439.8
	15°	439.4	439.8
0.25	0°	268.3	269.04
	15°	268.3	269.04
0.5	0°	211.1	210.3
	15°	211.1	210.3
0.75	0°	160.6	161.01
	15°	160.6	161.01
1	0°	95.8	96.2
	15°	95.8	96.2

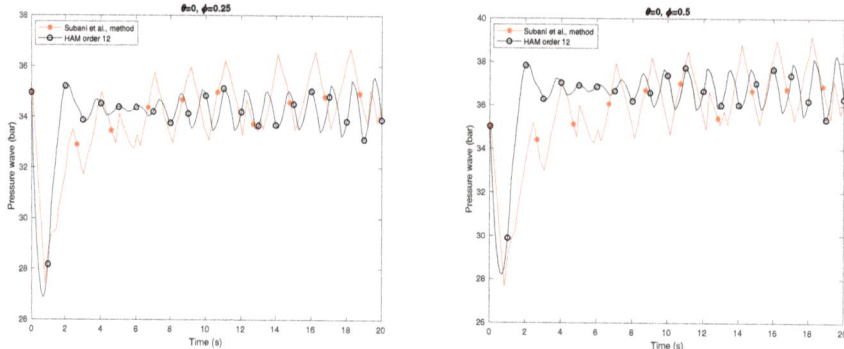

Figure 5. Transient pressure of hydrogen natural gas mixture for isothermal flow when leakage occurs at $X_L = L/3$ in horizontal pipeline when $\phi = 0.25$ and $\phi = 0.5$.

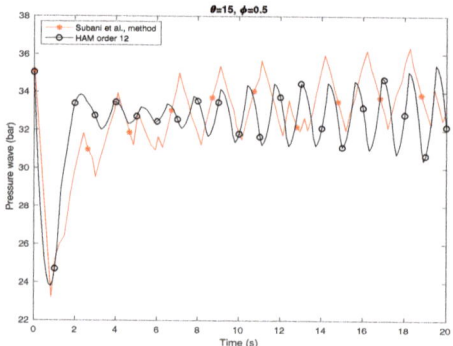

Figure 6. Transient pressure of hydrogen natural gas mixture for isothermal flow when leakage occurs at $X_L = L/3$ in an inclined pipeline when $\theta = 15°$ and $\phi = 0.5$.

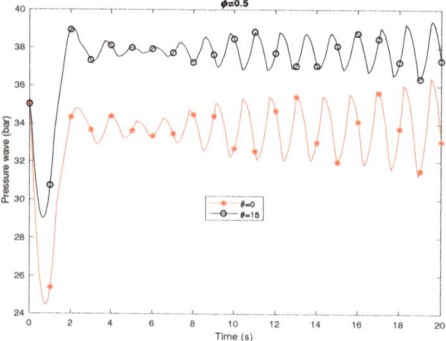

Figure 7. Transient pressure of hydrogen natural gas mixture with $\phi = 0.5$ for isothermal flow when leakage occurs at $X_L = L/3$ with different angles θ. HAM with order 12 and $\hbar = -0.5$.

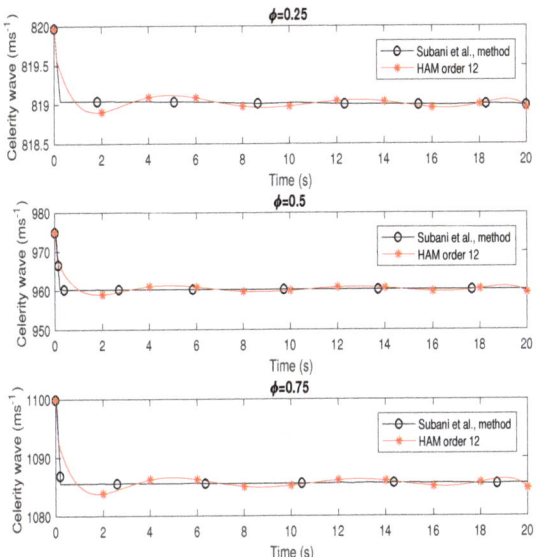

Figure 8. Celerity wave of hydrogen natural gas mixture for isothermal flow when leakage occurs at $X_L = L/3$ in horizontal pipeline with different mass ratio ϕ.

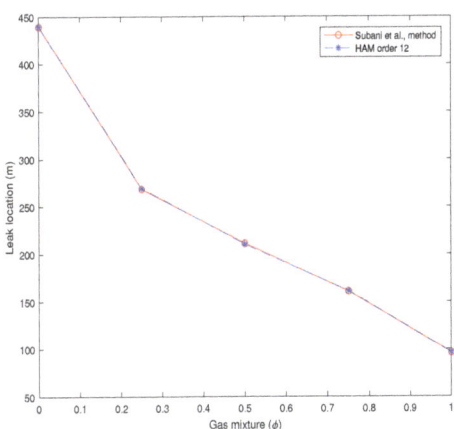

Figure 9. Leak location with respect to the gas mixture (ϕ).

In real (physical) pipelines, noise is expected to affect measurements [35,36]. The possible effects of noisy signals on the performance of the proposed method are Brownian motion or Wiener process or White noise, as the physical model of the stochastic procedure, as an indexed collection random variables. A Wiener process (notation $W = (W_t)_{t \geq 0}$) is named in the honor of Prof. Norbert Wiener; other name is the Brownian motion (notation $B = (B_t)_{t \geq 0}$). Wiener process is Gaussian process. As any Gaussian process, Wiener process is completely described by its expectation and correlation functions. A Brownian motion, also called a Wiener process, is obtained as the integral of a white noise signal as follows,

$$W(t) = \int_0^t \frac{dW(\tau)}{d\tau} d\tau. \tag{54}$$

The effects of noisy signals on the effectiveness of the proposed method and possible effects of noisy signals on the performance of leak locations will be proposed in the future works, by introducing white noise in the simulations.

For accurate pinpointing, we can use the zero-gradient control (ZGC) method which we have discussed in our recently published paper [6] about optimal mixture and controlling the pressure. In our next manuscript with title "Detecting Optimal Leak Locations using Delta Method and Zero Gradient Control for Non-isothermal Hydrogen/Natural Gas Mixture in an Inclined Pipeline" we used the delta method (DM) and zero gradient control (ZGC) method for detecting optimal leak locations. In our future works we will mixed the proposed methods with Artificial intelligence, Neural Network and Deep Learning [37] to predict and estimate the optimal mixture parameter for achieving more accurate pinpointing.

4. Conclusions

The homotopy analysis method used to solve the flow equations of hydrogen natural gas mixture in an inclined pipeline. To validate the approximation series for pressure compared with the Subani et al. method. The results in Figures 3, 5 and 6 show that the obtained results using proposed method are in good agreement with the reduced order modelling (ROM) proposed by Subani et al, in 2017. Then, homotopy analysis method is working as well as other methods and give the semi-analytical solutions.

The leak locations were detected using the homotopy analysis method for horizontal pipeline ($\theta = 0°$) and inclined pipeline ($\theta = 15°$) for gas mixture $\phi = 0, 0.25, 0.5, 0.27, 1$. Using the homotopy analysis method the celerity wave at leak point of the pipeline are 819.20 ms^{-1}, 964.60 ms^{-1} and 1086.60 ms^{-1} for $\phi = 0.25, 0.5$ and 0.75, respectively.

In an inclined pipeline $\theta = 15°$ the leak location for gas mixture $\phi = 0.5$ using the Subani et al. method (ROM) and homotopy analysis method respectively are 211.1 m and 210.3 m. Because of the real leak location is supposed at 200 m when the leak is located at $X_L = L/3$, the result of HAM method is more accurate than ROM method. As can be seen from Figure 9, with increases the gas mixture ϕ from 0 to 1 the leak location decreases and there is a steep slope for $\phi \in [0, 0.25] \cup [0.75, 1]$, and a mild slope for $\phi \in [0.25, 0.75]$.

The proposed HAM method is employed without using linearization, discretization, or transformation. It may be concluded that the HAM is very powerful and efficient in finding the analytical solutions for a wide class of gas transportation equations in a pipeline.

Author Contributions: Funding acquisition: N.A. and Z.I.; Methodology: S.S.C.; Software: S.S.C.; Supervision: N.A.; Writing(original draft): S.S.C.; Writing(review and editing): N.A. and Z.I. All authors have read and agreed to the published version of the manuscript.

Funding: This research received no external funding.

Acknowledgments: The authors gratefully acknowledge financial support from the Ministry of Higher Education, Malaysia and Universiti Teknologi Malaysia through Research and Innovation University Grant Scheme, UTMFR Grant (Vot No. Q.J130000.2554.21H48 and FRGS Grant (Vot No R.J130000.7854.5F255).

Conflicts of Interest: The authors declare that there was no conflict of interest regarding the publication of this paper.

References

1. Elaoud, S.; Hadj-Taieb, E. Transient flow in pipelines of high-pressure hydrogen-natural gas mixtures. *Int. J. Hydrogen Energy* **2008**, *33*, 4824–4832. [CrossRef]
2. Chaczykowski, M. Transient flow in natural gas pipeline the effect of pipeline thermal model. *Appl. Math. Model.* **2010**, *34*, 1051–1067. [CrossRef]

3. Karney, B.W.; Ruus, E. Charts for water hammer in pipelines resulting from valve closure from full opening only. *Can. Civ. Eng.* **1985**, *12*, 241–264. [CrossRef]
4. Kim, H.; Kim, S.; Kim, Y.; Kim, J. Optimization of Operation Parameters for Direct Spring Loaded Pressure Relief Valve in a Pipeline System. *J. Press. Vessel. Technol.* **2018**, *140*, 051603. [CrossRef]
5. Jalving, J.; Zavala, V.M. An Optimization-Based State Estimation Framework for Large-Scale Natural Gas Networks. *Ind. Eng. Chem. Res.* **2018**, *57*, 5966–5979. [CrossRef]
6. Seddighi Chahrborj, S.; Amin, N. Controlling the pressure of hydrogen-natural gas mixture in an inclined pipeline. *PLoS ONE* **2020**, *15*, e0228955.
7. Subani, N.; Amin, N.; Agaie, B.G. Leak detection of non-isothermal transient flow of hydrogennatural gas mixture. *J. Loss Prev. Process. Ind.* **2017**, *48*, 244–253. [CrossRef]
8. Subani, N.; Amin, N. Analysis of water hammer with different closing valve laws on transient flow of hydrogen-natural gas mixture. *Abstr. Appl. Anal.* **2015**, *2015*, 1–12. [CrossRef]
9. Subani, N.; Amin, N.; Agaie, B.G. Hydrogen-natural gas mixture leak detection using reduced order modelling. *Appl. Comput. Math.* **2015**, *4*, 135–144. [CrossRef]
10. Uilhoorn, F.E. Dynamic behaviour of non-isothermal compressible natural gases mixed with hydrogen in pipelines. *Int. J. Hydrogen Energy* **2009**, *34*, 6722–6729. [CrossRef]
11. Veziroglu, T.N.; Barbir, F. Hydrogen: The wonder fuel. *Int. J. Hydrogen Energy* **1992**, *17*, 391–404. [CrossRef]
12. Ebrahimzadeh, E.; Shahrak, M.N.; Bazooyar, B. Simulation of transient gas flow using the orthogonal collocation method. *Chem. Eng. Res. Des.* **2012**, *90*, 1701–1710. [CrossRef]
13. Elaoud, S.; Hadj-Taieb, E. Leak detection of hydrogen-natural gas mixtures in pipes using the pressure-time transient analysis. In Proceedings of the Conference and Exposition International of Ecologic Vehicles and Renewable Energy (EVRE), Monte Carlo, Monaco, 2019.
14. Elaoud, S.; Hadj-Taeb, L.; Hadj-Taeb, E. Leak detection of hydrogennatural gas mixtures in pipes using the characteristics method of specified time intervals. *J. Loss Prev. Process. Ind.* **2010**, *23*, 637–645. [CrossRef]
15. Turner, W.J.; Mudford, N.R. Leak detection, timing, location and sizing in gas pipelines. *Math. Comput. Model.* **1988**, *10*, 609–627. [CrossRef]
16. Wilkening, H.; Baraldi, D.C.F.D. CFD modelling of accidental hydrogen release from pipelines. *Int. J. Hydrogen Energy* **2007**, *32*, 2206–2215. [CrossRef]
17. Puust, R.; Kapelan, Z.; Savic, D.A.; Koppel, T. A review of methods for leakage management in pipe networks. *Urban Water J.* **2010**, *7*, 25–45. [CrossRef]
18. Diao, X.; Shen, G.; Jiang, J.; Chen, Q.; Wang, Z.; Ni, L.; Mebarki, A.; Dou, Z. Leak detection and location in liquid pipelines by analyzing the first transient pressure wave with unsteady friction. *J. Loss Prev. Process. Ind.* **2019**, *60*, 303–310. [CrossRef]
19. Li, S.; Zhang, J.; Yan, D.; Wang, P.; Huang, Q.; Zhao, X.; Cheng, Y.; Zhou, Q.; Xiang, N.; Dong, T. Leak detection and location in gas pipelines by extraction of cross spectrum of single non-dispersive guided wave modes. *J. Loss Prev. Process. Ind.* **2016**, *44*, 255–262. [CrossRef]
20. Silva, R.A.; Buiatti, C.M.; Cruz, S.L.; Pereira, J.A. Pressure wave behaviour and leak detection in pipelines. *Comput. Chem. Eng.* **1996**, *20*, S491–S496. [CrossRef]
21. Ge, C.; Wang, G.; Ye, H. Analysis of the smallest detectable leakage flow rate of negative pressure wave-based leak detection systems for liquid pipelines. *Comput. Chem.* **2008**, *32*, 1669–1680. [CrossRef]
22. Brunone, B.; Ferrante, M. Detecting leaks in pressurised pipes by means of transients. *J. Hydraul. Res.* **2001**, *39*, 539–547. [CrossRef]
23. Martini, A.; Rivola, A.; Troncossi, M. Autocorrelation analysis of vibro-acoustic signals measured in a test field for water leak detection. *Appl. Sci.* **2018**, *8*, 2450. [CrossRef]
24. Brennan, M.J.; Gao, Y.; Ayala, P.C.; Almeida, F.C.L.; Joseph, P.F.; Paschoalini, A.T. Amplitude distortion of measured leak noise signals caused by instrumentation: Effects on leak detection in water pipes using the cross-correlation method. *J. Sound Vib.* **2019**, *461*, 114905. [CrossRef]
25. Martini, A.; Troncossi, M.; Rivola, A. Leak detection in water-filled small-diameter polyethylene pipes by means of acoustic emission measurements. *Appl. Sci.* **2017**, *7*, 2. [CrossRef]
26. Liu, C.; Li, Y.; Fang, L.; Xu, M. New leak-localization approaches for gas pipelines using acoustic waves. *Measurement* **2019**, *134*, 54–65. [CrossRef]
27. Li, J.; Zheng, Q.; Qian, Z.; Yang, X. A novel location algorithm for pipeline leakage based on the attenuation of negative pressure wave. *Process. Saf. Environ.* **2019**, *123*, 309–316. [CrossRef]

28. Liao, S.J. The Proposed Homotopy Analysis Technique for the Solution of Nonlinear Problems. Ph.D. Thesis, Shanghai Jiao Tong University, Shanghai, China, 1992.
29. Rana, P.; Shukla, N.; Gupta, Y.; Pop, I. Homotopy analysis method for predicting multiple solutions in the channel flow with stability analysis. *Commun. Nonlinear Sci. Numer. Simul.* **2019**, *66*, 183–193. [CrossRef]
30. Yu, C.; Wang, H.; Fang, D.; Ma, J.; Cai, X.; Yu, X. Semi-analytical solution to one-dimensional advective-dispersive-reactive transport equation using homotopy analysis method. *J. Hydrol.* **2018**, *565*, 422–428. [CrossRef]
31. Seddighi Chahrborj, S.; Sadat Kiai, S.M.; Abu Bakar, M.R.; Ziaeian, I.; Gheisari, Y. Homotopy analysis method to study a quadrupole mass filter. *J. Mass Spectrom.* **2012**, *47*, 484–489. [CrossRef]
32. Seddighi Chahrborj, S.; Moameni, A. Spectral-homotopy analysis of MHD non-orthogonal stagnation point flow of a nanofluid. *J. Appl. Math. Comput. Mech.* **2018**, *17*. [CrossRef]
33. Liao, S. *Beyond Perturbation: Introduction to the Homotopy Analysis Method*; CRC Press: Boca Raton, FL, USA, 2003.
34. Mehmood, A.; Munawar, S.; Ali, A. Comments to: Homotopy analysis method for solving the MHD flow over a non-linear stretching sheet (Commun. Nonlinear Sci. Numer. Simul. 14 (2009) 26532663). *Commun. Nonlinear Sci. Numer.* **2010**, *15*, 4233–4240. [CrossRef]
35. Mandal, P.C. Gas leak detection in pipelines & repairing system of titas gas. *J. Appl. Eng.* **2014**, *2*.
36. Seddighi Chahrborj, S.; Kiai, S.M.S.; Arifina, N.M.; Gheisari, Y. Applications of Stochastic Process in the Quadrupole Ion traps. *Mass Spectrom. Lett.* **2015**, *6*, 91–98. [CrossRef]
37. Seddighi Chahrborj, S.; Chaharborj, S.S.; Mahmoudi, Y. Study of fractional order integro-differential equations by using Chebyshev Neural Network. *J. Math. Stat.* **2017**, *13*, 1–13. [CrossRef]

© 2020 by the authors. Licensee MDPI, Basel, Switzerland. This article is an open access article distributed under the terms and conditions of the Creative Commons Attribution (CC BY) license (http://creativecommons.org/licenses/by/4.0/).

Article

Error Estimation of the Homotopy Perturbation Method to Solve Second Kind Volterra Integral Equations with Piecewise Smooth Kernels: Application of the CADNA Library

Samad Noeiaghdam [1,2,*], Aliona Dreglea [1,3,*], Jihuan He [4], Zakieh Avazzadeh [5], Muhammad Suleman [6], Mohammad Ali Fariborzi Araghi [7], Denis N. Sidorov [1,3] and and Nikolai Sidorov [8]

1. Baikal School of BRICS, Irkutsk National Research Technical University, 664074 Irkutsk, Russia; dsidorov@isem.irk.ru
2. Department of Applied Mathematics and Programming, South Ural State University, Lenin Prospect 76, 454080 Chelyabinsk, Russia
3. Energy Systems Institute of Russian Academy of Science, 664033 Irkutsk, Russia
4. National Engineering Laboratory for Modern Silk, Soochow University, Suzhou 215021, China; hejihuan@suda.edu.cn
5. Department of Mathematical Sciences, Xi'an Jiaotong-Liverpool University, Suzhou 215123, China; zakiehavazzadeh@xjtlu.edu.cn
6. Department of Mathematics, Comsats Institute of Information Technology, Islamabad 45550, Pakistan; suleman@zju.edu.cn
7. Department of Mathematics, Central Tehran Branch, Islamic Azad University, Tehran 1955847881, Iran; m_fariborzi@iauctb.ac.ir
8. Institute of Mathematics and IT, Irkutsk State University, 664025 Irkutsk, Russia; sidorov@math.isu.runnet.ru
* Correspondence: noiagdams@susu.ru (S.N.); adreglea@isem.irk.ru (A.D.)

Received: 25 September 2020; Accepted: 15 October 2020; Published: 20 October 2020

Abstract: This paper studies the second kind linear Volterra integral equations (IEs) with a discontinuous kernel obtained from the load leveling and energy system problems. For solving this problem, we propose the homotopy perturbation method (HPM). We then discuss the convergence theorem and the error analysis of the formulation to validate the accuracy of the obtained solutions. In this study, the Controle et Estimation Stochastique des Arrondis de Calculs method (CESTAC) and the Control of Accuracy and Debugging for Numerical Applications (CADNA) library are used to control the rounding error estimation. We also take advantage of the discrete stochastic arithmetic (DSA) to find the optimal iteration, optimal error and optimal approximation of the HPM. The comparative graphs between exact and approximate solutions show the accuracy and efficiency of the method.

Keywords: stochastic arithmetic; homotopy perturbation method; CESTAC method; CADNA library; Volterra integral equation with piecewise continuous kernel

1. Introduction

The problem of finding approximate solution for linear Volterra IEs is one of the oldest problems in the applied mathematics researches. Specially, this problem with discontinuous kernel has many applications in the load leveling problems, energy storage with renewable and diesel generation, charge/discharge storages control and others [1–3].

There are various methods for solving linear and nonlinear problems [4–10] specially the Volterra IEs with discontinuous kernel. Muftahov et al. in [11] applied the Lavrentiev regularization and direct quadrature method, Sidorov in [12] used the successive approximations and Noeiaghdam et al. studied the Taylor-collocation method for solving Volterra IEs with discontinuous kernel [1,13]. Also, the nonlinear system of Volterra IE with applications was studied in [14,15]. Furthermore, the existence of a continuous solution depending on free parameters and sufficient conditions for the existence of a unique continuous solution of the system of Volterra IE with discontinuous kernels were derived in [16]. The class of integral operator equations of Volterra type with applications to p-Laplacian equations was illustrated in [17]. The problem of generalized solution (in the Sobolev-Schwartz sense) to the Volterra equations with piecewise continuous kernel was illustrated in [18]. Belbas and Bulka in [19] considered the multiple Volterra IEs. The problem of global solution's existence and blow-up of nonlinear Volterra IEs were discussed in [20]. For systematic study of the qualitative theory of Volterra IE with discontinuous kernels readers may refer to monograph [21] and part 1 in monograph [22].

The parametric continuation method for the first time was justified by Bernstein [23] for partial differential equations. Here readers may also refer to excellent review by Lusternik [24]. In community of numerical analysts the parametric continuation method is known as the HPM. This method is among of semi-analytical methods that was popularized by J.H. He [25–27]. Then, this method has been extended by many other researchers for solving different problems. The HPM was applied to find the approximate analytical solution of the Allen-Cahn equation in [28], to study the maximum power extraction from fractional order doubly fed induction generator based wind turbines in [29], dissipative nonplanar solitons in an electronegative complex plasma in [30] and others [31–33]. Convergence of the parameter continuation method in the homotopy method based on the theorem of V.A. Trenogin (see [34], Section 14, p. 146) will be global with respect to a parameter if there is an a priori estimate of the solution for all values of the parameter (this condition can be replaced with a more stringent requirement for the existence of a unique solution bounded for all values of the parameter). If there is no a priori estimate of the solution, then on the basis of the inverse operator theorem (see [34] p. 135), at least local convergence in the homotopy method can be guaranteed. Due to the models complexity, we addressed only some classes of the results in this field. Many other interesting results concerning nonlinear equations with discontinuous symmetric kernels with application of group symmetry have remained beyond this paper. Results of present paper in combination with methods of representation theory and group analysis in the bifurcation theory [35,36] make it possible to construct solutions of nonlinear models with discontinuous kernels using the HPM.

In the mentioned studies and many other researches, the numerical results have been obtained from the floating point arithmetic (FPA) and the accuracy of the method has been discussed using the traditional absolute error as follows

$$|w(t) - w_n(t)| < \varepsilon, \tag{1}$$

where $w(t)$ and $w_n(t)$ are the exact and approximate solutions. This condition depends on the existence of the exact solution and optimal value of ε. Also, based on condition (1) we will not be able to find the more accurate approximation because we do not have information about optimal ε and in some cases we do not know the exact solution. For small values of ε, the numerical algorithm can not be stopped and extra iterations will be produced without improving the accuracy. For large values of ε, the numerical algorithm will be stopped in initial steps without producing enough iterations. Moreover, in condition (1), researchers do not have any idea about optimal approximations, optimal errors or numerical instabilities. The aim of this study is to apply the HPM to solve the second kind linear Volterra IEs with jumping kernel and validate the numerical results using the CESTAC method [37–40]. In this method, instead of applying the condition (1), we need to produce other and better condition without having the disadvantages of (1) as follows:

$$|w_n(t) - w_{n+1}(t)| = @.0, \tag{2}$$

where @.0 is the informatical zero sign [41] and $w_n(t)$ and $w_{n+1}(t)$ are two successive approximations. Condition (2) is based on the DSA and Theorem 2 can support us to apply this condition theoretically. In this condition, not only we do not need to have the exact solution but also we would be able to identify the optimal approximation, optimal iteration and optimal error of numerical procedure.

Also, the CADNA library is applied as an important software for this validation. The CESTAC method and the CADNA library have been introduced and developed during decades by researchers from LIP6, the computer science laboratory in Sorbonne University in Paris, France (https://www-pequan.lip6.fr/). This principle was introduced in [38] and it was extended to various quadrature rules in [42–44] and others [45,46]. The CADNA library should be done on the LINUX operating system and its codes should be written using C, C++ or ADA codes [40,47–49]. The CESTAC method is based on the DSA and instead of applying the absolute error to show the precision of method, a termination criterion is applied based on two successive approximations [50–53]. Thus in this technique we do not need to have the exact solution to compare the results. Also, we will prove that number of common significant digits (NCSD) of two successive approximations are almost equal to the NCSD of exact and approximate solutions. So the new theorem gives the license to apply the new stopping condition instead of previous one. This technique has some advantages than other methods based on the FPA [37,39,50,52,53]. Due to the advantages of the CESTAC method we can find the optimal iteration of iterative and numerical methods, optimal approximation and optimal error. Furthermore, the extra iterations can be neglected and some of numerical instabilities can be detected too [13,54–56].

In recent years, this scheme was applied to estimate the round-off errors in different problems such as the numerical integration rules by Newton-Cotes and Gaussian rules [54,57–60], interpolation [61], solving IEs by Sinc-collocation method [55,62], homotopy analysis method for solving IEs [63] and Taylor-collocation method for discontinuous Volterra IEs [13]. Furthermore, this technique is applied for finding the optimal regularized parameter of the regularization method [56], solving ill-posed problems [56] and many other topics [64–66].

This paper is arranged as follows: In the next section, the preliminaries are described regarding to the HPM. In third section, the DSA and the CESTAC method are discussed. Also, algorithm of the CESTAC method and sample code of the CADNA library are presented. In forth section the main idea is described. Then using the HPM we solve the second kind linear Volterra IEs with jumping kernel. Furthermore, the convergence theorem is proved. Also, a theorem is presented which proves that instead of traditional absolute error which depends on the exact and approximate solutions, a termination criterion can be applied which depends on two successive approximations. Section five includes some examples. Also, several tables are presented to show the efficiency of method. The last section is conclusion.

2. Preliminaries

For operator F, given function g and prepared function x we get the following operator equation as

$$F(x) = g(z), \quad z \in \Gamma. \tag{3}$$

We can write the operator F in the following form

$$F(x) = \mathcal{L}(x) + \mathcal{N}(x), \tag{4}$$

where the remain part of F showed by $\mathcal{N}$ and $\mathcal{L}$ is the linear operator. Now, Equation (3) can be presented as

$$\mathcal{L}(x) + \mathcal{N}(x) = g(z), \quad z \in \Gamma. \tag{5}$$

According to the traditional homotopy [25–27], for parameter $\hat{a} \in [0,1]$, the homotopy operator H can be presented as

$$H(v, \hat{a}) = (1 - \hat{a})(\mathcal{L}(v) - \mathcal{L}(x_0)) + \hat{a}(F(v) - g(z)), \tag{6}$$

where $v(z, \hat{a})$ is defined on $\Gamma \times [0,1] \to R$ and x_0 is the initial guess of Equation (3). Now, by applying Equation (4) we get

$$H(v, \hat{a}) = \mathcal{L}(v) - \mathcal{L}(x_0) + \hat{a}\mathcal{L}(x_0) + \hat{a}(\mathcal{N}(v) - g(z)). \tag{7}$$

Putting $\hat{a} = 0$ in Equation (7) leads to $H(v, 0) = \mathcal{L}(v) - \mathcal{L}(x_0)$ and we get $\mathcal{L}(v) - \mathcal{L}(x_0) = 0$. Now, for $\hat{a} = 1$ we have $H(v, 1) = 0$ which it can produce the solution of Equation (3). Thus, when $\hat{a} : 0 \to 1$ we can change the solution v from x_0 to x. Now, the power series

$$v = \sum_{j=0}^{\infty} \hat{a}^j v_j, \tag{8}$$

can be applied to find the solution of $H(v, \hat{a}) = 0$. Then comparing the same powers of parameter $\hat{a}$ we can find the successive functions $v_j, j = 0, \cdots, n$.

Finally, applying

$$w = \lim_{\hat{a} \to 1} v = \sum_{j=0}^{\infty} v_j, \tag{9}$$

the solution of Equation (3) can be found and the n-th order approximation is in the following form

$$w_n = \sum_{j=0}^{n} v_j. \tag{10}$$

3. Stochastic Arithmetic and the CESTAC Method

In this section, the CESTAC method is described and the algorithm of this method is presented. Also, a sample program of the CADNA library is demonstrated and finally advantages of presented method based on the DSA are investigated in comparison with the traditional FPA [37–40,50].

Assume that some representable values are produced by computer and they are collected in set A. Then $W \in A$ can be produced for $w \in \mathbb{R}$ with $\mathcal{R}$ mantissa bits of the binary FPA in the following form

$$W = w - \chi 2^{E-\mathcal{R}} \xi, \tag{11}$$

where sign of w showed by χ, missing segment of the mantissa presented by $2^{-\mathcal{R}} \xi$ and the binary exponent of the result characterized by E. Moreover, there are single and double precisions by choosing $\mathcal{R} = 24, 53$ [40,50–53].

Assume ξ is the casual variable that uniformly distributed on $[-1, 1]$. After making perturbation on final mantissa bit of w we will have (μ) and (σ) as mean and standard deviation for results of W which they have important rule in precision of W. Repeating this process J times for $W_i, i = 1, \ldots, J$ we will have quasi Gaussian distribution for results. It means that μ for these data equals to the exact w. It is clear that we should find μ and σ based on $W_i, i = 1, \ldots, J$. For more consideration, the following Algorithm 1 is presented where τ_δ is the value of T distribution as the confidence interval is $1 - \delta$ with $J - 1$ freedom degree [52].

Usually, in order to find the numerical results we need to apply the usual packages like Mathematica and Matlab. Here, instead of them we introduce the CADNA library and the CESTAC method to validate the numerical results [1,55,56,62].

This library should run on LINUX operating system and all commands should be written by C, C++, FORTRAN or ADA codes [13,54,59,60,63].

We have many advantages to apply the CESTAC method and the CADNA library instead of traditional schemes using the FPA. In this method, a novel criterion independence of absolute error and tolerance value like ε is presented. Applying the CADNA library, we can find the optimal iteration, approximation and error of numerical methods. Moreover, the numerical instabilities can be identified [13,54–56]. A sample program of the CADNA library is presented as

Algorithm 1:

Step 1- Make the perturbation of the last bit of mantissa to produce J samples of W as $\Phi = \{W_1, W_2, ..., W_J\}$.

Step 2- Find $W_{ave} = \dfrac{\sum_{i=1}^{J} W_i}{J}$.

Step 3- Compute $\sigma^2 = \dfrac{\sum_{i=1}^{J}(W_i - W_{ave})^2}{J-1}$.

Step 4- Find the NCSDs of w and W_{ave} applying $C_{W_{ave},w} = \log_{10} \dfrac{\sqrt{J}|W_{ave}|}{\tau_\delta \sigma}$.

Step 5- Print $W = @.0$ if $W_{ave} = 0$, or $C_{W_{ave},W} \leq 0$.

```
#include <cadna.h>
cadna_init(-1);
main()
{
double_st Parameter;
do
{
Write the main codes here;
printf(" %s ",strp(Parameter));
}
while(u[n]-u[n-1]!=0);
cadna_end();
}
```

4. Main idea

Consider the following second kind linear IE

$$w(t) = g(t) + \int_0^t k(t,s)w(s)ds, a = 0 \leq t \leq T \leq b, \tag{12}$$

where $k(t,s)$ is discontinuous along continuous curves $\gamma_i, i = 0, 1, \cdots, m-1$ and it can be written in the following form

$$w(t) = g(t) + \int_{\gamma_0(t)}^{\gamma_1(t)} k_1(t,s)w(s)ds + \int_{\gamma_1(t)}^{\gamma_2(t)} k_2(t,s)w(s)ds + \cdots + \int_{\gamma_{m-1}(t)}^{\gamma_m(t)} k_m(t,s)w(s)ds, \tag{13}$$

and finally for brief form we get

$$w(t) = g(t) + \sum_{i=1}^{m} \int_{\gamma_{i-1}(t)}^{\gamma_i(t)} k_i(t,s)w(s)ds. \tag{14}$$

Indeed, the kernel is the principal part of the IE (14). One may think about considered Volterra IE as generalization of classic Duhamel integral. So, the kernel can be understood as instrumental response function (IF, or spectral sensitivity, transmission function, point spread function, frequency response), see e.g., [67]. In this study, we do not focus on specific physical problems, but more on numerical aspects of solutions only.

Based on the HPM and applying Equations (4) and (5) for solving Equation (14), operators $\mathcal{L}(v)$ and $\mathcal{N}(v)$ should be defined as follows

$$\mathcal{L}(v) = v,$$

and
$$\mathcal{N}(v) = \sum_{i=1}^{m} \int_{\gamma_{i-1}(t)}^{\gamma_i(t)} k_i(t,s) w(s) ds. \tag{15}$$

For next step, using Equation (7) the homotopy map can be constructed as follows

$$H(v,\hat{a}) = v(t) - w_0(t) + \hat{a}\left[w_0(t) - \sum_{i=1}^{m} \int_{\gamma_{i-1}(t)}^{\gamma_i(t)} k_i(t,s) w(s) ds - g(t)\right], \tag{16}$$

and we have

$$\sum_{j=0}^{\infty} \hat{a}^j v_j(t) = w_0(t) + \hat{a}[g(t) - w_0(t)] + \sum_{j=1}^{\infty} \hat{a}^j \sum_{i=1}^{m} \int_{\gamma_{i-1}(t)}^{\gamma_i(t)} k_i(t,s) v_{j-1}(s) ds. \tag{17}$$

Now, Equation (17) can be written in the following form

$$\sum_{j=0}^{\infty} \hat{a}^j v_j(t) = w_0(t) + \hat{a}[g(t) - w_0(t)] + \sum_{j=1}^{\infty} \hat{a}^j A_{j-1}(t), \tag{18}$$

where

$$A_{j-1}(t) = \sum_{i=1}^{m} \int_{\gamma_{i-1}(t)}^{\gamma_i(t)} k_i(t,s) v_{j-1}(s) ds.$$

By disjointing the different powers of $\hat{a}$ in both sides of Equation (18) the following successive iterations can be obtained as

$$\begin{aligned}
\hat{a}^0 &: v_0(t) = w_0(t), \\
\hat{a}^1 &: v_1(t) = g(t) - w_0(t) + A_0(t) \\
&= g(t) - w_0(t) + \sum_{i=1}^{m} \int_{\gamma_{i-1}(t)}^{\gamma_i(t)} k_i(t,s) v_0(s) ds, \\
\hat{a}^2 &: v_2(t) = A_1(t) = \sum_{i=1}^{m} \int_{\gamma_{i-1}(t)}^{\gamma_i(t)} k_i(t,s) v_1(s) ds, \\
&\vdots \quad \vdots \qquad \vdots \\
\hat{a}^n &: v_n(t) = A_{n-1}(t) = \sum_{i=1}^{m} \int_{\gamma_{i-1}(t)}^{\gamma_i(t)} k_i(t,s) v_{n-1}(s) ds.
\end{aligned} \tag{19}$$

Applying Equation (10) and successive iterations (19), the approximate solution of Equation (14) can be obtained.

Theorem 1. *Assume that functions $k_i(t,s)$ and $g(t)$ of Equation (14) are continuous in $\eta_1 = [a,b] \times [a,b]$ and $\eta = [a,b]$ respectively where these functions are bounded. If*

$$\exists \alpha_i, N_1; |k_i(t,s)| \leq \alpha_i, |g(t)| \leq N_1, \forall s, t \in \eta, i = 1, 2, \cdots, m,$$

then for initial approximation w_0 which is continuous in $[a,b]$, the series solution (9) will be uniformly convergent to the exact solution for each $\hat{a} \in [0,1]$.

Proof. Assume $w_0(t) \in C[a,b]$, then we have a positive number N_0 such that $|w_0(t)| \leq N_0$. Therefore, we can write

$$|v_0(t)| = |w_0(t)| \leq N_0,$$

$$|v_1(t)| = \left| g(t) - w_0(t) + \int_{\gamma_0(t)}^{\gamma_1(t)} k_1(t,s)v_0(s)ds + \int_{\gamma_1(t)}^{\gamma_2(t)} k_2(t,s)v_0(s)ds \right.$$

$$\left. + \cdots + \int_{\gamma_{m-1}(t)}^{\gamma_m(t)} k_m(t,s)v_0(s)ds \right|$$

$$\leq |g(t)| + |w_0(t)| + \int_{\gamma_0(t)}^{\gamma_1(t)} |k_1(t,s)||v_0(s)|ds + \int_{\gamma_1(t)}^{\gamma_2(t)} |k_2(t,s)||v_0(s)|ds$$

$$+ \cdots + \int_{\gamma_{m-1}(t)}^{\gamma_m(t)} |k_m(t,s)||v_0(s)|ds|$$

$$\leq N_1 + N_0 + \alpha_1 N_0 (\gamma_1 - \gamma_0) + \alpha_2 N_0 (\gamma_2 - \gamma_1) + \cdots + \alpha_m N_0 (\gamma_m - \gamma_{m-1}) = \beta,$$

$$|v_2(t)| = \left| \int_{\gamma_0(t)}^{\gamma_1(t)} k_1(t,s)v_1(s)ds + \int_{\gamma_1(t)}^{\gamma_2(t)} k_2(t,s)v_1(s)ds \right.$$

$$\left. + \cdots + \int_{\gamma_{m-1}(t)}^{\gamma_m(t)} k_m(t,s)v_1(s)ds \right|,$$

$$|v_2(t)| \leq \int_{\gamma_0(t)}^{\gamma_1(t)} |k_1(t,s)||v_1(s)|ds + \int_{\gamma_1(t)}^{\gamma_2(t)} |k_2(t,s)||v_1(s)|ds$$

$$+ \cdots + \int_{\gamma_{m-1}(t)}^{\gamma_m(t)} |k_m(t,s)||v_1(s)|ds$$

$$\leq \alpha_1(\gamma_1 - \gamma_0)\beta + \alpha_2(\gamma_2 - \gamma_1)\beta + \cdots + \alpha_m(\gamma_m - \gamma_{m-1})\beta$$

$$= \beta \sum_{i=1}^{m} \alpha_i (\gamma_i - \gamma_{i-1}).$$

Accordingly, we obtain the following general form

$$|v_j(t)| \leq \beta \left(\sum_{i=1}^{m} \alpha_i^{j-1} \frac{(\gamma_i - \gamma_{i-1})^{j-1}}{(j-1)!} \right), \quad s,t \in [a,b], j \geq 2. \tag{20}$$

Finally, for series solution (8) and for any $\hat{a} \in [0,1]$ we can write

$$\sum_{j=0}^{\infty} \hat{a}^j v_j(t) \leq \sum_{j=0}^{\infty} |v_j(t)| \leq \sum_{j=0}^{\infty} a_j = N_0 + \beta + \beta \exp\left(\sum_{i=1}^{m} \alpha_i (\gamma_i - \gamma_{i-1}) \right),$$

where $a_0 = N_0, a_1 = \beta, a_j = \beta \left(\sum_{i=1}^{m} \alpha_i^{j-1} \frac{(\gamma_i - \gamma_{i-1})^{j-1}}{(j-1)!} \right), j \geq 2$. It means that series solution (8) for any $\hat{a} \in [0,1]$ is uniformly convergent in interval $[a,b]$. □

From Equation (20), the following remark can be deduced:

Remark 1. Based on the n-th order approximate solution (10), the error function $E_n = \sup_{t \in [a,b]} |w(t) - w_n(t)|$ can be approximated as follows:

$$|w(t) - w_n(t)| = \left| \sum_{j=0}^{\infty} v_j(t) - \sum_{j=0}^{n} v_j(t) \right| = \left| \sum_{j=n+1}^{\infty} v_j(t) \right| \leq \sum_{j=n+1}^{\infty} |v_j(t)|$$

$$\leq \beta \sum_{j=n+1}^{\infty} \left(\sum_{i=1}^{m} \alpha_i^{j-1} \frac{(\gamma_i - \gamma_{i-1})^{j-1}}{(j-1)!} \right).$$

Order of error E_n can be obtained in the following form:

$$E_n = \mathcal{O}\left[\sum_{j=n+1}^{\infty} \frac{1}{(j-1)!} \left(\sum_{i=1}^{m} \alpha_i^j (\gamma_i - \gamma_{i-1})^j \right) \right] = \mathcal{O}\left(\frac{L^n}{n!} \right),$$

where L is a positive real number.

Definition 1 ([38]). For numbers $z_1, z_2 \in \mathbb{R}$, the NCSDs can be computed as follows:

(1) for $z_1 \neq z_2$,
$$C_{z_1,z_2} = \log_{10}\left| \frac{z_1 + z_2}{2(z_1 - z_2)} \right| = \log_{10}\left| \frac{z_1}{z_1 - z_2} - \frac{1}{2} \right|, \qquad (21)$$

(2) for all real numbers z_1, $C_{z_1,z_1} = +\infty$.

Theorem 2. Let $w(t)$ and $w_n(t)$ be the exact and numerical solutions of problem (12) which $w_n(t)$ is obtained by using the HPM and Equation (10). Based on assumptions of Theorem 1 and Remark 1 for n enough large we have

$$C_{w_n(t),w_{n+1}(t)} \simeq C_{w_n(t),w(t)}, \qquad (22)$$

where $C_{w_n(t),w(t)}$ shows the NCSDs of $w_n(t), w(t)$ and $C_{w_n(t),w_{n+1}(t)}$ is the NCSDs of two successive iterations $w_n(t), w_{n+1}(t)$.

Proof. Using Definition 1 and Remark 1 we get

$$C_{w_n(t),w_{n+1}(t)} = \log_{10}\left| \frac{w_n(t)}{w_n(t) - w_{n+1}(t)} - \frac{1}{2} \right|$$

$$= \log_{10}\left| \frac{w_n(t)}{w_n(t) - w_{n+1}(t)} \right| + \log_{10}\left| 1 - \frac{1}{2w_n(t)}(w_n(t) - w_{n+1}(t)) \right|$$

$$= \log_{10}\left| \frac{w_n(t)}{w_n(t) - w_{n+1}(t)} \right| + \mathcal{O}(w_n(t) - w_{n+1}(t))$$

$$= \log_{10}\left| \frac{w_n(t)}{(w_n(t) - w(t)) - (w_{n+1}(t) - w(t))} \right| + \mathcal{O}\left[(w_n(t) - w(t)) - (w_{n+1}(t) - w(t))\right] \qquad (23)$$

$$= \log_{10}\left| \frac{w_n(t)}{(w_n(t) - w(t))\left[1 - \frac{w_{n+1}(t) - w(t)}{w_n(t) - w(t)}\right]} \right| + \mathcal{O}(E_n) + \mathcal{O}(E_{n+1})$$

$$= \log_{10}\left| \frac{w_n(t)}{w_n(t) - w(t)} \right| - \log_{10}\left| 1 - \frac{w_{n+1}(t) - w(t)}{w_n(t) - w(t)} \right| + \mathcal{O}\left(\frac{L^n}{n!} \right)$$

$$= \log_{10}\left| \frac{w_n(t)}{w_n(t) - w(t)} \right| - \log_{10}\left| 1 - \frac{w_{n+1}(t) - w(t)}{w_n(t) - w(t)} \right| + \mathcal{O}\left(\frac{L^n}{n!} \right).$$

Also,
$$C_{w_n(t),w(t)} = \log_{10}\left|\frac{w_n(t)}{w_n(t)-w(t)} - \frac{1}{2}\right|$$
$$= \log_{10}\left|\frac{w_n(t)}{w_n(t)-w(t)}\right| + \mathcal{O}(w_n(t) - w(t)) \qquad (24)$$
$$= \log_{10}\left|\frac{w_n(t)}{w_n(t)-w(t)}\right| + \mathcal{O}\left(\frac{L^n}{n!}\right).$$

Applying Equations (23) and (24) we have
$$C_{w_n(t),w_{n+1}(t)} = C_{w_n(t),w(t)} - \log_{10}\left|1 - \frac{w_{n+1}(t)-w(t)}{w_n(t)-w(t)}\right| + \mathcal{O}\left(\frac{L^n}{n!}\right).$$

From Remark 1, we can write $\frac{w_{n+1}(t)-w(t)}{w_n(t)-w(t)} = \frac{\mathcal{O}\left(\frac{L^{n+1}}{(n+1)!}\right)}{\mathcal{O}\left(\frac{L^n}{n!}\right)} = \mathcal{O}\left(\frac{1}{n}\right)$. Thus for n enough large we get $\mathcal{O}\left(\frac{1}{n}\right) \ll 1$ and consequently
$$C_{w_n(t),w_{n+1}(t)} \simeq C_{w_n(t),w(t)}.$$

□

Theorem 2 shows that when n increases, the NCSDs between two sequential results obtained from the algorithm is almost equal to the common significant digits of the n-th iteration and the exact solution at the given point t which means that for an optimal index like $n = n_o pt$, when $w_n(t) - w_{n+1}(t) = @.0$ then $w_n(t) - w(t) = @.0$.

5. Numerical Results

In this section, several examples of second kind linear Volterra IEs with discontinuous kernel are presented. The numerical process is based on the HPM that we discussed in previous sections. Also, using the CESTAC method and the CADNA library for all examples we will arrange some numerical procedures based on the following algorithm to find the optimal approximation, optimal error and optimal step of the HPM for solving linear Volterra IEs with jumping kernels. Having the exact solution in the examples is only to compare the numerical results based on both conditions (1) and (2). Some comparative graphs between exact and approximate solutions are plotted to show the accuracy and efficiency of the method.

Algorithm 2:

Step 1- Let $n = 1$.
Step 2- Do the following steps while $|w_n(t) - w_{n+1}(t)| \neq @.0$
{
Step 2-1- Produce $w_n(t)$ using Equations (10) and (19).
Step 2-2- Print $n, w_n(x), |w(t) - w_n(t)|, |w_n(t) - w_{n+1}(t)|$.
Step 2-3- $n = n+1$.
}

Example 1. *Consider the following second kind Volterra IE with discontinuous kernel*
$$w(t) = -t + \frac{13t^2}{9} - \frac{41t^3}{162} - \frac{5t^4}{324} + \int_0^{\frac{t}{3}}(s+t)w(s)ds + \int_{\frac{t}{3}}^t w(s)ds, \qquad (25)$$

with exact solution $w(t) = t^2 - t$. Applying the homotopy map (16) and relations (17), (18) and successive iterations (19) we get

$$\hat{a}^0 : v_0(t) = w_0(t) = g(t) = -t + \frac{13t^2}{9} - \frac{41t^3}{162} - \frac{5t^4}{324},$$

$$\hat{a}^1 : v_1(t) = g(t) - w_0(t) + \int_0^{\frac{t}{3}} (s+t) v_0(s) ds + \int_{\frac{t}{3}}^{t} v_0(s) ds$$

$$= -\frac{4t^2}{9} + \frac{577t^3}{1458} - \frac{1055t^4}{26244} - \frac{3199t^5}{787320} - \frac{23t^6}{1417176},$$

$$\hat{a}^2 : v_2(t) = \int_0^{\frac{t}{3}} (s+t) v_1(s) ds + \int_{\frac{t}{3}}^{t} v_1(s) ds$$

$$= \frac{104t^3}{729} + \frac{5365t^4}{59049} - \frac{411953t^5}{63772920} - \frac{1237231t^6}{1721868840} - \frac{761899t^7}{216955473840} - \frac{713t^8}{520693137216},$$

$$\hat{a}^n : v_n(t) = \int_0^{\frac{t}{3}} (s+t) v_{n-1}(s) ds + \int_{\frac{t}{3}}^{t} v_{n-1}(s) ds, \quad n \geq 2,$$

and finally using series solution (10), the approximate solution for $n = 5$ can be obtained as follows

$$w_5(t) = -t + t^2 + 0.000166742 \, t^7 - 0.0000302385 \, t^8 - 1.52193 \times 10^{-6} \, t^9 - 5.93206 \times 10^{-9} \, t^{10}$$

$$- 2.63436 \times 10^{-12} \, t^{11} - 1.3852 \times 10^{-16} \, t^{12} - 8.13306 \times 10^{-22} \, t^{13} - 4.25383 \times 10^{-28} \, t^{14}.$$

In this example, in order to show the accuracy of method, the CESTAC method and the CADNA library are applied according to Algorithm 2. Also, instead of applying the termination criterion (1) and using the traditional absolute error, the stopping condition (2) is applied. This condition is based on two successive approximations $w_n(t)$ and $w_{n+1}(t)$. When the difference of these terms is @.0 the CESTAC algorithm will be stopped. It shows that the NCSDs of the difference between two successive iterates is zero. The numerical results using the DSA are presented in Table 1 for $t = 0.2$ in double precision. According to this table the optimal step of iterations for the HPM is $n_{opt} = 10$, the optimal approximation is -0.16 and the optimal absolute error is 0.231×10^{-13}. Figure 1, shows the comparison between the exact and approximate solutions for optimal value $n_{opt} = 10$ obtained from the CESTAC method.

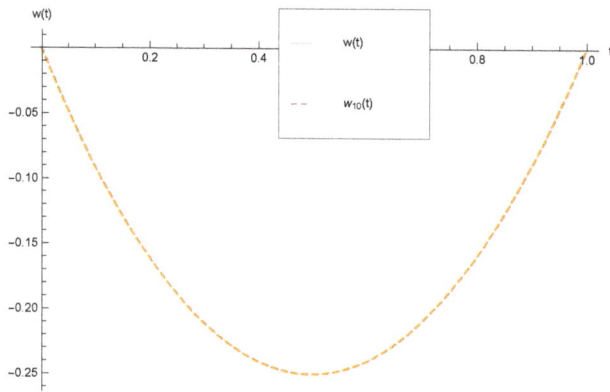

Figure 1. Comparison between the exact and optimal approximate solutions for $n_{opt} = 10$.

Table 1. Applying Algorithm 2 for Example 1 with $t = 0.2$.

n	$w_n(t)$	$|w_n(t) - w_{n+1}(t)|$	$|w(t) - w_n(t)|$
1	−0.158949024126688	0.158949024126688	$0.1050975873312 \times 10^{-2}$
2	−0.159947054328260	$0.9980302015726 \times 10^{-3}$	$0.529456717393 \times 10^{-4}$
3	−0.159997865982876	$0.508116546153 \times 10^{-4}$	$0.21340171239 \times 10^{-5}$
4	−0.159999928407226	$0.2062424350 \times 10^{-5}$	$0.715927734 \times 10^{-7}$
5	−0.159999997943235	$0.695360085 \times 10^{-7}$	0.2056764×10^{-8}
6	−0.159999999946935	$0.20037004 \times 10^{-8}$	0.530644×10^{-10}
7	−0.159999999999848	0.52913×10^{-10}	0.151×10^{-12}
8	−0.159999999999976	0.128×10^{-12}	0.231×10^{-13}
9	−0.159999999999999	0.23×10^{-13}	@.0
10	−0.160000000000000	@.0	@.0

Example 2. *Consider the following Volterra IE [11]*

$$w(t) = \frac{1}{8}t^3 - \frac{271}{8192}t^4 - \frac{1099}{20480}t^5 - \frac{31}{40960}t^6 + \int_0^{\frac{1}{4}}(1+t+s)w(s)ds + \int_{\frac{1}{4}}^{\frac{1}{2}}(2+ts)w(s)ds + \int_{\frac{1}{2}}^{t}(1+t+s)w(s)ds, \quad (26)$$

where the exact solution is $w(t) = \frac{t^3}{8}$. Applying the homotopy map (16) and relations (17), (18) and successive iterations (19) we can find the approximate solution in the following form

$$w_5(t) = 0.125\,t^3 - 2.60209 \times 10^{-18}\,t^8 - 2.32004 \times 10^{-6}\,t^9 - 0.0000188571\,t^{10} - 0.0000644212\,t^{11}$$

$$-0.000118496\,t^{12} - 0.000123946\,t^{13} - 0.0000701359\,t^{14} - 0.0000170219\,t^{15} - 2.00669 \times 10^{-7}\,t^{16}$$

$$-2.8387 \times 10^{-10}\,t^{17} - 5.39533 \times 10^{-14}\,t^{18} - 1.37516 \times 10^{-18}\,t^{19} - 4.46435 \times 10^{-24}\,t^{20}$$

$$-1.5236 \times 10^{-30}\,t^{21}$$

In this example, the DSA and the CADNA library are applied to validate the numerical approximations. Also, using the stopping condition (2) we do not need to have the exact solution to show that accuracy of presented method. The numerical results are presented in Table 2 for $t = 0.3$ by applying Algorithm 2. Using these results, the optimal approximation is $0.337499999999999 \times 10^{-2}$ and the optimal absolute error is 0.26×10^{-15} and $n_{opt} = 11$ is the optimal step of iteration for HPM method for solving Example 2. Comparison between the exact and approximate solutions for $n_{opt} = 11$ is demonstrated in Figure 2.

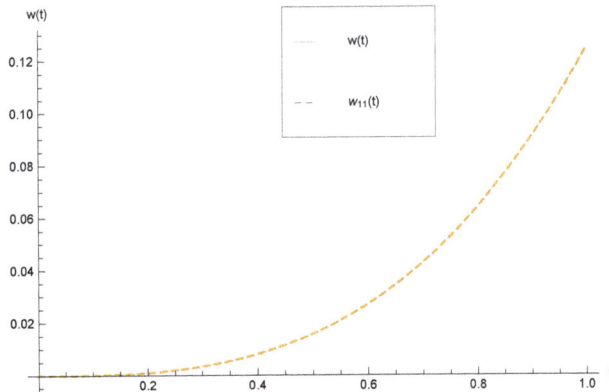

Figure 2. Comparison between the exact and optimal approximate solutions for $n_{opt} = 11$.

Table 2. Applying Algorithm 2 for Example 2 with $t = 0.3$.

n	$w_n(t)$	$\|w_n(t) - w_{n+1}(t)\|$	$\|w(t) - w_n(t)\|$
1	$0.333953193046557 \times 10^{-2}$	$0.333953193046557 \times 10^{-2}$	$0.35468069534420 \times 10^{-4}$
2	$0.337246339699770 \times 10^{-2}$	$0.3293146653212 \times 10^{-4}$	$0.253660300229 \times 10^{-5}$
3	$0.337484790802514 \times 10^{-2}$	$0.238451102744 \times 10^{-5}$	$0.15209197485 \times 10^{-6}$
4	$0.337499213815335 \times 10^{-2}$	$0.14423012820 \times 10^{-6}$	$0.7861846644 \times 10^{-8}$
5	$0.337499968103952 \times 10^{-2}$	$0.7542886169 \times 10^{-8}$	$0.31896047 \times 10^{-9}$
6	$0.337499999258881 \times 10^{-2}$	$0.31154929 \times 10^{-9}$	0.741118×10^{-11}
7	$0.337499999979560 \times 10^{-2}$	0.720678×10^{-11}	0.20439×10^{-12}
8	$0.337499999998390 \times 10^{-2}$	0.18830×10^{-12}	0.1609×10^{-13}
9	$0.337499999999973 \times 10^{-2}$	0.1582×10^{-13}	0.26×10^{-15}
10	$0.337499999999998 \times 10^{-2}$	0.25×10^{-15}	@.0
11	$0.337499999999999 \times 10^{-2}$	@.0	@.0

Table 3. Numerical approximations for Example 3 with $t = 0.2$.

n	$w_n(t)$	$\|w_n(t) - w_{n+1}(t)\|$	$\|w(t) - w_n(t)\|$
1	$0.997548914719302 \times 10^{-5}$	$0.997548914719302 \times 10^{-5}$	$0.245108528069 \times 10^{-7}$
2	$0.100010710205707 \times 10^{-4}$	$0.255818733777 \times 10^{-7}$	$0.10710205707 \times 10^{-8}$
3	$0.999995689009746 \times 10^{-5}$	$0.111413047328 \times 10^{-8}$	$0.4310990253 \times 10^{-10}$
4	$0.100000016025571 \times 10^{-4}$	$0.4471245970 \times 10^{-10}$	$0.16025571 \times 10^{-11}$
5	$0.999999994481069 \times 10^{-5}$	$0.16577464 \times 10^{-11}$	$0.5518930 \times 10^{-13}$
6	$0.100000000017666 \times 10^{-4}$	$0.5695594 \times 10^{-13}$	0.17666×10^{-14}
7	$0.999999999994729 \times 10^{-5}$	0.18193×10^{-14}	0.527×10^{-16}
8	$0.100000000000014 \times 10^{-4}$	0.5418×10^{-16}	0.14×10^{-17}
9	$0.999999999999996 \times 10^{-5}$	0.15×10^{-17}	0.3×10^{-19}
10	$0.100000000000000 \times 10^{-4}$	@.0	@.0

Example 3. *Consider the following linear Volterra IE of the second kind*

$$w(t) = t^5 - \frac{811201}{1572864}t^6 + \frac{38249}{14680064}t^7 - \frac{3938545}{939524096}t^8$$
$$+ \int_0^{\frac{t}{8}} (1 - 3t - s)w(s)ds + \int_{\frac{t}{8}}^{\frac{t}{2}} (2 + s^3 - t)w(s)ds + \int_{\frac{t}{2}}^{\frac{3t}{4}} (2t^2 s + 1)w(s)ds - 4\int_{\frac{3t}{4}}^{t} w(s)ds, \quad (27)$$

where the exact solution is $w(t) = t^5$.

The numerical results are presented in Table 3. The optimal iteration of the HPM for solving this example is $n_{opt} = 10$, the optimal approximation is 0.1×10^{-4} and the optimal error is 0.3×10^{-19}. To validate the results, the CADNA library is applied based on the termination criterion (2). Theorem 2 is able to permit us to apply the stopping condition instead of the traditional condition (1). In Figure 3, the graph of exact and approximate solutions for optimal value $n_{opt} = 10$ is studied.

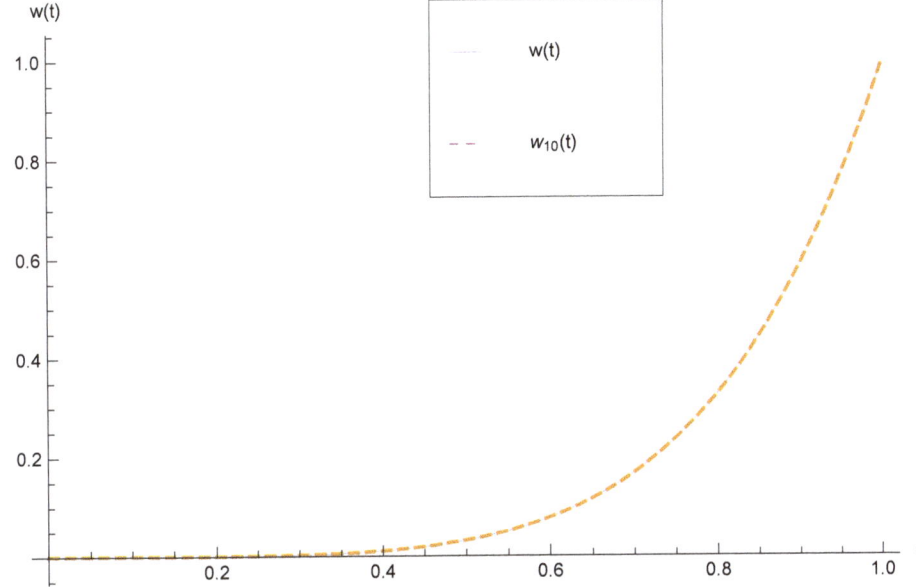

Figure 3. Comparison between exact and optimal approximate solutions for $n_{opt} = 10$.

6. Conclusions

Volterra IEs with discontinuous kernel are among applicable problems in power engineering and especially in load leveling problems. In this study, we applied the HPM while the CESTAC method and the CADNA library used to examine the numerical results. Applying this method not only the optimal iteration of the HPM, the optimal approximation and the optimal error can be found but also some of numerical instabilities can be detected. Furthermore, the substantial theorem is provided which approves the appropriateness of the termination criterion (2) instead of traditional absolute error. We will focus on validating the nonlinear Volterra IEs with discontinuous kernel in fuzzy and crisp forms using the CESTAC method for our future works.

Author Contributions: Conceptualization, D.N.S., J.H., N.S. and S.N.; methodology, S.N., D.N.S.; software, S.N.; validation, D.N.S., A.D. and N.S.; formal analysis, Z.A., N.S.; data curation, D.N.S., A.D.; writing—original draft preparation, S.N., D.N.S.; writing—review and editing, A.D., M.S., J.H., Z.A. and M.A.F.A.; supervision, D.N.S.; project administration, D.N.S.; funding acquisition, D.N.S. All authors have read and agreed to the published version of the manuscript.

Funding: This research received no external funding.

Acknowledgments: This work was partially supported by the base part of the Government Assignment for Scientific Research from the Ministry of Science and Higher Education of Russia (project code: FZZS-2020-0039).

Conflicts of Interest: The authors declare no conflict of interest.

References

1. Noeiaghdam, S.; Sidorov, D.; Muftahov, I.; Zhukov, A.V. Control of Accuracy on Taylor-Collocation Method for Load Leveling Problem. The Bulletin of Irkutsk State University. *Ser. Math.* **2019**, *30*, 59–72. [CrossRef]
2. Sidorov, D.; Muftahov, I.; Tomin, N.; Karamov, D.; Panasetsky, D.; Dreglea, A.; Liu, F.; Foley, A. A Dynamic Analysis of Energy Storage with Renewable and Diesel Generation using Volterra Equations. *IEEE Trans. Ind.* **2019**, *14*, 3451–3459. [CrossRef]
3. Sidorov, D.; Zhukov, A.; Foley, A.; Tynda, A.; Muftahov, I.; Panasetsky, D.; Li, Y. Volterra Models in Load Leveling Problem. *E3S Web Conf.* **2018**, *69*, 01015. [CrossRef]

4. Fariborzi Araghi, M.A.; Noeiaghdam, S. Homotopy analysis transform method for solving generalized Abel's fuzzy integral equations of the first kind. In Proceedings of the 4-th Iranian Joint Congress on Fuzzy and Intelligent Systems (CFIS), Zahedan, Iran, 9–11 September 2015. [CrossRef]
5. Fariborzi Araghi, M.A.; Noeiaghdam, S. Homotopy regularization method to solve the singular Volterra integral equations of the first kind. *Jordan J. Math. Stat.* **2018**, *10*, 1–12.
6. Srivastava, H.M.; Günerhan, H.; Ghanbari, B. Exact traveling wave solutions for resonance nonlinear Schrödinger equation with intermodal dispersions and the Kerr law nonlinearity. *Math. Methods Appl. Sci.* **2019**, *42*, 7210–7221. [CrossRef]
7. Sabir, Z.; Günerhan, H.; Guirao, J.L.G. On a new model based on third-order nonlinear multisingular functional differential equations. *Math. Probl. Eng.* **2020**, *2020*. [CrossRef]
8. Gao, W.; Ghanbari, B.; Günerhan, H.; Baskonus, H.M. Some mixed trigonometric complex soliton solutions to the perturbed nonlinear Schrödinger equation. *Mod. Phys. Lett. B* **2020**, *34*, 2050034. [CrossRef]
9. Sidorov, N.A.; Leontev, R.Y.; Dreglya, A.I. On small solutions of nonlinear equations with vector parameter in sectorial neighborhoods. *Math. Notes* **2012**, *91*, 90–104. [CrossRef]
10. El-Nabulsi, R.A. Nonlocal Effects to Neutron Diffusion Equation in a Nuclear Reactor. *J. Comput. Theor. Transp.* **2020**, *49*, 267–281. [CrossRef]
11. Muftahov, I.; Tynda, A.; Sidorov, D. Numeric solution of Volterra integral equations of the first kind with discontinuous kernels. *J. Comput. Appl. Math.* **2017**, *313*, 119–128. [CrossRef]
12. Sidorov, D.N. On Parametric Families of Solutions of Volterra Integral Equations of the First Kind with Piecewise Smooth Kernel. *Differ. Equ.* **2013**, *49*, 210–216. [CrossRef]
13. Noeiaghdam, S.; Sidorov, D.; Sizikov, V.; Sidorov, N. Control of accuracy on Taylor-collocation method to solve the weakly regular Volterra integral equations of the first kind by using the CESTAC method. *Appl. Comput. Math.* **2020**, *19*, 87–105.
14. Sidorov, D.; Tynda, A.; Muftahov, I.; Dreglea, A.; Liu, F. Nonlinear Systems of Volterra Equations with Piecewise Smooth Kernels: Numerical Solution and Application for Power Systems Operation. *Mathematics* **2020**, *8*, 1257. [CrossRef]
15. Raffou, Y.N. Classification of positive solutions of nonlinear system of Volterra integral equations. *Ann. Funct. Anal.* **2011**, *2*, 34–41. [CrossRef]
16. Sidorov, D. Solvability of system of integral Volterra equations of the first kind with piecewise continuous kernels. *Russ. Math. (Iz.VUZ)* **2013**, *57*, 54–63. [CrossRef]
17. Goodrich, C.S. Perturbed Integral Operator Equations of Volterra Type with Applications to *p*-Laplacian Equations. *Mediterr. J. Math.* **2018**, *15*. [CrossRef]
18. Sidorov, D.N. Generalized Solution to the Volterra Equations with Piecewise Continuous Kernels. *Bull. Malays. Math. Sci. Soc.* **2014**, *37*, 757–768.
19. Belbas, S.A.; Bulka, Y. Numerical solution of multiple nonlinear Volterra integral equations. *Appl. Math. Comput.* **2011**, *217*, 4791–4804. [CrossRef]
20. Sidorov, D.N. Existence and blow-up of Kantorovich principal continuous solutions of nonlinear integral equations. *Differ. Equ.* **2014**, *50*, 1217–1224. [CrossRef]
21. Sidorov, D. *Integral Dynamical Models: Singularities, Signals And Control*; World Scientific Series on Nonlinear Sciences Series A; Chua, L.O., Ed.; World Scientific Press: Singapore, 2015; Volume 87.
22. Sidorov, N.; Sidorov, D.; Sinitsyn, A. *Toward General Theory of Differential-Operator and Kinetic Models*; World Scientific Series on Nonlinear Sciences Series A; Chua, L.O., Ed.; World Scientific Press: Singapore, 2020; Volume 97.
23. Bernstein, S. Sur la nature analytique des solutions des certaines equations aux derivees partielles du second ordre. *C. R. Acad. Sci. Paris* **1903**, *137*, 778–781.
24. Lyusternik, L.A. Certain questions in non-linear functional analysis. *Uspekhi Mat. Nauk* **1956**, *11*, 145–168.
25. He, J.H. Homotopy perturbation technique. *Comput. Meth. Appl. Mech. Engrg.* **1999**, *178*, 257–262. [CrossRef]
26. He, J.H. A coupling method of a homotopy technique and a perturbation technique for non-linear problems. *Internat. J. Non- Mech.* **2000**, *35*, 37–43. [CrossRef]
27. He, J.H. Homotopy perturbation method: A new non-linear analytical technique. *Appl. Math. Comput.* **2003**, *135*, 73–79.
28. Hussain, S.; Shah, A.; Ayub, S.; Ullah, A. An approximate analytical solution of the Allen-Cahn equation using homotopy perturbation method and homotopy analysis method. *Heliyon* **2019**, *5*, e03060. [CrossRef]

29. Abolvafaei, M.; Ganjefar, S. Maximum power extraction from fractional order doubly fed induction generator based wind turbines using homotopy singular perturbation method. *Int. J. Electr. Power Energy Syst.* **2020**, *119*, 105889. [CrossRef]
30. Kashkari, B.S.; El-Tantawy, S.A.; Salas, A.H.; El-Sherif, L.S. Homotopy perturbation method for studying dissipative nonplanar solitons in an electronegative complex plasma. *Chaos Solitons Fractals* **2020**, *130*, 109457. [CrossRef]
31. Bota, C.; Caruntu, B. Approximate analytical solutions of nonlinear differential equations using the Least Squares Homotopy Perturbation Method. *J. OfMath. Anal. Appl.* **2017**, *448*, 401–408. [CrossRef]
32. Eshkuvatov, Z.K.; Samihah Zulkarnain, F.; Long, N.M.A.N.; Muminov, Z. Homotopy perturbation method for the hypersingular integral equations of the first kind. *Ain Shams Eng. J.* **2018**, *9*, 3359–3363. [CrossRef]
33. Javeed, S.; Baleanu, D.; Waheed, A.; Shaukat Khan, M.; Affan, H. Analysis of Homotopy Perturbation Method for Solving Fractional Order Differential Equations. *Mathematics* **2019**, *7*, 40. [CrossRef]
34. Trenogin, V.A. *Functional Analysis*; Fizmatlit: Moscow, Russia, 2007.
35. Sidorov, N.; Loginov, B.; Sinitsyn, A.; Falaleev, M. *Lyapunov-Schmidt Methods in Nonlinear Analysis and Applications*; Kluwer Academic Publisher: Dordrecht, The Netherlands; Boston, UK; London, UK, 2002.
36. Trenogin, V.A.; Sidorov, N.A.; Loginov, B.V. Potentiality, group symmetry and bifurcation in the theory of branching equation. *Differ. Integral Equ.* **1990**, *3*, 145–154.
37. Alt, R.; Lamotte, J.-L.; Markov, S. Stochastic arithmetic, Theory and experiments. *Serdica J. Comput.* **2010**, *4*, 1–10.
38. Chesneaux, J.M.; Jézéquel, F. Dynamical control of computations using the Trapezoidal and Simpson's rules. *J. Univers. Comput. Sci.* **1998**, *4*, 2–10.
39. Chesneaux, J.M. Stochastic arithmetic properties. *IMACS Comput. Appl. Math.* **1992**, 81–91.
40. Chesneaux, J.M. *CADNA, an ADA Tool for Round-Off Error Analysis and for Numerical Debugging*; ADA in Aerospace: Barcelone, Spain, 1990.
41. Vignes, J. Zéro mathématique et zéro informatique, in: La Vie des Sciences. *Comptes Rendus De L'Académie De Sci.* **1987**, *4*, 1–13.
42. Jézéquel, F.; Rico, F.; Chesneaux, J.-M.; Charikhi, M. Reliable computation of a multiple integral involved in the neutron star theory. *Math. Comput. Simul.* **2006**, *71*, 44–61. [CrossRef]
43. Jézéquel, F.; Chesneaux, J.-M. Computation of an infinite integral using Romberg's method. *Numer. Algorithms* **2004**, *36*, 265–283. [CrossRef]
44. Scott, N.S.; Jézéquel, F.; Denis, C.; Chesneaux, J.-M. Numerical 'health check' for scientific codes: The CADNA approach. *Comput. Phys. Commun.* **2007**, *176*, 507–521. [CrossRef]
45. Jézéquel, F. Dynamical control of converging sequences computation. *Appl. Numer. Math.* **2004**, *50*, 147–164. [CrossRef]
46. Jézéquel, F. A dynamical strategy for approximation methods. *C. R. Acad. Sci. Paris-Mécanique* **2006**, *334*, 362–367. [CrossRef]
47. Jézéquel, F.; Chesneaux, J.-M. CADNA: A library for estimating round-off error propagation. *Comput. Phys. Commun.* **2008**, *178*, 933–955. [CrossRef]
48. Lamotte, J.-L.; Chesneaux, J.-M.; Jézéquel, F. CADNA_C: A version of CADNA for use with C or C++ programs. *Comput. Phys. Commun.* **2010**, *181*, 1925–1926. [CrossRef]
49. Eberhart, P.; Brajard, J.; Fortin, P.; Jézéquel, F. High Performance Numerical Validation using Stochastic Arithmetic. *Reliab. Comput.* **2015**, *21*, 35–52.
50. Graillat, S.; Jézéquel, F.; Wang, S.; Zhu, Y. Stochastic arithmetic in multi precision. *Math. Comput. Sci.* **2011**, *5*, 359–375. [CrossRef]
51. Graillat, S.; Jézéquel, F.; Picot, R. Numerical Validation of Compensated Summation Algorithms with Stochastic Arithmetic. *Electron. Notes Theor. Comput. Sci.* **2015**, *317*, 55–69. [CrossRef]
52. Vignes, J. Discrete Stochastic Arithmetic for Validating Results of Numerical Software. *Spec. Issue Numer. Algorithms* **2004**, *37*, 377–390. [CrossRef]
53. Vignes, J. A stochastic arithmetic for reliable scientific computation. *Math. Comput. Simul.* **1993**, *35*, 233–261. [CrossRef]
54. Noeiaghdam, S.; Fariborzi Araghi, M.A. Finding optimal step of fuzzy Newton-Cotes integration rules by using the CESTAC method. *J. Fuzzy Set Valued Anal.* **2017**, *2017*, 62–85. [CrossRef]

55. Noeiaghdam, S.; Fariborzi Araghi, M.A.; Abbasbandy, S. Valid implementation of Sinc-collocation method to solve the fuzzy Fredholm integral equation. *J. Comput. Appl. Math.* **2020**, *370*, 112632. [CrossRef]
56. Noeiaghdam, S.; Fariborzi Araghi, M.A. A Novel Approach to Find Optimal Parameter in the Homotopy-Regularization Method for Solving Integral Equations. *Appl. Math. Inf. Sci.* **2020**, *14*, 1–8.
57. Abbasbandy, S.; Fariborzi Araghi, M.A. Numerical solution of improper integrals with valid implementation. *Math. Comput. Appl.* **2002**, *7*, 83–91. [CrossRef]
58. Abbasbandy, S.; Fariborzi Araghi, M.A. A stochastic scheme for solving definite integrals. *Appl. Numer. Math.* **2005**, *55*, 125–136. [CrossRef]
59. Fariborzi Araghi, M.A.; Noeiaghdam, S. A Valid Scheme to Evaluate Fuzzy Definite Integrals by Applying the CADNA Library. *Int. Fuzzy Syst. Appl.* **2017**, *6*, 1–20. [CrossRef]
60. Fariborzi Araghi, M.A.; Noeiaghdam, S. Dynamical control of computations using the Gauss-Laguerre integration rule by applying the CADNA library. *Adv. Appl. Math.* **2016**, *16*, 1–18.
61. Abbasbandy, S.; Fariborzi Araghi, M.A. The use of the stochastic arithmetic to estimate the value of interpolation polynomial with optimal degree. *Appl. Numer. Math.* **2004**, *50*, 279–290. [CrossRef]
62. Fariborzi Araghi, M.A.; Noeiaghdam, S. Valid implementation of the Sinc-collocation method to solve linear integral equations by the CADNA library. *J. Math. Model.* **2019**, *7*, 63–84.
63. Noeiaghdam, S.; Fariborzi Araghi, M.A.; Abbasbandy, S. Finding optimal convergence control parameter in the homotopy analysis method to solve integral equations based on the stochastic arithmetic. *Numer. Algorithms* **2019**, *81*, 237–267. [CrossRef]
64. Khojasteh Salkuyeh, D.; Toutounian, F. Optimal iterate of the power and inverse iteration methods. *Appl. Numer. Math.* **2009**, *59*, 1537–1548. [CrossRef]
65. Khojasteh Salkuyeh, D.; Toutounian, F. Numerical accuracy of a certain class of iterative methods for solving linear system. *Appl. Math. Comput.* **2006**, *176*, 727–738.
66. Graillat, S.; Jézéquel, F.; Ibrahim, M.S. Dynamical Control of Newton's Method for Multiple Roots of Polynomials. *Reliab. Comput.* **2016**, *21*, 117–139.
67. Sizikov, V.S.; Sidorov, D.N. Discrete Spectrum Reconstruction Using Integral Approximation Algorithm. *Appl. Spectrosc.* **2017**, *71*, 1640–1651. [CrossRef] [PubMed]

Publisher's Note: MDPI stays neutral with regard to jurisdictional claims in published maps and institutional affiliations.

© 2020 by the authors. Licensee MDPI, Basel, Switzerland. This article is an open access article distributed under the terms and conditions of the Creative Commons Attribution (CC BY) license (http://creativecommons.org/licenses/by/4.0/).

Article

On (ϕ, ψ)-Metric Spaces with Applications

Eskandar Ameer [1], Hassen Aydi [2,3,4,*], Hasanen A. Hammad [5,*], Wasfi Shatanawi [4,6,7] and Nabil Mlaiki [6]

1. Department of Mathematics, Taiz University, Taiz, Yemen; eskandar.msma154@iiu.edu.pk
2. Nonlinear Analysis Research Group, Ton Duc Thang University, Ho Chi Minh City, Vietnam
3. Faculty of Mathematics and Statistics, Ton Duc Thang University, Ho Chi Minh City, Vietnam
4. China Medical University Hospital, China Medical University, Taichung 40402, Taiwan; wshatanawi@psu.edu.sa
5. Department of Mathematics, Faculty of Science, Sohag University, Sohag 82524, Egypt
6. Department of Mathematics and General Sciences, Prince Sultan University, P. O. Box 66833, Riyadh 11586, Saudi Arabia; nmlaiki@psu.edu.sa
7. Department of Mathematics, Hashemite University, Zarqa 13133, Jordan
* Correspondence: hassen.aydi@tdtu.edu.vn (H.A.); hassanein_hamad@science.sohag.edu.eg (H.A.H.)

Received: 11 August 2020; Accepted: 29 August 2020; Published: 5 September 2020

Abstract: The aim of this article is to introduce the notion of a (ϕ, ψ)-metric space, which extends the metric space concept. In these spaces, the symmetry property is preserved. We present a natural topology $\tau_{(\phi,\psi)}$ in such spaces and discuss their topological properties. We also establish the Banach contraction principle in the context of (ϕ, ψ)-metric spaces and we illustrate the significance of our main theorem by examples. Ultimately, as applications, the existence of a unique solution of Fredholm type integral equations in one and two dimensions is ensured and an example in support is given.

Keywords: (ϕ, ψ)-metric space; topological property; fixed point; Fredholm integral equation

MSC: 46T99; 47H10; 54H25

1. Introduction

Fixed-point technique offers a focal concept with many diverse applications in nonlinear analysis. It is an important theoretical tool in many fields and various disciplines such as topology, game theory, optimal control, artificial intelligence, logic programming, dynamical systems (and chaos), functional analysis, differential equations, and economics.

Recently, many important extensions (or generalizations) of the metric space notion have been investigated (as examples, see References [1–5]). In 1989, the class of of b-metric spaces has been introduced by Bakhtin [6], that is, the classical triangle inequality is relaxed in the right-hand term by a parameter $s \geq 1$. This class was formally defined by Czerwik [7] (see also References [8,9]) in 1993 with a view of generalizing the Banach contraction principle (BCP). The above class has been generalized by Mlaiki et al. [10] and Abdeljawad et al. [11], by introduction of control functions (see also Reference [12]). Fagin et al. [13] presented the notion of an s-relaxed metric. A 2-metric introduced by Gahler [14] is a function defined on $\Im \times \Im \times \Im$ (where $\Im$ is a nonempty set), and verifies some particular conditions. Gahler showed that a 2-metric generalizes the classical concept of a metric. While, different authors established that no relations exist between these two notions (see Reference [15]). Mustafa and Sims [16] initiated the class of G-metric spaces. Branciari [17] gave a new generalization of the metric concept by replacing the triangle inequality with a more general one involving four points. Partial metric spaces have been introduced by Matthews [18] (for related works, see References [19–21]) as a part of the discussion of denotational semantics in dataflow networks. Jleli and Samet [22] introduced the notion of a JS-metric, where the triangle inequality is replaced by a lim sup-condition. Very recently, Jleli and Samet [23] also introduced the concept of F-metric spaces. For this, denote by Ξ the set of functions $F : (0, \infty) \to (-\infty, \infty)$ satisfying the following conditions:

(F_1) F is non-decreasing;
(F_2) for each sequence $\{t_n\} \subset (0, \infty)$;

$$\lim_{n \to +\infty} F(t_n) = -\infty \text{ if and only if } \lim_{n \to +\infty} t_n = 0.$$

Definition 1 ([23]). *Let $\Im$ be a nonempty set and $D : \Im \times \Im \to [0, \infty)$ be a function. Assume that there exist a function $F \in \Xi$ and $\alpha \in [0, \infty)$ such that for $\sigma, \varsigma \in \Im$,*

(D_1) $D(\sigma, \varsigma) = 0$ if and only if $\sigma = \varsigma$;
(D_2) $D(\sigma, \varsigma) = D(\varsigma, \sigma)$;
(D_3) *for each $n \in \mathbb{N}$ with $n \geq 2$, and for each $\{u_i\}_{i=1}^n \subset \Im$ with $(u_1, u_n) = (\sigma, \varsigma)$, we have,*

$$D(\sigma, \varsigma) > 0 \Rightarrow F(D(\sigma, \varsigma)) \leq F\left(\sum_{i=1}^{n-1} D(u_i, u_{i+1})\right) + \alpha.$$

Then D is said to be a F-metric on $\Im$. The pair $(\Im, D)$ is said to be a F-metric space.

In this paper, we present a new generalization of the concept of metric spaces, namely, a (ϕ, ψ)−metric space. We compare our concept with the existing generalizations in the literature. Next, we give a natural topology $\tau_{\phi,\psi}$ on these spaces, and study their topological properties. Moreover, we establish the BCP in the setting of (ϕ, ψ)-metric spaces. As applications, we ensure the existence of a unique solution of two Fredholm type integral equations.

2. On (ϕ, ψ)−Metric Spaces

Definition 2. *Let $\mathcal{D}$ be the set of functions $\phi : (0, \infty) \to (0, \infty)$ such that:*

(ϕ_1) ϕ is non-decreasing;
(ϕ_2) *for each positive sequence $\{t_n\}$,*

$$\lim_{n \to \infty} \phi(t_n) = 0 \text{ if and only if } \lim_{n \to \infty} t_n = 0.$$

Let $\psi : (0, \infty) \to (0, \infty)$ be such that:

(i) ψ is monotone increasing, that is, $\sigma < \varsigma \Rightarrow \psi(\sigma) \leq \psi(\varsigma)$;
(ii) $\psi(t) \leq t$ for every $t > 0$.

We denote by Ψ the set of functions satisfying (i)–(ii).

Now, we introduce the notion of (ϕ, ψ)-metric spaces.

Definition 3. *Let $\Im$ be a nonempty set and $d : \Im \times \Im \to [0, \infty)$ be a function. Assume that there exist two functions $\psi \in \Psi$ and $\phi \in \mathcal{D}$ such that for all $\sigma, \varsigma \in \Im$, the following hold:*

(d_1) $d(\sigma, \varsigma) = 0$ if and only if $\sigma = \varsigma$;
(d_2) $d(\sigma, \varsigma) = d(\varsigma, \sigma)$;
(d_3) *for each $n \in \mathbb{N}$, $n \geq 2$, and for each $\{\omega_i\}_{i=1}^n \subset \Im$ with $(\omega_1, \omega_n) = (\sigma, \varsigma)$, we have*

$$d(\sigma, \varsigma) > 0 \Rightarrow \phi(d(\sigma, \varsigma)) \leq \psi\left(\phi\left(\sum_{i=1}^{n-1} d(\omega_i, \omega_{i+1})\right)\right).$$

Then d is named as a (ϕ, ψ)-metric on $\Im$. The pair $(\Im, d)$ is called a (ϕ, ψ)-metric space. It is known that property (d_2) states that this metric should measure the distances symmetrically.

Remark 1. *Any metric on $\Im$ is a (ϕ, ψ)-metric on $\Im$. Indeed, if d is a metric on $\Im$, then it satisfies (d_2) and (d_2). On the other hand, by the triangle inequality, for every $(\sigma, \varsigma) \in \Im \times \Im$, for each integer $n \geq 2$, and for each $\{\omega_i\}_{i=1}^n \subset \Im$ with $(\omega_1, \omega_n) = (\sigma, \varsigma)$,*

$$d(\sigma, \varsigma) \leq \sum_{i=1}^{n-1} d(\omega_i, \omega_{i+1}).$$

It yields that

$$d(\sigma, \varsigma) > 0 \Rightarrow e^{d(\sigma, \varsigma)} \leq e^{\left[\sum_{i=1}^{n-1} d(\omega_i, \omega_{i+1})\right]}.$$

That is,

$$d(\sigma, \varsigma) e^{d(\sigma, \varsigma)} \leq \sum_{i=1}^{n-1} d(\omega_i, \omega_{i+1}) \left(e^{\left[\sum_{i=1}^{n-1} d(\omega_i, \omega_{i+1})\right]} \right).$$

Thus,

$$\phi(d(\sigma, \varsigma)) \leq \psi \left(\phi \left(\sum_{i=1}^{n-1} d(\omega_i, \omega_{i+1}) \right) \right).$$

Then (d_3) holds with $\phi(t) = te^t$ and $\psi(t) = t$.

Example 1. *Let $\Im = \mathbb{N}$ and let $d : \Im \times \Im \to [0, \infty)$ be defined by*

$$d(\sigma, \varsigma) = \begin{cases} |\sigma - \varsigma|, & \text{if } (\sigma, \varsigma) \notin [0,2] \times [0,2], \\ \frac{(\sigma - \varsigma)^2}{9}, & \text{if } (\sigma, \varsigma) \in [0,2] \times [0,2], \end{cases}$$

for all $\sigma, \varsigma \in \Im$. It is easy to see that d satisfies (d_1) and (d_2). But, d does not verify the triangle inequality.
Indeed,

$$d(0,2) = \frac{4}{9} > \frac{2}{9} = \frac{1}{9} + \frac{1}{9} = d(0,1) + d(1,2).$$

Hence, d is not a metric on $\Im$. Further, let $\sigma, \varsigma \in \Im$ such that $d(\sigma, \varsigma) > 0$. Let $\{\omega_i\}_{i=1}^n \subset \Im$ where $n \geq 2$ and $(\omega_1, \omega_n) = (\sigma, \varsigma)$. Consider,

$$I = \{1, 2, 3, ..., n-1 : (\omega_i, \omega_{i+1}) \in [0,2] \times [0,2]\},$$

and

$$J = \{1, 2, 3, ..., n-1\} \setminus I.$$

Hence, we have

$$\sum_{i=1}^{n-1} d(\omega_i, \omega_{i+1}) = \sum_{i \in I} d(\omega_i, \omega_{i+1}) + \sum_{j \in J} d(\omega_j, \omega_{j+1})$$

$$= \sum_{i \in I} \frac{(\omega_{i+1} - \omega_i)^2}{9} + \sum_{j \in J} |\omega_{j+1} - \omega_j|.$$

Now, we have two cases:

Case 1: If $(\sigma, \varsigma) \notin [0,2] \times [0,2]$, we have

$$
\begin{aligned}
d(\sigma, \varsigma) &= |\sigma - \varsigma| \leq \sum_{i=1}^{n-1} |\omega_{i+1} - \omega_i| \leq \sum_{i=1}^{n-1} \frac{4}{3} |\omega_{i+1} - \omega_i| \\
&= \sum_{i \in I} \frac{4 |\omega_{i+1} - \omega_i|}{3} + \sum_{j \in J} \frac{4}{3} |\omega_{j+1} - \omega_j| \\
&\leq \sum_{i \in I} \frac{4 |\omega_{i+1} - \omega_i|}{3} + \sum_{j \in J} 4 |\omega_{j+1} - \omega_j|.
\end{aligned}
$$

Observe that

$$
\frac{|\omega_{i+1} - \omega_i|}{3} \leq \frac{(\omega_{i+1} - \omega_i)^2}{9}.
$$

Thus, we get that

$$
\begin{aligned}
d(\sigma, \varsigma) &\leq 4 \left[\sum_{i \in I} \frac{(\omega_{i+1} - \omega_i)^2}{9} + \sum_{j \in J} |\omega_{j+1} - \omega_j| \right] \\
&= 4 \sum_{i=1}^{n-1} d(\omega_i, \omega_{i+1}).
\end{aligned}
$$

Case 2: If $(\sigma, \varsigma) \in [0,2] \times [0,2]$, we have

$$
\begin{aligned}
d(\sigma, \varsigma) &= \frac{|\sigma - \varsigma|^2}{9} \leq \frac{|\sigma - \varsigma|}{3} \\
&= \sum_{i \in I} \frac{|\omega_{i+1} - \omega_i|}{3} + \sum_{j \in J} \frac{|\omega_{j+1} - \omega_j|}{3} \\
&\leq \sum_{i \in I} \frac{|\omega_{i+1} - \omega_i|}{3} + \sum_{j \in J} 3 |\omega_{j+1} - \omega_j| \\
&\leq \sum_{i \in I} \frac{|\omega_{i+1} - \omega_i|^2}{3} + \sum_{j \in J} 3 |\omega_{j+1} - \omega_j| \\
&= \frac{1}{3} \left[\sum_{i \in I} \frac{|\omega_{i+1} - \omega_i|^2}{9} + \sum_{j \in J} |\omega_{j+1} - \omega_j| \right] \\
&= \frac{1}{3} \sum_{i=1}^{n-1} d(\omega_i, \omega_{i+1}).
\end{aligned}
$$

By combining the above, we conclude that for all $\sigma, \varsigma \in \Im$, for each integer $n \geq 2$, and for each $\{\omega_i\}_{i=1}^n \subset \Im$ with $(\omega_1, \omega_n) = (\sigma, \varsigma)$, we have

$$
d(\sigma, \varsigma) > 0 \Rightarrow d(\sigma, \varsigma) \leq \frac{1}{3} \sum_{i=1}^{n-1} d(\omega_i, \omega_{i+1}).
$$

Therefore,

$$
\begin{aligned}
d(\sigma, \varsigma) e^{d(\sigma, \varsigma)} &\leq \frac{1}{3} \sum_{i=1}^{n-1} d(\omega_i, \omega_{i+1}) e^{\left[\frac{1}{3} \sum_{i=1}^{n-1} d(\omega_i, \omega_{i+1})\right]} \\
&\leq \frac{1}{3} \sum_{i=1}^{n-1} d(\omega_i, \omega_{i+1}) e^{\left[\sum_{i=1}^{n-1} d(\omega_i, \omega_{i+1})\right]}.
\end{aligned}
$$

It further implies that

$$d(\sigma,\varsigma)e^{d(\sigma,\varsigma)} \leq \frac{1}{3}\sum_{i=1}^{n-1} d(\omega_i,\omega_{i+1})e^{\left[\sum_{i=1}^{n-1} d(\omega_i,\omega_{i+1})\right]}.$$

Therefore, d is a (ϕ,ψ)-metric.

Remark 2. It should be noted that the class of (ϕ,ψ)-metric spaces is effectively larger than the set of F-metric spaces. Indeed, a (ϕ,ψ)-metric is a F-metric by considering $\phi(t) = e^{f(t)}$ and $\psi(t) = e^{-\alpha}t$. We present an easy example to show that a (ϕ,ψ)-metric need not be a F-metric.

Example 2. Let $\Im = [0,1]$. Define $d : \Im \times \Im \to [0,\infty)$ as

$$d(\sigma,\varsigma) = \left(\frac{\sigma-\varsigma}{6}\right)^2.$$

Clearly, d is a (ϕ,ψ)-metric on $\Im$ with $\phi(t) = t$ and $\psi(t) = \frac{t}{36}$. Assume that there are $F \in \Xi$ and $\alpha \in [0,\infty)$. Let $n \in \mathbb{N}$ and $\omega_i = \frac{i}{n}$ for $i = 0,2,...,n$. Using (D_3), we obtain

$$f(d(0,1)) \leq f(d(0,\omega_1) + d(\omega_1,\omega_2) + ... + d(\omega_{n-1},1)) + \alpha, \ n \in \mathbb{N}.$$

Thus,

$$f(\frac{1}{36}) \leq f(\frac{1}{36n}) + \alpha, \ n \in \mathbb{N}.$$

Using (F_2), we get

$$\lim_{n \to \infty} f(\frac{1}{36n}) + \alpha = -\infty,$$

which is a contradiction. Therefore, d is not a F-metric space on $\Im$.

3. Topology of (ϕ,ψ)-Metric Spaces

Here, we study the natural topology defined on (ϕ,ψ)-metric spaces.

Definition 4. Let $(\Im,d)$ be a (ϕ,ψ)-metric space and M be a subset of $\Im$. M is said to be (ϕ,ψ)-open if for each $\sigma \in M$, there is $r > 0$ so that $B(\sigma,r) \subset M$, where

$$B(\sigma,r) = \{\varsigma \in \Im : d(\sigma,\varsigma) < r\}.$$

A subset Z of $\Im$ is called (ϕ,ψ)-closed if $\Im \setminus Z$ is (ϕ,ψ)-open. We denote by $\tau_{(\phi,\psi)}$ the set of all (ϕ,ψ)-open subsets of $\Im$.

Proposition 1. Let $(\Im,d)$ be a (ϕ,ψ)-metric space. Then $\tau_{(\phi,\psi)}$ is a topology on $\Im$.

Proposition 2. Let $(\Im,d)$ be a (ϕ,ψ)-metric space. Then, for each nonempty subset C of $\Im$, we have equivalence of the following assertions:

(i) C is (ϕ,ψ)-closed.
(ii) For any sequence $\{\sigma_n\} \subset \Im$, we have

$$\lim_{n \to \infty} d(\sigma_n,\sigma) = 0, \ \sigma \in \Im \Rightarrow \sigma \in C.$$

Proof. Suppose that C is (ϕ,ψ)-closed. Let $\{\sigma_n\}$ be a sequence in C such that

$$\lim_{n \to \infty} d(\sigma_n,\sigma) = 0, \tag{1}$$

where $\sigma \in \Im$. Assume that $\sigma \in \Im \backslash C$. Since C is (ϕ, ψ)-closed, $\Im \backslash C$ is (ϕ, ψ)-open. Hence, there is $r > 0$ so that $B(\sigma, r) \subset \Im \backslash C$, that is, $B(\sigma, r) \cap C = \emptyset$. Also, by (1), there is $N \in \mathbb{N}$ so that

$$d(\sigma_n, \sigma) < r, n \geq N.$$

That is, $\sigma_n \in B(\sigma, r), n \geq N$. Hence, $\sigma_N \in B(\sigma, r) \cap C$. It is a contradiction, and so $\sigma \in C$. That is, $(i) \Rightarrow (ii)$ is proved. Conversely, assume that (ii) is verified. Let $\sigma \in \Im \backslash C$. We now show that there is some $r > 0$ so that $B(\sigma, r) \subset \Im \backslash C$. We argue by contradiction. assume that for each $r > 0$, there is $\sigma_r \in B(\sigma, r) \cap C$. Thus, for each $n \in \mathbb{N}$, there is $\sigma_n \in B(\sigma, \frac{1}{n}) \cap C$. Then $\{\sigma_n\} \subset C$ and

$$\lim_{n \to \infty} d(\sigma_n, \sigma) = 0.$$

By (ii), we get $\sigma \in C$, which is a contradiction with $\sigma \in \Im \backslash C$. Thus, C is (ϕ, ψ)-closed and so $(ii) \Rightarrow (i)$. □

Proposition 3. *Let $(\Im, d)$ be a (ϕ, ψ)-metric space, $\alpha \in \Im$ and $r > 0$. Let $B(\alpha, r)$ be the subset of $\Im$ given as*

$$B(\alpha, r) = \{\sigma \in \Im : d(\alpha, \sigma) \leq r\}.$$

Assume that for each sequence $\{\sigma_n\} \subset \Im$, we have

$$\lim_{n \to \infty} d(\sigma_n, \sigma) = 0, \sigma \in \Im \Rightarrow d(\sigma, \varsigma) \leq \limsup_{n \to \infty} d(\sigma, \varsigma), \varsigma \in \Im. \quad (2)$$

Then $B(\alpha, r)$ is (ϕ, ψ)-closed.

Proof. Let $\{\sigma_n\} \subset B(\alpha, r)$ be a sequence so that

$$\lim_{n \to \infty} d(\sigma_n, \sigma) = 0, \sigma \in \Im.$$

From Proposition 2, we show that $\sigma \in B(\alpha, r)$. By using the definition of $B(\alpha, r)$, we obtain $d(\sigma_n, \sigma) \leq r, n \in \mathbb{N}$. Taking $\limsup_{n \to \infty}$, by (2), we get

$$d(\sigma, \varsigma) \leq \limsup_{n \to \infty} d(\sigma_n, \varsigma) \leq r,$$

which yields that $\sigma \in B(\alpha, r)$. Consequently, $B(\alpha, r)$ is (ϕ, ψ)-closed. □

Remark 3. *Proposition 3 gives only a sufficient condition ensuring that $B(\alpha, r)$ is (ϕ, ψ)-closed. An interesting problem is devoted to get a sufficient and necessary condition under which $B(\alpha, r)$ is (ϕ, ψ)-closed.*

Definition 5. *Let $(\Im, d)$ be a (ϕ, ψ)-metric space. Let C be a nonempty subset of $\Im$. Let $\overline{C}$ be the closure of C with respect to the topology $\tau_{(\phi, \psi)}$, that is, $\overline{C}$ is the intersection of all (ϕ, ψ)-closed subsets of $\Im$ containing C. Obviously, $\overline{C}$ is the smallest (ϕ, ψ)-closed subset containing C.*

Proposition 4. *Let $(\Im, d)$ be a (ϕ, ψ)-metric space. Let C be a nonempty subset of $\Im$. If $\sigma \in \overline{C}$, then $B(\sigma, r) \cap C \neq \emptyset$ for $r > 0$.*

Proof. Let $\psi \in \Psi$ and $\phi \in D$ be such that (d_3) holds. Define

$$C' = \{\sigma \in \Im : \text{for every } r > 0, \text{ there is } c \in C : d(\sigma, \varsigma) < r\}.$$

By ($d1$), it is easy to see that $C \subset C'$. Next, we will show that C' is (ϕ, ψ)-closed. Let $\{\sigma_n\}$ be a sequence in C' such that
$$\lim_{n \to \infty} d(\sigma_n, \sigma) = 0, \ \sigma \in \Im. \tag{3}$$

By (3), there are some $\delta > 0$ and $N \in \mathbb{N}$ so that
$$d(\sigma_n, \sigma) < \frac{\delta}{2}, \text{ for } n \geq N.$$

Since $\sigma_N \in C$, there is $\alpha \in C$ so that
$$d(\sigma_N, \alpha) < \frac{\delta}{2}, \text{ for } n \geq N.$$

If $d(\sigma, \alpha) > 0$, by (d_3), we have
$$\phi(d(\sigma, \alpha)) \leq \psi[\phi(d(\sigma_N, \sigma) + d(\sigma_N, \alpha))] \leq \psi[\phi(\delta)]$$
$$< \phi(\delta).$$

Hence,
$$\phi(d(\sigma, \alpha)) < \phi(\delta).$$

Using ($\phi 1$), we get
$$d(\sigma, \alpha) < \delta.$$

Hence, in all cases, we obtain $d(\sigma, \alpha) < \delta$, which yields that $\sigma \in C'$. Then by Proposition 2, C' is (ϕ, ψ)-closed, which contains C. Then $\overline{C} \subset C'$. □

Definition 6. *Let $(\Im, d)$ be a (ϕ, ψ)-metric space. Let $\{\sigma_n\}$ be a sequence in $\Im$. We say that $\{\sigma_n\}$ is (ϕ, ψ)-convergent to $\sigma \in \Im$ if $\{\sigma_n\}$ is convergent to σ with respect to the topology $\tau_{(\phi, \psi)}$, that is, for each (ϕ, ψ)-open subset $\grave{O}_\sigma$ of $\Im$ containing σ, there is $N \in \mathbb{N}$ so that $\sigma_n \in \grave{O}_\sigma$ for any $n \geq N$. Here, σ is called the limit of $\{\sigma_n\}$.*

The next result comes directly by combining the above definition and the definition of $\tau_{(\phi, \psi)}$.

Proposition 5. *Let $(\Im, d)$ be a (ϕ, ψ)-metric space. Let $\{\sigma_n\}$ be a sequence in $\Im$ and $\sigma \in \Im$. We have equivalence of the following assertions:*

(i) *$\{\sigma_n\}$ is (ϕ, ψ)-convergent to σ.*
(ii) *$\lim_{n \to \infty} d(\sigma_n, \sigma) = 0$.*

In the following, the limit of a (ϕ, ψ)-convergent sequence is unique.

Proposition 6. *Let $(\Im, d)$ be a (ϕ, ψ)-metric space. Let $\{\sigma_n\}$ be a sequence in $\Im$. Then*
$$\lim_{n \to \infty} d(\sigma_n, \sigma) = \lim_{n \to \infty} d(\sigma_n, \varsigma) = 0 \Rightarrow \sigma = \varsigma.$$

Proof. Let $\sigma, \varsigma \in \Im$ be so that
$$\lim_{n \to \infty} d(\sigma_n, \sigma) = \lim_{n \to \infty} d(\sigma_n, \varsigma) = 0.$$

Assume that $\sigma \neq \varsigma$. By (d_1), $d(\sigma, \varsigma) > 0$. Using (d_3), there are $\psi \in \Psi$ and $\phi \in D$ such that
$$\phi(d(\sigma, \varsigma)) \leq \psi[\phi(d(\sigma_n, \sigma) + d(\sigma_n, \varsigma))]$$
$$< \phi(d(\sigma_n, \sigma) + d(\sigma_n, \varsigma)),$$

for every n. Next, in view of (d_2) and $(\phi 2)$,

$$\lim_{n \to \infty} \phi\left(d\left(\sigma_n, \sigma\right) + d\left(\sigma_n, \varsigma\right)\right) = 0,$$

and so $\phi\left(d\left(\sigma, \varsigma\right)\right) = 0$, which is a contradiction, and so $\sigma = \varsigma$. □

Definition 7. *Let $(\Im, d)$ be a (ϕ, ψ)-metric space. Let $\{\sigma_n\}$ be a sequence in $\Im$. Then,*

(i) *$\{\sigma_n\}$ is (ϕ, ψ)-Cauchy if $\lim\limits_{n,m \to \infty} d\left(\sigma_n, \sigma_m\right) = 0$.*

(ii) *$(\Im, d)$ is (ϕ, ψ)-complete, if any (ϕ, ψ)-Cauchy sequence in $\Im$ is (ϕ, ψ)-convergent to some element in $\Im$.*

Proposition 7. *Let $(\Im, d)$ be a (ϕ, ψ)-metric space. If $\{\sigma_n\} \subset \Im$ is (ϕ, ψ)-convergent, then it is (ϕ, ψ)-Cauchy.*

Proof. Let $\psi \in \Psi$ and $\phi \in D$ be such that (d_3) holds. Let $\sigma \in \Im$ be so that

$$\lim_{n \to \infty} d\left(\sigma_n, \sigma\right) = 0.$$

For any $\delta > 0$, there is $N \in \mathbb{N}$ such that

$$d\left(\sigma_n, \sigma\right) + d\left(\sigma_m, \sigma\right) < \delta, \ n, m \geq N. \tag{4}$$

Let $m, n \geq N$. We consider the two following cases.

Case 1: If $\sigma_n = \sigma_m$. Here, by (d_1),

$$d\left(\sigma_n, \sigma_m\right) = 0 < \delta.$$

Case 2: If $\sigma_n \neq \sigma_m$. Here, from (4),

$$0 < d\left(\sigma_n, \sigma\right) + d\left(\sigma_m, \sigma\right) < \delta.$$

One writes

$$\phi\left(d\left(\sigma_n, \sigma\right) + d\left(\sigma_m, \sigma\right)\right) < \phi(\delta).$$

It implies that

$$\psi\left(d\left(\sigma_n, \sigma\right) + d\left(\sigma_m, \sigma\right)\right) < \psi\left(\phi(\delta)\right).$$

Now, using (d_3), we obtain

$$\phi\left(d\left(\sigma_n, \sigma_m\right)\right) \leq \psi\left(\phi d\left(\sigma_n, \sigma\right) + d\left(\sigma_m, \sigma\right)\right) < \psi\left(\phi(\delta)\right)$$
$$< \phi(\delta),$$

which implies from $(\phi 1)$ that

$$d\left(\sigma_n, \sigma_m\right) < \delta.$$

Hence,

$$d\left(\sigma_n, \sigma_m\right) < \delta, \ n, m \geq N.$$

Consequently,

$$\lim_{n,m \to \infty} d\left(\sigma_n, \sigma_m\right) = 0,$$

that is, $\{\sigma_n\}$ is (ϕ, ψ)-Cauchy. □

Now, we study the compactness on (ϕ, ψ)-metric spaces.

Definition 8. *Let $(\Im, d)$ be a (ϕ, ψ)-metric space. Let C be a nonempty subset of $\Im$. then C is called (ϕ, ψ)-compact if C is compact with respect to the topology $\tau_{(\phi, \psi)}$ on $\Im$.*

Proposition 8. Let $(\Im, d)$ be a (ϕ, ψ)-metric space. Let C be a nonempty subset of $\Im$. Then, we have equivalent of the following assertions:

(i) C is (ϕ, ψ)-compact.
(ii) For each sequence $\{\sigma_n\} \subset C$, there is a subsequence $\{\sigma_{n(k)}\}$ of $\{\sigma_n\}$ so that
$$\lim_{k \to \infty} d\left(\sigma_{n(k)}, \sigma\right) = 0.$$

Proof. Assume that C is (ϕ, ψ)-compact. Note that the set of decreasing sequences of nonempty (ϕ, ψ)-closed subsets of C has a nonempty intersection. Let $\{\sigma_n\}$ be a sequence in C. For any $n \in \mathbb{N}$, let $Z_n = \{\sigma_m : m \geq n\}$. Clearly, $Z_{n+1} \subset Z_n$ for each $n \in \mathbb{N}$. This implies that $\{\overline{Z}_n\}_{n \in \mathbb{N}}$ is a decreasing sequence of nonempty (ϕ, ψ)-closed subsets of Z. Thus, there is $\sigma \in \cap_{n \in \mathbb{N}} \overline{Z}_n$. Given an arbitrary element $\varepsilon > 0$. Since $\sigma \in \overline{Z}_0$, by Proposition 4, there are $n_0 \geq 0$ and $\sigma_{n_0} \in C$ so that $d(\sigma_{n_0}, \sigma) < \varepsilon$. Continuing in this direction, for any $k \in \mathbb{N}$, there are $n(k) \geq k$ and $\sigma_{n(k)} \in C$ so that
$$d\left(\sigma_{n(k)}, \sigma\right) < \varepsilon.$$

Consequently,
$$\lim_{k \to \infty} d\left(\sigma_{n(k)}, \sigma\right) = 0.$$

Since C is (ϕ, ψ)-compact, one says that C is (ϕ, ψ)-closed, and $\sigma \in C$. Hence, we established that $(i) \Rightarrow (ii)$. Conversely, suppose that (ii) is satisfied. Let $\psi \in \Psi$ and $\phi \in D$ such that (d_3) is satisfied.

First, we claim that
$$\forall r > 0, \exists (\sigma_0), i = 1, \ldots, n \subset C : C \subset \bigcup_{i=1,\ldots,n} B(\sigma_i, r). \tag{5}$$

We argue by contradiction. Suppose there is $r > 0$ so that for any finite number of elements $(\sigma_0), i = 1, \ldots, n \subset C$,
$$C \subsetneq \bigcup_{i=1,\ldots,n} B(\sigma_i, r).$$

Let $\sigma_1 \in C$ be a fixed element. Then
$$C \subsetneq B(\sigma_1, r).$$

That is, there is $\sigma_2 \in C$ so that $d(\sigma_1, \sigma_2) \geq r$. Also,
$$C \subsetneq B(\sigma_1, r) \cup B(\sigma_2, r).$$

So there is $\sigma_3 \in C$ so that $d(\sigma_i, \sigma_3) \geq r$ for $i = 1, \ldots, n$. Continuing in this direction and by induction, we build a sequence $\{\sigma_n\} \subset C$ so that $d(\sigma_n, \sigma_m) \geq r$, $n, m \in \mathbb{N}$. Note that we could bot extract from $\{\sigma_n\}$ any (ϕ, ψ)-Cauchy subsequence, and so (from Proposition 7), any (ϕ, ψ)-convergent subsequence. We get so a contradiction with (ii), which proves (5). Next, let $\{\grave{O}_i\}_{i \in I}$ be an arbitrary family of (ϕ, ψ)-open subsets of $\Im$ so that
$$C \subset \cup_{i \in I} \grave{O}_i. \tag{6}$$

We claim that
$$\forall r_0 > 0 : \forall \sigma \in C, \exists i \in I : B(\sigma, r_0) \subset \grave{O}_i. \tag{7}$$

We argue by contradiction. Assume that for every $r > 0$, there is $\sigma_r \in C$ so that $B(\sigma_r, r) \subsetneq \dot{O}_i$, for all $i \in I$. Particularly, for all $n \in \mathbb{N}$, there is $\sigma_n \in C$ so that $B(\sigma_n, \frac{1}{n}) \subsetneq \dot{O}_i$ for all $i \in I$. By (ii), we build a subsequence $\{\sigma_{n(k)}\}$ from $\{\sigma_n\}$ so that

$$\lim_{k \to \infty} d\left(\sigma_{n(k)}, \sigma\right) = 0, \tag{8}$$

for some $\sigma \in C$. Moreover, using (6), there is $j \in I$ so that $\sigma \in \Im$. In view of the fact that $\dot{O}_j$ is a (ϕ, ψ)-open subset of $\Im$, there is $r_0 > 0$ so that $B(\sigma, r_0) \subset \dot{O}_j$. Now, for each $n(k) \in \mathbb{N}$ and for every $q \in B(\sigma_{n(k)}, \frac{1}{n(k)})$, one writes

$$\begin{aligned} d(\sigma, q) &> 0 \Rightarrow \phi(d(\sigma, q)) \le \psi\left(\phi\left(d\left(\sigma, \sigma_{n(k)}\right) + d\left(\sigma_{n(k)}, q\right)\right)\right) \\ &< \psi\left(\phi\left(d\left(\sigma, \sigma_{n(k)}\right) + \frac{1}{n(k)}\right)\right) \\ &\phi\left(d\left(\sigma, \sigma_{n(k)}\right) + \frac{1}{n(k)}\right). \end{aligned}$$

Using (8) and ($\phi2$), there is $K \in \mathbb{N}$ so that

$$\phi\left(d\left(\sigma, \sigma_{n(k)}\right) + \frac{1}{n(k)}\right) < \phi(r_0)$$

for each $k \ge K$. It yields that

$$d(\sigma, q) > 0 \Rightarrow \phi(d(\sigma, q)) < \phi(r_0).$$

Consequently, by ($\phi1$), we find that $d(\sigma, q) < r_0$. Hence, we get

$$B(\sigma_{n(k)}, \frac{1}{n(k)}) \subset B(\sigma, r_0),$$

for $n(k) \in \mathbb{N}$. Thus,

$$B(\sigma_{n(k)}, \frac{1}{n(k)}) \subset \dot{O}_j, n(k) \in \mathbb{N}.$$

We get a contradiction with respect to

$$B(\sigma_{n(k)}, \frac{1}{n(k)}) \subsetneq \dot{O}_i, n(k) \in \mathbb{N}.$$

for all $i \in I$. Then (7) holds. Further, by (5), there is $\{\sigma_p\}_{p=1,\ldots,n} \subset C$ so that

$$C \subset \bigcup_{p=1,\ldots,n} B(\sigma_p, r_0).$$

But by (7), for any $p = 1, \ldots, n$, there exists $i(p) \in I$ such that $B(\sigma_p, r_0) \subset \dot{O}_{i(p)}$, which yields

$$C \subset \bigcup_{p=1,\ldots,n} \dot{O}_{i(p)}.$$

Thus, C is (ϕ, ψ)-compact, and so (ii)$\Rightarrow$(i). $\square$

Definition 9. Let $(\Im, d)$ be a (ϕ, ψ)-metric space. Let C be a nonempty subset of $\Im$. The subset C is said to be sequentially (ϕ, ψ)-compact, if for each sequence, there are a subsequence $\{\sigma_{n(k)}\}$ of $\{\sigma_n\}$ and $\sigma \in C$ so that

$$\lim_{k \to \infty} d\left(\sigma_{n(k)}, \sigma\right) = 0.$$

Definition 10. Let $(\Im, d)$ be a (ϕ, ψ)-metric space. Let C be a nonempty subset of $\Im$. The subset C is called (ϕ, ψ)-totally bounded if

$$\forall r > 0, \exists (\sigma_0), i = 1, ..., n \subset C : C \subset \bigcup_{i=1,...,n} B(\sigma_i, r).$$

Due to the proof of Proposition 8, we may state the following proposition.

Proposition 9. Let $(\Im, d)$ be a (ϕ, ψ)-metric space. Let C be a nonempty subset of $\Im$.

(i) C is (ϕ, ψ)-compact if and only if C is sequentially (ϕ, ψ)-compact.
(ii) If C is (ϕ, ψ)-compact, then C is (ϕ, ψ)-totally bounded.

4. Banach Contraction Principle on (ϕ, ψ)-Metric Spaces

In this section, we prove a new version of the BCP in the context of (ϕ, ψ)-metric spaces.

Theorem 1. Let $(\Im, d)$ be a complete (ϕ, ψ)-metric space and $T : \Im \to \Im$ be a self-mapping. Suppose that there exists $\lambda \in (0, 1)$ such that for all $\sigma, \varsigma \in \Im$,

$$d(T(\sigma), T(\varsigma)) \leq \lambda d(\sigma, \varsigma). \qquad (9)$$

Then T has a unique fixed point in $\Im$.

Proof. Let $\sigma_0 \in \Im$. Define the sequence $\{\sigma_n\}$ in $\Im$ by

$$\sigma_{n+1} = T(\sigma_n), \text{ where } n \in \mathbb{N}.$$

If for some n, $d(\sigma_n, \sigma_{n+1}) = 0$, then σ_n is a fixed point of T. Without restriction of the generality, we may suppose that $d(\sigma_n, \sigma_{n+1}) > 0$ for all n. Using (9), we get

$$\begin{aligned} d(\sigma_n, \sigma_{n+1}) &\leq \lambda d(\sigma_{n-1}, \sigma_n) \leq \lambda^2 d(\sigma_{n-2}, \sigma_{n-1}) \\ &\leq ... \leq \lambda^n d(\sigma_0, \sigma_1), \end{aligned}$$

for all $n \in \mathbb{N}$. Thus,

$$\sum_{i=n}^{m-1} d(\sigma_i, \sigma_{i+1}) \leq \frac{\lambda^n}{1-\lambda} d(\sigma_0, \sigma_1), \quad m > n.$$

Hence, by $(\phi 1)$, we have

$$\phi\left(\sum_{i=n}^{m-1} d(\sigma_i, \sigma_{i+1})\right) \leq \phi\left(\frac{\lambda^n}{1-\lambda} d(\sigma_0, \sigma_1)\right), \quad m > n.$$

Since ψ is monotone increasing, we obtain for $m > n$,

$$\begin{aligned} \psi\left(\phi\left(\sum_{i=n}^{m-1} d(\sigma_i, \sigma_{i+1})\right)\right) &\leq \psi\left(\phi\left(\frac{\lambda^n}{1-\lambda} d(\sigma_0, \sigma_1)\right)\right) \\ &< \phi\left(\frac{\lambda^n}{1-\lambda} d(\sigma_0, \sigma_1)\right). \end{aligned}$$

Since
$$\lim_{n \to \infty} \frac{\lambda^n}{1-\lambda} d(\sigma_0, \sigma_1) = 0,$$
by ($\phi 2$), we have
$$\lim_{n \to \infty} \phi \left(\frac{\lambda^n}{1-\lambda} d(\sigma_0, \sigma_1) \right) = 0. \tag{10}$$

Using (d_3), we obtain
$$d(\sigma_n, \sigma_m) > 0, m > n \Rightarrow \phi(d(\sigma_n, \sigma_m)) \leq \psi \left(\phi \left(\sum_{i=n}^{m-1} d(\sigma_i, \sigma_{i+1}) \right) \right)$$
$$< \phi \left(\frac{\lambda^n}{1-\lambda} d(\sigma_0, \sigma_1) \right).$$

It implies that
$$\phi(d(\sigma_n, \sigma_m)) < \phi \left(\frac{\lambda^n}{1-\lambda} d(\sigma_0, \sigma_1) \right).$$

By using (10), we obtain
$$\lim_{n,m \to \infty} \phi(d(\sigma_n, \sigma_m)) = 0.$$

Then from ($\phi 2$), we have
$$\lim_{n,m \to \infty} d(\sigma_n, \sigma_m) = 0.$$

Therefore, $\{\sigma_n\}$ is a (ϕ, ψ)-Cauchy sequence in $\Im$. Since $\Im$ is (ϕ, ψ)-complete, we can find $\sigma^* \in \Im$ such that
$$\lim_{n \to \infty} d(\sigma_n, \sigma^*) = 0. \tag{11}$$

Next, we prove that $T(\sigma^*) = \sigma^*$. We argue by contradiction. Assume that $d(T(\sigma^*), \sigma^*) > 0$. By using ($d_3$), we obtain
$$\phi(d(T(\sigma^*), \sigma^*)) \leq \psi(\phi(d(T(\sigma^*), T(\sigma_n)) + d(T(\sigma_n), \sigma^*)))$$
$$< \phi(d(T(\sigma^*), T(\sigma_n)) + d(T(\sigma_n), \sigma^*)),$$
for $n \in \mathbb{N}$. By (9) and ($\phi 1$),
$$d(T(\sigma^*), \sigma^*) < \lambda d(\sigma^*, \sigma_n) + d(\sigma_{n+1}, \sigma^*).$$

By using ($\phi 2$) and (11), we get
$$\lim_{n \to \infty} \phi(\lambda d(\sigma^*, \sigma_n) + d(\sigma_{n+1}, \sigma^*)) = 0,$$
which is a contradiction. Therefore, $d(T(\sigma^*), \sigma^*) = 0$ and $T(\sigma^*) = \sigma^*$. Thus, T has a fixed point $\sigma^* \in \Im$. Next, we prove that T has at most one fixed point. Assume that σ^* and ς^* are two fixed points of T such that $\sigma^* \neq \varsigma^*$. Then from (9), we have
$$0 < d(\sigma^*, \varsigma^*) = d(T(\sigma^*), T(\varsigma^*)) \leq \lambda d(\sigma^*, \varsigma^*) < d(\sigma^*, \varsigma^*).$$

It is a contradiction. Hence, T has a unique fixed point in $\Im$. □

Corollary 1. Let $(\Im, d)$ be a (ϕ, ψ)-metric space. Suppose there exist a continuous comparison function $\psi \in \Psi$ and $\phi \in D$ so that (d_3) holds. Let $S : B(\sigma_0, r) \to \Im$ be a given mapping, where $\sigma_0 \in \Im$ and $r > 0$. Assume that:

(i) Suppose that for each sequence $\{\sigma_n\} \subset \Im$, we have

$$\lim_{n \to \infty} d(\sigma_n, \sigma) = 0 \Rightarrow d(\sigma, \varsigma) \leq \lim_{n \to \infty} \sup d(\sigma_n, \varsigma), \varsigma \in \Im;$$

(ii) $(\Im, d)$ is (ϕ, ψ)-complete;
(iii) There exists $\lambda \in (0, 1)$ such that

$$d(S(\sigma), S(\varsigma)) \leq \lambda d(\sigma, \varsigma), \ (\sigma, \varsigma) \in B(\sigma_0, r) \times B(\sigma_0, r);$$

(iv) There exists $0 < \varepsilon < r$ such that

$$\phi(\lambda \varepsilon + d(S\sigma_0, \sigma_0)) \leq \phi(\varepsilon).$$

Then S has a fixed point.

Proof. Consider $0 < \varepsilon < r$ such that (iv) is satisfied. First, we will show that

$$S(B(\sigma_0, \varepsilon)) \subset B(\sigma_0, \varepsilon).$$

Let $\sigma \in B(\sigma_0, \varepsilon)$, that is, $d(\sigma_0, \sigma) \leq \varepsilon$. Assume that $d(S\sigma, \sigma_0) > 0$. By (d_3),

$$\phi(d(S\sigma, \sigma_0)) \leq \psi(\phi(d(S\sigma, S\sigma_0) + d(S\sigma_0, \sigma_0))).$$

Using (iii), we obtain

$$\begin{aligned}
\phi(d(S\sigma, \sigma_0)) &\leq \psi(\phi(d(S\sigma, S\sigma_0) + d(S\sigma_0, \sigma_0))) \\
&\leq \psi(\phi(\lambda d(\sigma, \sigma_0) + d(S\sigma_0, \sigma_0))) \\
&\leq \psi(\phi(\lambda \varepsilon + d(S\sigma_0, \sigma_0))) \\
&< \phi(\lambda \varepsilon + d(S\sigma_0, \sigma_0)) \\
&\leq \phi(\varepsilon).
\end{aligned}$$

Hence, by $(\phi 1)$, we have $d(S\sigma, \sigma_0) \leq \varepsilon$, which yields $S(\sigma) \in B(\sigma_0, \varepsilon)$. Therefore,

$$S(B(\sigma_0, \varepsilon)) \subset B(\sigma_0, \varepsilon).$$

Further, the mapping $S : B(\sigma_0, \varepsilon) \to B(\sigma_0, \varepsilon)$ is well-defined, and the Banach contraction condition holds. Next, since the condition of Proposition 3 is satisfied, it is known that $B(\sigma_0, \varepsilon)$ is (ϕ, ψ)-closed, so from (i), it is (ϕ, ψ)-complete. Finally, the result is deduced by using Theorem 1. □

5. Solving a Nonlinear Fredholm Integral Equation

This section is devoted to discusses the existence and uniqueness of a solution of a Fredholm type integral equation of the 2nd kind [24–29]. Consider the equation below:

$$\sigma(\mu) = \beta(\mu) + \int_u^v \Omega(\mu, \ell) \Re(\mu, \ell, \sigma(\ell)) \, d\ell, \ \mu \in [u, v]. \tag{12}$$

Let $\Theta = C[u, v]$ be the set of all continuous functions defined on $[u, v]$. For $\sigma, \zeta \in \Theta$ and $q > 1$, define $d : \Theta \times \Theta \to [0, \infty)$ by

$$d(\sigma, \zeta) = \left(\frac{1}{6} \sup_{\mu \in [u,v]} |\sigma(\mu) - \zeta(\mu)| \right)^q.$$

Then (Θ, d) is a complete (ϕ, ψ)−metric space with $\phi(\rho) = \rho$ and $\psi(\rho) = \frac{\rho}{6^q}$.

To study the existence of a solution for the problem (12), we state and prove the theorem below.

Theorem 2. *Consider the problem (12) via the assumptions below:*

($\dagger_1$) $\Re : [u, v] \times [u, v] \times \mathbb{R} \to \mathbb{R}$, $\Omega : [u, v] \times [u, v] \to \mathbb{R}$, and $\beta : [u, v] \to \mathbb{R}$ are continuous functions;

($\dagger_2$) *For $\mu \in [u, v]$, we have*

$$\sup_{\mu \in [u,v]} \int_u^v \Omega(\mu, \ell) d\ell \leq 1;$$

($\dagger_3$) *For $q > 1$, consider*

$$|\Re(\mu, \ell, \sigma(\ell)) - \Re(\mu, \ell, \zeta(\ell))| \leq \frac{1}{\sqrt[q]{3}} |\sigma(\ell) - \zeta(\ell)|.$$

Then the nonlinear integral equation (12) has a unique solution in Θ.

Proof. Define the operator $T : C[u, v] \to C[u, v]$ by

$$T\sigma(\mu) = \beta(\mu) + \int_u^v \Omega(\mu, \ell) \Re(\mu, \ell, \sigma(\ell)) d\ell, \ \mu \in [u, v]. \tag{13}$$

The solution of problem (12) is a fixed point for the operator (13). By hypotheses ($\dagger_1$) − ($\dagger_3$), we have

$$d(T\sigma(\mu), T\zeta(\mu))$$
$$= \left(\frac{1}{6} \sup_{\mu \in [u,v]} |T\sigma(\mu) - T\zeta(\mu)| \right)^q$$
$$= \frac{1}{6^q} \left(\sup_{\mu \in [u,v]} \left| \int_u^v \Omega(\mu, \ell) \Re(\mu, \ell, \sigma(\ell)) d\ell - \int_u^v \Omega(\mu, \ell) \Re(\mu, \ell, \zeta(\ell)) d\ell \right| \right)^q$$
$$\leq \frac{1}{6^q} \left(\sup_{\mu \in [u,v]} \int_u^v \Omega(\mu, \ell) |\Re(\mu, \ell, \sigma(\ell)) - \Re(\mu, \ell, \zeta(\ell))| dv \right)^q$$
$$\leq \frac{1}{6^q} \left(\sup_{\mu \in [u,v]} \int_u^v \Omega(\mu, \ell) \right)^q \times \sup_{\ell \in [u,v]} \left(\frac{1}{\sqrt[q]{3}} |\sigma(\ell) - \zeta(\ell)| \right)^q$$
$$\leq \frac{1}{3} \sup_{\ell \in [u,v]} \left(\frac{1}{6} |\sigma(\ell) - \zeta(\ell)| \right)^q$$
$$= \lambda d(\sigma(\mu), \zeta(\mu)).$$

Thus, the condition (9) of Theorem 1 holds with $\lambda = \frac{1}{3}$. Therefore, all hypotheses of Theorem 1 are fulfilled. So the problem (12) has a unique solution in Θ. □

The example below supports Theorem 2.

Example 3. *The following problem:*

$$\sigma(\mu) = \frac{1}{36} \int_0^1 \ell^2 \sigma(\ell) d\ell, \ \mu \in [0, 1], \tag{14}$$

has a solution in $C[0, 1]$.

Proof. Define the operator $T : C[0,1] \to C[0,1]$ by $T\sigma(\mu) = \frac{1}{36} \int_0^1 \ell^2 \sigma(\ell) d\ell$. Customize $\Omega(\mu, \ell) = \frac{\ell}{6}$, $\beta(\mu) = 0$ and $\Re(\mu, \ell, \sigma(\ell)) = \frac{\ell\sigma(\ell)}{6}$ in Theorem 2. Note that

- $\Re$ and Ω are continuous functions;
- For $\mu \in [0,1]$, we have

$$\sup_{\mu \in [u,v]} \int_u^v \Omega(\mu, \ell) d\ell = \sup_{\mu \in [0,1]} \int_0^1 \frac{\ell}{6} d\ell = \frac{1}{12} < 1;$$

- Take $q = 2$. For $\ell \in [0,1]$, we get

$$\begin{aligned}
|\Re(\mu, \ell, \sigma(\ell)) - \Re(\mu, \ell, \zeta(\ell))| &= \left|\frac{\ell\sigma(\ell)}{6} - \frac{\ell\zeta(\ell)}{6}\right| \\
&= \frac{\ell}{6}|\sigma(\ell) - \zeta(\ell)| \\
&\leq \frac{1}{\sqrt{3}}|\sigma(\ell) - \zeta(\ell)|.
\end{aligned}$$

Therefore, the stipulations of Theorem 2 are justified, hence the mapping T has a unique fixed point in $C[0,1]$, which is the unique solution of the equation (14). □

6. Solving a Two-Dimensional Nonlinear Fredholm Integral Equation

In many problems in engineering and mechanics under a suitable transformation, two-dimensional Fredholm integral equations of the second kind appear. For example, in the calculation of plasma physics, it is usually required to solve some Fredholm integral equations, see References [30–32].

Now, consider the two-dimensional Fredholm integral equation of the shape:

$$\zeta(r,j) = e(r,j) + \int_0^1 \int_0^1 \Omega(r,j,f,g) \daleth(r,j,\zeta(f,g)) df dg; \quad (r,j) \in [0,1]^2, \tag{15}$$

where e, Ω and $\daleth$ are given continuous functions defined on $L^2(C([0,1] \times [0,1]))$ and ζ is a function in $L^2(C([0,1] \times [0,1]))$.

Let $\nabla = C([0,1])$ be the set of all real valued continuous functions on $[0,1]$. Consider the same distance of the above section, then for $\sigma, \zeta \in \nabla$, the pair (∇, d) is a complete (ϕ, ψ)-metric space with $\phi(\rho) = \rho$ and $\psi(\rho) = \frac{\rho}{6^q}$.

Now, we consider the problem (15) under the hypotheses below:

($\ddagger_1$) $\Omega : [0,1]^4 \to \mathbb{R}$, and $\daleth : [0,1]^2 \times \mathbb{R} \to \mathbb{R}$ and $e : [0,1]^2 \to \mathbb{R}$ are continuous functions;
($\ddagger_2$) for all $\sigma, \zeta \in \nabla$, there is a constant $\kappa < 1$ such that

$$|\daleth(r,j,\sigma(f,g)) - \daleth(r,j,\zeta(f,g))| \leq \frac{1}{\sqrt[q]{2\kappa}} |\sigma(h,g) - \zeta(h,g)|, \ q > 1;$$

($\ddagger_3$) we have $\int_0^1 \int_0^1 \Omega(r,j,f,g) df dg \leq \kappa$.

Our related theorem in this part is listed as follows.

Theorem 3. *The problem (15) has a unique solution in $L^2(C([0,1] \times [0,1]))$ if the hypotheses ($\ddagger_1$) – ($\ddagger_3$) hold.*

Proof. Define the operator $T: \nabla \to \nabla$ by

$$T(\zeta(\tau,\mu)) = e(r,j) + \int_0^1 \int_0^1 \Omega(r,j,f,g) \daleth(r,j,\zeta(f,g)) df dg, \ (a,b) \in [0,1] \times [0,1], \tag{16}$$

then for $q > 1$, we get

$$\frac{1}{6^q} |T(\sigma(r,j)) - T(\zeta(r,j)))|^q$$

$$= \frac{1}{6^q} \left| \int_0^1 \int_0^1 \Omega(r,j,f,g) \daleth(r,j,\sigma(f,g)) df dg - \int_0^1 \int_0^1 \Omega(r,j,f,g) \daleth(r,j,\zeta(f,g)) df dg \right|^q$$

$$\leq \frac{1}{6^q} \left(\int_0^1 \int_0^1 \Omega(r,j,f,g) \, |\daleth(r,j,\sigma(f,g)) - \daleth(r,j,\zeta(f,g))| df dg \right)^q$$

$$\leq \frac{1}{6^q} \left(\int_0^1 \int_0^1 \Omega(r,j,f,g) df dg \right)^q (|\daleth(r,j,\sigma(f,g)) - \daleth(r,j,\zeta(f,g))|)^q$$

$$\leq \frac{1}{6^q} \kappa^q \left(\frac{1}{\sqrt[q]{2\kappa}} |\sigma(h,g) - \zeta(h,g)| \right)^q$$

$$= \frac{1}{2} \left(\frac{1}{6} |\sigma(h,g) - \zeta(h,g)| \right)^q.$$

Taking the supremum, we get

$$d(T\sigma, T\zeta) = \left(\frac{1}{6} \sup_{\mu \in [u,v]} |T(\sigma(r,j)) - T(\zeta(r,j)))| \right)^q$$

$$\leq \frac{1}{2} \left(\frac{1}{6} \sup_{\mu \in [u,v]} |\sigma(h,g) - \zeta(h,g)| \right)^q$$

$$= \lambda d(\sigma, \zeta).$$

Thus, from Theorem 1, the operator (16) has a unique fixed point in $L^2(C([0,1] \times [0,1]))$, which is considered as the unique solution of the problem (15). □

7. Conclusions

In this manuscript, we initiated the concept a (ϕ, ψ)-metric space. It is a generalization of the metric space setting. We also presented its topological structure natural topology. The Banach contraction principle in this class has been established. Moreover, we gave some examples and applications in support of the introduced new concepts and presented results. As perspectives, it is an open problem to treat the cases of Kannan, Chatterjea, Hardy-Rogers, Ćirić and Suzuki type contractions. Also, it would be interesting to investigate the case of common fixed points.

Author Contributions: E.A. contributed in conceptualization and writing some of the original draft; H.A. contributed methodology, revision and reduced the similarity and typos; H.A.H. contributed in conceptualization, investigation, validation, writing some original draft and approve the final version of this manuscript; W.S. contributed in the revision, supervision, methodology; N.M. contributed in funding acquisition, methodology, project administration, visualization. All authors have read and agreed to the published version of the manuscript.

Funding: Prince Sultan University for funding this work through research group Nonlinear Analysis Methods in Applied Mathematics (NAMAM) group number RG-DES-2017-01-17.

Acknowledgments: The last two authors would like to thank Prince Sultan University for funding this work through research group Nonlinear Analysis Methods in Applied Mathematics (NAMAM) group number RG-DES-2017-01-17.

Conflicts of Interest: The authors declare that they have no competing interests concerning the publication of this article.

References

1. Shatanawi, W.; Gupta, V.; Kanwar, A. New results on modified intuitionistic generalized fuzzy metric spaces by employing E.A property and common E.A. property for coupled maps. *J. Intell. Fuzzy Syst.* **2020**, *38*, 3003–3010. [CrossRef]
2. Aydi, H.; Lakzian, H.; Mitrović, Z.D.; Radenović, S. Best proximity points of MF-cyclic contractions with property UC. *Numer. Funct. Anal. Optim.* **2020**, *41*, 871–882. [CrossRef]
3. Gupta, V.; Shatanawi, W.; Kanwar, A. Coupled fixed point theorems employing CLR_Ω-Property on V-fuzzy metric spaces. *Mathematics* **2020**, *8*, 404. [CrossRef]
4. Ameer, E.; Aydi, H.; Arshad, M.; De la Sen, M. Hybrid Ćirić type graphic (Y, Λ)-contraction mappings with applications to electric circuit and fractional differential equations. *Symmetry* **2020**, *12*, 467. [CrossRef]
5. Patle, P.; Patel, D.; Aydi, H.; Radenović, S. On H^+-type multivalued contractions and applications in symmetric and probabilistic spaces. *Mathematics* **2019**, *7*, 144. [CrossRef]
6. Bakhtin, I.A. The contraction mapping in almost metric spaces. *Funct. Anal. Gos. Ped. Inst. Unianowsk* **1989**, *24*, 26–37.
7. Czerwik, S. Contraction mappings in b-metric spaces. *Acta Math. Univ. Ostrav.* **1993**, *1*, 5–11.
8. Qawaqneh, H.; Noorani, M.S.M.; Shatanawi, W.; Aydi, H.; Alsamir, H. Fixed point results for multi-valued contractions in b-metric spaces and an application. *Mathematics* **2018**, *7*, 132. [CrossRef]
9. Karapinar, E.; Czerwik, S.; Aydi, H. (α, ψ)-Meir-Keeler contraction mappings in generalized b-metric spaces. *J. Funct. Spaces* **2018**, *2018*, 3264620. [CrossRef]
10. Mlaiki, N.; Aydi, H.; Souayah, N.; Abdeljawad, T. Controlled metric type spaces and the related contraction principle. *Mathematics* **2018**, *6*, 194. [CrossRef]
11. Abdeljawad, T.; Mlaiki, N.; Aydi, H.; Souayah, N. Double controlled metric type spaces and some fixed point results. *Mathematics* **2018**, *6*, 320. [CrossRef]
12. Alamgir, N.; Kiran, Q.; Isik, H.; Aydi, H. Fixed point results via a Hausdorff controlled type metric. *Adv. Differ. Equ.* **2020**, *2020*, 24. [CrossRef]
13. Fagin, R.; Kumar, R.; Sivakumar, D.; Comparing top k lists. *SIAM J. Discret. Math.* **2003**, *17*, 134–160. [CrossRef]
14. Gahler, V.S. 2-metrische Raume und ihre topologische struktur. *Math. Nachr.* **1964**, *26*, 115–118. [CrossRef]
15. Ha, K.S.; Cho, Y.J.; White, A. Strictly convex and 2-convex 2-normed spaces. *Math. Jpn.* **1988**, *33*, 375–384.
16. Mustafa, Z.; Sims, B.; A new approach to generalized metric spaces. *J. Nonlinear Convex Anal.* **2006**, *7*, 289–297.
17. Branciari, A. A fixed point theorem of Banach–Caccioppoli type on a class of generalized metric spaces. *Publ. Math. Debr.* **2000**, *57*, 31–37.
18. Matthews, S.G. Partial metric topology. *Ann. N. Y. Acad. Sci.* **1994**, *728*, 183–197. [CrossRef]
19. Aydi, H.; Felhi, A.; Karapinar, E.; Sahmim, S. A Nadler-type fixed point theorem in dislocated spaces and applications. *Miscolc Math. Notes* **2018**, *19*, 111–124. [CrossRef]
20. Aydi, H.; Karapinar, E.; Shatanawi. Coupled fixed point results for (ψ, φ)-weakly contractive condition in ordered partial metric spaces. *Comput. Math. Appl.* **2011**, *62*, 4449–4460. [CrossRef]
21. Aydi, H.; Abbas, M.; Vetro, C. Partial Hausdorff metric and Nadler's fixed point theorem on partial metric spaces. *Topol. Its Appl.* **2012**, *159*, 3234–3242. [CrossRef]
22. Jleli, M.; Samet, B. A generalized metric space and related fixed point theorems. *Fixed Point Theory Appl.* **2015**, *14*, 1–14. [CrossRef]
23. Jleli, M.; Samet, B., On a new generalization of metric spaces. *J. Fixed Point Theory Appl.* **2018**, *20*, 128. [CrossRef]
24. Fredholm, E.I. Sur une classe d'equations fonctionnelles. *Acta Math.* **1903**, *27*, 365–390. [CrossRef]

25. Rus, M.D. A note on the existence of positive solution of Fredholm integral equations. *Fixed Point Theory* **2004**, *5*, 369–377.
26. Berenguer, M.I.; Munoz, M.V.F.; Guillem, A.I.G.; Galan, M.R. Numerical treatment of fixed point applied to the nonlinear fredholm integral equation. *Fixed Point Theory Appl.* **2009**, 735–638. [CrossRef]
27. Hammad, H.A.; De la Sen, M. A coupled fixed point technique for solving coupled systems of functional and nonlinear integral equations. *Mathematics* **2019**, *7*, 634. [CrossRef]
28. Hammad, H.A.; De la Sen, M. A solution of Fredholm integral equation by using the cyclic η_s^q-rational contractive mappings technique in b-metric-like spaces. *Symmetry* **2019**, *11*, 1184. [CrossRef]
29. Hammad, H.A.; De la Sen, M. Solution of nonlinear integral equation via fixed point of cyclic α_s^q-rational contraction mappings in metric-like spaces. *Bull. Braz. Math. Soc. New Ser.* **2020**, *51*, 81–105. [CrossRef]
30. Farengo, R.; Lee, Y.C.; Guzdar, P.N. An electromagnetic integral equation: Application to microtearing modes. *Phys. Fluids* **1983**, *26*, 3515–3523. [CrossRef]
31. Sidorov, N.A.; Sidorov, D.N. Solving the Hammerstein integral equation in the irregular case by successive approximations. *Sib. Math. J.* **2010**, *51*, 325–329. [CrossRef]
32. Sidorov, D.N. Existence and blow-up of Kantorovich principal continuous solutions of nonlinear integral equations. *Differ. Equat.* **2014**, *50*, 1217–1224. [CrossRef]

© 2020 by the authors. Licensee MDPI, Basel, Switzerland. This article is an open access article distributed under the terms and conditions of the Creative Commons Attribution (CC BY) license (http://creativecommons.org/licenses/by/4.0/).

MDPI
St. Alban-Anlage 66
4052 Basel
Switzerland
Tel. +41 61 683 77 34
Fax +41 61 302 89 18
www.mdpi.com

Symmetry Editorial Office
E-mail: symmetry@mdpi.com
www.mdpi.com/journal/symmetry

www.ingramcontent.com/pod-product-compliance
Lightning Source LLC
LaVergne TN
LVHW070733100526
838202LV00013B/1225